Lecture Notes in Bioinformatics 9859

Subseries of Lecture Notes in Computer Science

Ezio Bartocci · Pietro Lio
Nicola Paoletti (Eds.)

Computational Methods in Systems Biology

14th International Conference, CMSB 2016
Cambridge, UK, September 21–23, 2016
Proceedings

 Springer

Editors
Ezio Bartocci
TU Wien
Vienna
Austria

Pietro Lio
Computer Laboratory
University of Cambridge
Cambridge
UK

Nicola Paoletti
Department of Computer Science
University of Oxford
Oxford
UK

ISSN 0302-9743   ISSN 1611-3349   (electronic)
Lecture Notes in Bioinformatics
ISBN 978-3-319-45176-3   ISBN 978-3-319-45177-0   (eBook)
DOI 10.1007/978-3-319-45177-0

Library of Congress Control Number: 2016948626

LNCS Sublibrary: SL8 – Bioinformatics

Printed on acid-free paper

This Springer imprint is published by Springer Nature
The registered company is Springer International Publishing AG
The registered company address is: Gewerbestrasse 11, 6330 Cham, Switzerland

Preface

This volume contains the papers presented at CMSB 2016, the 14th Conference on Computational Methods in Systems Biology, held on September 21–23, 2016 at the Computer Laboratory, University of Cambridge (UK).

The CMSB annual conference series, initiated in 2003, provides a unique forum of discussion for computer scientists, biologists, mathematicians, engineers, and physicists interested in a system-level understanding of biological processes. Topics of interest include formalisms for modelling biological processes; models and their biological applications; frameworks for model verification, validation, analysis, and simulation of biological systems; high-performance computational systems biology and parallel implementations; model inference from experimental data; model integration from biological databases; multi-scale modelling and analysis methods; and computational approaches for synthetic biology. Case studies in systems and synthetic biology are especially encouraged.

There were 37 regular submissions, 3 tools papers, and 9 poster submissions. Each regular submission and tool paper submission was reviewed by at least 4 Program Committee members. The committee decided to accept 17 regular papers, 3 tool papers, and all submitted posters. On average, regular and tool papers received 4.2 reviews each, while each poster submissions received 2 reviews. To complement the contributed papers, we also included in the program four invited lectures: Luca Cardelli (Microsoft Research, UK), Joëlle Despeyroux (Inria Sophia Antipolis, France), Radu Grosu (TU Wien, Austria), and Jane Hillston (University of Edinburgh, UK).

As program co-chairs, we have many people to thank. We are extremely grateful to the members of the Program Committee and the external reviewers for their peer reviews and the valuable feedback they provided to the authors. We thank also the authors of the accepted papers for revising the papers according to the suggestions of the program committee and for their responsiveness on providing the camera-ready copies within the deadline. Our special thanks goes to François Fages and all the members of the CMSB Steering Committee for their advice on organizing and running the conference. We acknowledge the support of the EasyChair conference system during the reviewing process and the production of these proceedings. We thank Kaushik Chowdhury and the IEEE Computer Society Technical Committee on Simulation for supporting the best student paper award and the best poster award. We thank NVIDIA for providing their equipment as the best paper award. Our gratitude also goes to the tool track chair, Claudio Angione, and the local organization chair, Max Conway, for their help, support, and spirited participation before, during, and after the conference. We are also really grateful to Paolo Zuliani for having organized a minisymposium on *Automated Reasoning for Systems Biology,* which was held a day before the conference. It is our pleasant duty to acknowledge the financial support from our sponsor, Microsoft Research, and the support of the Computer Laboratory at the University of Cambridge, where this year's event was hosted. Finally, we would like to thank all the participants of the conference. It was the quality of their presentations and their contribution to the discussions that made the meeting a scientific success.

September 2016

Ezio Bartocci
Pietro Lio
Nicola Paoletti

Organization

Program Committee Co-chairs

Ezio Bartocci   TU Wien, Austria
Pietro Lio   University of Cambridge, UK
Nicola Paoletti   Oxford University, UK

Tools Track Chair

Claudio Angione   Teesside University, UK

Local Organization Chair

Max Conway   University of Cambridge, UK

Program Committee

Claudio Angione   Teesside University, UK
Gianluca Ascolani   University of Cambridge, UK
Julio Banga   IIM-CSIC, Spain
Ezio Bartocci   TU Wien, Austria
Gregory Batt   Inria Paris-Rocquencourt, France
Luca Bortolussi   University of Trieste, Italy
Jérémie Bourdon   Nantes University, France
Andrea Bracciali   University of Stirling, UK
Luca Cardelli   Microsoft Research, UK
Milan Češka   Oxford University, UK
Vincent Danos   University of Edinburgh, UK
Joëlle Despeyroux   Inria Sophia Antipolis, France
Diego Di Bernardo   University of Naples Federico II, Italy
François Fages   Inria Paris-Rocquencourt, France
Flavio H. Fenton   Georgia Tech, USA
Jérôme Feret   Inria/Ecole Normale Supérieure, France
Calin Guet   IST Austria
Monika Heiner   Brandenburg University of Technology, Germany
Lila Kari   University of Western Ontario, Canada
Heinz Köppl   TU Darmstadt, Germany
Hillel Kugler   Bar-Ilan University, Israel
Marta Kwiatkowska   University of Oxford, UK
Pietro Lio   University of Cambridge, UK

Oded Maler	CNRS-VERIMAG, France
Giancarlo Mauri	University of Milano Bicocca, Italy
Pedro Mendes	University of Manchester, UK/University of Connecticut Health Center, USA
Nicola Paoletti	University of Oxford, UK
Tatjana Petrov	IST Austria
Andrew Phillips	Microsoft Research Cambridge, UK
Carla Piazza	University of Udine, Italy
Ovidiu Radulescu	University of Montpellier 2, France
Blanca Rodriguez	University of Oxford, UK
Olivier Roux	École Centrale de Nantes, France
David Šafránek	Masaryk University, Czech Republic
Guido Sanguinetti	University of Edinburgh, UK
Scott A. Smolka	Stony Brook University, USA
Oliver Stegle	EBI, UK
Jörg Stelling	ETH Zurich, Switzerland
Carolyn Talcott	SRI International, USA
P.S. Thiagarajan	National University of Singapore, Singapore
Adelinde Uhrmacher	University of Rostock, Germany
Verena Wolf	Saarland University, Germany
Boyan Yordanov	Microsoft Research Cambridge, UK
Paolo Zuliani	Newcastle University, UK

Tool Evaluation Committee

Claudio Angione	Teesside University, UK
Liu Bing	Carnegie Mellon University, USA
Pierre Boutillier	Harvard Medical School, USA
Giulio Caravagna	University of Edinburgh, UK
Tommaso Dreossi	UC Berkeley, USA
Maxime Folschette	University of Nice Sophia-Antipolis, France
Fabian Fröhlich	Helmholtz Zentrum München, Germany
Attila Gabot	Aachen University, Germany
Emanuel Goncalves	EBI, UK
Benjamin Gyori	Harvard Medical School, USA
Ariful Islam	Carnegie Mellon University, USA
Luca Laurenti	University of Oxford, UK
Curtis Madsen	Boston University, USA
Dimitrios Milios	University of Edinburgh, UK
Niall Murphy	Microsoft Research Cambridge, UK
Abhishek Murthy	Philips Research, USA
Aurélien Naldi	Université de Montpellier, France
Rasmus Petersen	Queen Mary University of London, UK
Ly Kim Quyen	Ecole Normale Supérieure, France
Giselle Reis	Inria-Saclay, France
Satya Swarup Samal	University of Bonn, Germany

Fedor Shmarov   Newcastle University, UK
Elisabeth Yaneske   Teesside University, UK

Steering Committee

Jérémie Bourdon   Nantes University, France
Finn Drablos   NTNU, Norway
François Fages   Inria Paris-Rocquencourt, France
David Harel   Weizmann Institute of Science, Israel
Monika Heiner   Brandenburg University of Technology, Germany
Tommaso Mazza   IRCCS Casa Sollievo della Sofferenza - Mendel, Italy
Pedro Mendes   University of Manchester, UK/University of Connecticut
                             Health Center, USA
Satoru Miyano   University of Tokyo, Japan
Gordon Plotkin   University of Edinburgh, UK
Corrado Priami   CoSBi/Microsoft Research, University of Trento, Italy
Olivier Roux   École Centrale de Nantes, France
Carolyn Talcott   SRI International, USA
Adelinde Uhrmacher   University of Rostock, Germany

Additional Reviewers

Ahmad, Jamil
Asarin, Eugene
Barbot, Benoit
Beica, Andreea
Bueno-Orovio, Alfonso
Casagrande, Alberto
Červený, Jan
Cinquemani, Eugenio
Daca, Przemyslaw
Dannenberg, Frits
Eyassu, Filmon
Forets, Marcelo

Fränzle, Martin
Galpin, Vashti
Gilbert, David
Herajy, Mostafa
Islam, Md. Ariful
Krivine, Jean
Kyriakopoulos,
    Charalampos
Lück, Alexander
Magnin, Morgan
Magron, Victor
Niehren, Joachim

Nobile, Marco
Patanè, Andrea
Paulevé, Loïc
Ramanathan, S.
Rohr, Christian
Ruet, Paul
Schnoerr, David
Soliman, Sylvain
Srivastav, Abhinav
Tschaikowski, Max

Contents

Tool Papers

Invited Paper

(Mathematical) Logic for Systems Biology
(Invited Paper)

Joëlle Despeyroux[✉]

Inria and CNRS, I3S, Sophia-Antipolis, France
joelle.despeyroux@inria.fr

Abstract. We advocates here the use of (mathematical) logic for systems biology, as a unified framework well suited for both modeling the dynamic behaviour of biological systems, expressing properties of them, and verifying these properties. The potential candidate logics should have a traditional proof theoretic pedigree (including a sequent calculus presentation enjoying cut-elimination and focusing), and should come with (certified) proof tools. Beyond providing a reliable framework, this allows the adequate encodings of our biological systems. We present two candidate logics (two modal extensions of linear logic, called HyLL and SELL), along with biological examples. The examples we have considered so far are very simple ones - coming with completely formal (interactive) proofs in Coq. Future works includes using automatic provers, which would extend existing automatic provers for linear logic. This should enable us to specify and study more realistic examples in systems biology, biomedicine (diagnosis and prognosis), and eventually neuroscience.

1 Introduction

We consider here the question of reasoning about biological systems in (mathematical) logic. We show that two new logics, both modal extensions of linear logic [12] (LL), are particularly well-suited to this purpose. The first logic, called Hybrid Linear Logic (HyLL), has been developed by the author in joint work with K. Chaudhuri [8]. The second logic, an extension of Subexponential Linear Logic (SELL$^{\cap}$), has been independently proposed by C. Olarte, E. Pimentel and V. Nigam [15]. Both HyLL and SELL provides a unified framework to encode biological systems, to express temporal properties of their dynamic behaviour, and to prove these properties. By constructing proofs in the logics, we directly witness reachability as logical entailment [13,17]. This approach is in contrast to most current approaches to applying formal methods to systems biology, which generally encode biological systems either in a dedicated programming language [6,10,19], or in differential equations [5], express properties in a temporal logic, and then verify these properties against some form of traces (model-checking), eventually built using an external simulator.

In a joint work with E. De Maria and A. Felty, we presented some first applications of HyLL to systems biology [13]. In these first experiments, we focused on

E. Bartocci et al. (Eds.): CMSB 2016, LNBI 9859, pp. 3–12, 2016.
DOI: 10.1007/978-3-319-45177-0_1

Boolean systems and in this case a time unit corresponds to a transition in the system. We believe that discrete modeling is crucial in systems biology because it allows taking into account some phenomena that have a very low chance of happening (and could thus be neglected by differential approaches), but which may have a strong impact on system behavior.

In a recent joint work with C. Olarte and E. Pimentel [9], we compared HyLL and SELL, providing two encodings. The first enoding is from HyLL's logical rules into LL with the highest level of adequacy, hence showing that HyLL is as expressive as LL. We also proposed an encoding of HyLL into SELL$^{\cap}$ (SELL plus quantification over locations) that gives better insights about the meaning of worlds in HyLL. This shows that SELL is more expressive than HyLL. However, the simplicity of HyLL might be of interest, both from the user point of view and as far as proof search is concerned (a priori easier and more efficient in HyLL than in SELL). In this joint work, we furthermore encoded temporal operators of Computational Tree Logic (CTL) into linear logic with fixed point operators.

We first recall here these two previous works. Then we briefly mention our current joint work with P. Lio, on formalizing the evolution of cancer cells, concluding with some future work.

This note is thus based on joint works with K. Chaudhuri (INRIA Saclay), A. Felty (Univ. of Ottawa), P. Lio (Cambridge Univ.), and C. Olarte and E. Pimentel (Universidade Federal do Rio Grande do Norte, Brazil).

2   Preliminaries

Although we assume that the reader is familiar with linear logic [12] (LL), we review some of its basic proof theory in the following sections. First, let us gently introduce linear logic by means of an example.

2.1   Linear Logic for Biology

Linear Logic (LL) [12] is particularly well suited for describing state transition systems. LL has been successfully used to model such diverse systems as: the π-calculus, concurrent ML, security protocols, multiset rewriting, and games.

In the area of biology, a rule of activation (e.g., a protein activates a gene or the transcription of another protein) can be modeled by the following LL axiom:

$$\texttt{active}(a,b) \overset{\text{def}}{=} \texttt{pres}(a) \multimap (\texttt{pres}(a) \otimes \texttt{pres}(b)).$$

The formula $\texttt{active}(a,b)$ describes the fact that a state where a is present ($\texttt{pres}(a)$ is true) can evolve into a state where both $\texttt{pres}(a)$ and $\texttt{pres}(b)$ are true.

Propositions such as $\texttt{pres}(a)$ are called *resources*, and a rule in the logic can be viewed as a rewrite rule from a set of resources into another set of resources, where a set of resources describes a state of the system. Thus, a particular state transition system can be modeled by a set of rules of the above shape. The rules of the logic then allow us to prove some desired properties of the system, such as, for example, the existence of a stable state. However, linear implication is timeless. Linear implication can be used to model one event occurring after another, but it cannot be precise about how many steps or how long the delay is without explicitly encoding time. In a domain where resources have lifetimes and state changes have temporal, probabilistic or stochastic *constraints*, then the logic will allow inferences that may not be realizable in the system being modeled. This was the motivation of the development of HyLL, which was designed to represent constrained transition systems.

2.2   Linear Logic and Focusing

Literals are either atomic formulas (p) or their negations ($p^{\perp}$). The connectives $\otimes$ and $\bindnasrepma$ and their units 1 and $\perp$ are *multiplicative*; the connectives $\oplus$ and $\&$ and their units 0 and $\top$ are *additive*; $\forall$ and $\exists$ are (first-order) quantifiers; and ! and ? are the exponentials (called bang and question-mark, respectively).

First proposed by Andreoli [1] for linear logic, focused proof systems provide normal form proofs for cut-free proofs. The connectives of linear logic can be divided into two classes. The *negative* connectives have invertible introduction rules: these connectives are $\bindnasrepma$, $\perp$, $\&$, $\top$, $\forall$, and ?. The *positive* connectives $\otimes$, 1, $\oplus$, **0**, $\exists$, and ! are the de Morgan duals of the negative connectives. A formula is *positive* if it is a negated atom or its top-level logical connective is positive. Similarly, a formula is *negative* if it is an atom or its top-level logical connective is negative.

Focused proofs are organized into two *phases*. In the *negative* phase, all the invertible inference rules are eagerly applied. The *positive* phase begins by choosing a positive formula F on which to focus. Positive rules are applied to F until either 1 or a negated atom is encountered (and the proof must end by applying the initial rules), the promotion rule (!) is applied, or a negative subformula is encountered and the proof switches to the negative phase.

This change of phases on proof search is particularly interesting when the focused formula is a *bipole* [1]. Focusing on a bipole will produce a single positive and a single negative phase. This two-phase decomposition enables us to adequately capture the application of object-level inference rules by the meta-level linear logic, as shown in [9].

2.3   Hybrid Linear Logic

Hybrid Linear Logic (HyLL) is a conservative extension of Intuitionistic first-order Linear Logic (ILL) [12] where the truth judgments are labelled by worlds representing constraints on states and state transitions. Instead of the ordinary judgment "A is true", for a proposition A, judgments of HyLL are of the form "A is true at world w", abbreviated as $A @ w$. Particular choices of worlds produce particular instances of HyLL. Typical examples are "A is true at time t", or "A is true with probability p". HyLL was first proposed in [8] and it has been used as a logical framework for specifying biological systems [13].

Formally, worlds are defined as follows.

Definition 1 (HyLL worlds). A *constraint domain* $\mathcal{W}$ is a monoid structure $\langle W, ., \iota \rangle$. The elements of W are called *worlds* and its *reachability relation* $\preceq$: $W \times W$ is defined as $u \preceq w$ if there exists $v \in W$ such that $u.v = w$.

The identity world ι is $\preceq$-initial and is intended to represent the lack of any constraints. Thus, the ordinary first-order linear logic is embeddable into any instance of HyLL by setting all world labels to the identity. A typical, simple example of constraint domain is $\mathcal{T} = \langle I\!N, +, 0 \rangle$, representing instants of time.

Atomic propositions $(p, q, \ldots)$ are applied to a sequence of terms $(s, t, \ldots)$, which are drawn from an untyped term language containing constants $(c, d, \ldots)$, term variables $(x, y, \ldots)$ and function symbols $(f, g, \ldots)$ applied to a list of terms (t). Non-atomic propositions are constructed from the connectives of first-order intuitionistic linear logic and the two hybrid connectives *satisfaction* (at), which states that a proposition is true at a given world $(w, \iota, u.v, \ldots)$, and *localization* ($\downarrow$), which binds a name for the (current) world the proposition is true at. The following grammar summarizes the syntax of HyLL.

$$t \quad ::= c \mid x \mid f(t)$$
$$A, B ::= p(t) \mid A \otimes B \mid 1 \mid A \rightarrow B \mid A \,\&\, B \mid \top \mid A \oplus B \mid 0 \mid !A \mid$$
$$\forall x.\, A \mid \exists x.\, A \mid (A \text{ at } w) \mid \downarrow u.\, A \mid \forall u.\, A \mid \exists u.\, A$$

Note that world u is bounded in the propositions $\downarrow u.\ A$, $\forall u.\ A$ and $\exists u.\ A$. World variables cannot be used in terms, and neither can term variables occur in worlds. This restriction is important for the modular design of HyLL because it keeps purely logical truth separate from constraint truth. We note that $\downarrow$ and at commute freely with all non-hybrid connectives [8].

The sequent calculus [11] presentation of HyLL uses sequents of the form $\Gamma; \Delta \vdash C @ w$ where Γ (*unbounded context*) is a set and Δ (*linear context*) is a multiset of judgments of the form $A @ w$. Note that in a judgment $A @ w$ (as in a proposition A at w), w can be any expression in $\mathcal{W}$, not only a variable.

The inference rules dealing with the new hybrid connectives are depicted below (the complete set of rules can be found in [8]).

$$\frac{\Gamma; \Delta \vdash A@u}{\Gamma; \Delta \vdash (A \text{ at } u)@w} \text{at}R \qquad \frac{\Gamma; \Delta, A@u \vdash C@w}{\Gamma; \Delta, (A \text{ at } u)@v \vdash C@w} \text{at}L$$

$$\frac{\Gamma; \Delta \vdash A[w/u]@w}{\Gamma; \Delta \vdash \downarrow u.A@w} \downarrow R \qquad \frac{\Gamma; \Delta, A[v/u]@v \vdash C@w}{\Gamma; \Delta, \downarrow u.A@v \vdash C@w} \downarrow L$$

Note that $(A$ at $u)$ is a *mobile* proposition: it carries with it the world at which it is true. Weakening and contraction are admissible rules for the unbounded context.

The most important structural properties are the admissibility of the general identity (i.e. over any formulas, not only atomic propositions) and cut theorems. While the first provides a syntactic completeness theorem for the logic, the latter guarantees consistency (i.e. that there is no proof of $.; . \vdash 0 @ w$).

Theorem 1 (Identity/Cut).

1. $\Gamma; A @ w \vdash A @ w$
2. If $\Gamma; \Delta \vdash A @ u$ and $\Gamma; \Delta', A @ u \vdash C @ w$, then $\Gamma; \Delta, \Delta' \vdash C @ w$
3. If $\Gamma; . \vdash A @ u$ and $\Gamma, A @ u; \Delta \vdash C @ w$, then $\Gamma; \Delta \vdash C @ w$.

Moreover, HyLL is conservative with respect to intuitionistic linear logic: as long as no hybrid connectives are used, the proofs in HyLL are identical to those in ILL. It is worth noting that HyLL is more expressive than S5, as it allows direct manipulation of the worlds using the hybrid connectives and HyLL's δ connective (see Sect. 5) is not definable in S5. We also note that HyLL admits a complete focused [1] proof system. The interested reader can find proofs and further meta-theoretical theorems about HyLL in [8].

Modal Connectives. We can define modal connectives in HyLL as follows:

Definition 2 (Modal connectives).

$$\Box A \overset{\text{def}}{=} \downarrow u. \forall w. (A \text{ at } u.w) \qquad \Diamond A \overset{\text{def}}{=} \downarrow u. \exists w. (A \text{ at } u.w) \qquad \delta_v A \overset{\text{def}}{=} \downarrow u. (A \text{ at } u.v)$$

$\Box A$ [resp. $\Diamond A$] represents all [resp. some] state(s) satisfying A and reachable from now. The connective δ represents a form of delay.

2.4 Subexponentials in Linear Logic

Linear logic with subexponentials (SELL) shares with LL all its connectives except the exponentials: instead of having a single pair of exponentials ! and ?, SELL may contain as many *subexponentials* [7,18], written $!^a$ and $?^a$, as one needs. The grammar of formulas in SELL is as follows:

$$F ::= \mathbf{0} \mid 1 \mid \top \mid \bot \mid p(t) \mid F_1 \otimes F_2 \mid F_1 \oplus F_2 \mid F_1 \,\invamp\, F_2 \mid F_1 \,\&\, F_2 \mid$$
$$\exists x. F \mid \forall x. F \mid !^a F \mid ?^a F$$

The proof system for SELL is specified by a *subexponential signature* $\Sigma = \langle I, \preceq, U \rangle$, where I is a set of labels, $U \subseteq I$ is a set specifying which subexponentials allow weakening and contraction, and $\preceq$ is a pre-order among the elements of I. We shall use $a, b, \ldots$ to range over elements in I and we will assume that $\preceq$ is upwardly closed with respect to U, *i.e.*, if $a \in U$ and $a \preceq b$, then $b \in U$.

The system SELL is constructed by adding all the rules for the linear logic connectives except for the exponentials. The rules for subexponentials are dereliction and promotion of the subexponential labelled with $a \in I$

$$\frac{\vdash ?^{a_1} F_1, \ldots ?^{a_n} F_n, G}{\vdash ?^{a_1} F_1, \ldots ?^{a_n} F_n, !^a G} \, !^a \qquad \frac{\vdash \Gamma, G}{\vdash \Gamma, ?^a G} \, ?^a$$

Here, the rule $!^a$ has the side condition that $a \preceq a_i$ for all i. That is, one can only introduce a $!^a$ on the right if all other formulas in the sequent are marked with indices that are greater or equal than a. Moreover, for all indices $a \in U$, we add the usual rules for weakening and contraction.

We can enhance the expressiveness of SELL with the subexponential quantifiers $\cap$ and $\cup$ ([15,18]) given by the rules (omitting the subexponential signature)

$$\frac{\vdash \Gamma, G[l_e/l_x]}{\vdash \Gamma, \cap l_x : a.G} \cap \qquad \frac{\vdash \Gamma, G[l/l_x]}{\vdash \Gamma, \cup l_x : a.G} \cup$$

where l_e is fresh. Intuitively, subexponential variables play a similar role as eigenvariables. The generic variable $l_x : a$ represents any subexponential, constant or variable in the ideal of a. Hence l_x can be substituted by any subexponential l of type b (i.e., $l : b$) if $b \preceq a$. We call the resulting system SELL$^{\cap}$.

As shown in [15,18], SELL$^{\cap}$ admits a cut-free and also a complete focused proof system.

Theorem 2. *SELL$^{\cap}$ admits cut-elimination for any subexponential signature.*

Modal connectives. We can define modal connectives in SELL as follows:

$$\Box_u A \stackrel{\text{def}}{=} \forall l : u.\, !^l A \qquad \Diamond_u A \stackrel{\text{def}}{=} \exists l : u.\, !^l A \qquad \Box A \stackrel{\text{def}}{=} \forall t : \infty.\, !^t A \qquad \Diamond A \stackrel{\text{def}}{=} \exists t : \infty.\, !^t A$$

3   First Experiments with HyLL

In a joint work with E. De Maria and A. Felty, we presented some first applications of HyLL to systems biology [13]. In these first experiments, we focused on Boolean systems and in this case a time unit corresponds to a transition in the system.

The activation rule seen in LL (Sect. 2.1) can be written in HyLL as

$$\texttt{active}(a,b) \stackrel{\text{def}}{=} \texttt{pres}(a) \multimap \delta_1 \,(\texttt{pres}(a) \otimes \texttt{pres}(b)).$$

We chosed a simple yet representative biological example concerning the DNA-damage repair mechanism based on proteins p53 and Mdm2, and present and proved several properties of this system. All these properties were reachability properties or the existence of an invariant. Most interesting proofs require induction or case analysis, that we borrowed from the meta-level (Coq). We fully formalized these proofs in the Coq Proof Assistant [3]. In Coq, we can both reason in HyLL and formalize meta-theoretic properties about it.

We discussed the merits and eventual drawbacks of this new approach compared to approaches using temporal logic and model checking. To better illustrate the correspondence with such approaches, which all use temporal logic to reason about (simulations of models of) the biological systems described, we also presented, informally but in some detail, the encoding of temporal logic operators in HyLL.

4 Relative Expressiveness Power of HyLL and SELL

We observe that, while linear logic has only seven logically distinct prefixes of bangs and question-marks, SELL allows for an unbounded number of such prefixes, *e.g.*, $!^i$, or $!^i?^j$. Hence, by using different prefixes, we allow for the specification of richer systems where subexponentials are used to mark different modalities/states. For instance, subexponentials can be used to represent contexts of proof systems [16]; to specify systems with temporal, epistemic and spatial modalities [18] and to specify and verify biological systems [17]. An inhibition rule can be written in (classical) SELL as

$$\text{inhib}(a,b) \overset{\text{def}}{=} !^t a \multimap !^{t+1}(a \otimes b^\perp).$$

HyLL and Linear Logic. One may wonder whether the use of worlds in HyLL increases also the expressiveness of LL. In a joint work with C. Olarte and E. Pimentel [9], we proved that this is not the case, by showing that HyLL rules can be directly encoded into LL by using the methods proposed in [14]. Moreover, the encoding of HyLL into LL is *adequate* in the sense that a focused step in LL corresponds *exactly* to the application of one inference rule in HyLL.

HyLL and SELL. Linear logic allows for the specification of two kinds of context maintenance: both weakening and contraction are available (classical context) or neither is available (linear context). That is, when we encode (linear) judgments in HyLL belonging to different worlds, the resulting meta-level atomic formulas will be stored in the same (linear) LL context. The same happens with classical HyLL judgments and the classical LL context.

Although this is perfectly fine, encoding HyLL into SELL$^\cap$ allows for a better understanding of worlds in HyLL. For that, we use subexponentials to represent worlds, having each world as a linear context. A HyLL judgment of the shape $F@w$ in the (left) linear context is encoded as the SELL$^\cap$ formula $?^w \lfloor F@w \rfloor$. Hence, HyLL judgments that hold at world w are stored at the w linear context of SELL$^\cap$. A judgment of the form $G@w$ in the classical HyLL context is encoded as the SELL$^\cap$ formula $?^c?^w \lfloor G@w \rfloor$. Then, the encoding of $G@w$ is stored in the unbounded (classical) subexponential context c.

We showed that our encoding is indeed adequate. Moreover, as before, the adequacy of the encodings is on the *level of derivations*.

Information Confinement. One of the features needed to specify spatial modalities is information *confinement*: a space/world can be inconsistent and this does not imply the inconsistency of the whole system. We showed in [9] that information confinement cannot be specified in HyLL. The authors in [15] exploit the combination of subexponentials of the form $!^w?^w$ in order to specify information confinement in SELL$^\cap$. More precisely, note that the sequents (in a 2-sided presentation of SELL) $!^w?^w \mathbf{0} \nvdash \mathbf{0}$ and $!^w?^w \mathbf{0} \nvdash !^v?^v \mathbf{0}$, representing "inconsistency is local" and "inconsistency is not propagated" respectively hold in SELL.

5   Computation Tree Logic (CTL) in Linear Logic

Hybrid linear logic is expressive enough to encode some forms of modal operators, thus allowing for the specification of properties of transition systems. As mentioned in [13], it is possible to encode CTL temporal operators into HyLL considering existential (E) and bounded universal (A) path quantifiers. We extended these encodings in [9], showing how to fully capture E and A CTL quantifiers in linear logic with fixed points. For that, we used the system μMALL [2] that extends MALL (multiplicative, additive linear logic) with fixed point operators. In [13], proofs of (encodings of) properties involving CTL quantifiers use induction borrowed from the (Coq) meta-level. In [9], we could directly use fixed points in linear logic.

6   Concluding Remarks and Future Work

Concerning related work, it is worth noticing that there are some other logical frameworks that are extensions of LL, for example, HLF [20]. Being a logic in the LF family, HLF is based on natural deduction, hence having a complex notion of $(\beta\eta)$ normal forms. Thus adequacy (of encodings of systems) results are often much harder to prove in HLF than in (focused) HyLL/SELL. HLF seems to have been later abandonned in favour of Hybridized Intuitionistic Linear Logic (HILL) [4] - a type theory based on a subpart of HyLL.

Both HyLL and SELL have been used for formalizing and analyzing biological systems [13,17]. SELL proved to be a broader framework for handling such systems (in particular localities). However, the simplicity of HyLL may be of interest for specific purposes, such as building tools for diagnosis in biomedicine.

Formal proofs in HyLL were implemented in [13], in the Coq [3] proof assistant. It would be interesting to extend the implementations of HyLL given there to SELL. Such an interactive proof environment would enable both formal studies of encoded systems in SELL and formal meta-theoretical study of SELL itself.

We may pursue the goal of using HyLL/SELL for further applications. That might include neuroscience, a young and promising science where many hypotheses are provided and need to be verified. Indeed, logic is a general tool whose area of potential applications are not restricted per se. This is in contrast to most of the other approaches, which are valid only in a restricted area (typically inside or outside the cell).

In an ongoing joint work with P. Lio, we are formalizing the evolution of cancer cells, acquiring driver or passenger mutations. A rule describing an intravasating Circulating Tumour Cell, for example, might be:

$$C(n, breast, f, [\text{EPCAM}]) \multimap \delta_d\, C(n, blood, 1, [\text{EPCAM}])$$

where f is a fitness parameter, here in $\{0, 1\}$. Our long term goal here is the design of a Logical Framework for disease diagnosis and therapy prognosis. This requires the development of automatic tools for proof search in our logics.

These tools should benefit both from current research on proof search in linear logic and from current developments of automatic provers for SELL.

References

1. Andreoli, J.: Logic programming with focusing proofs in linear logic. J. Log. Comput. **2**(3), 297–347 (1992)
2. Baelde, D.: Least and greatest fixed points in linear logic. ACM Trans. Comput. Log. **13**(1), 2 (2012)
3. Bertot, Y., Castéran, P.: Interactive Theorem Proving and Program Development. Coq'Art: The Calculus of Inductive Constructions. Springer, Heidelberg (2004)
4. Caires, L., Perez, J., Pfenning, F.: Logic-based domain-aware session types (2014) (submitted)
5. Campagna, D., Piazza, C.: Hybrid automata in systems biology: how far can we go? Electron. Notes TCS **229**(1), 93–108 (2009)
6. Danos, V.: Agile modelling of cellular signalling (invited paper). In: Proceedings of the 5th Workshop on Structural Operational Semantics (SOS). Electronic Notes in TCS, vol. 229, pp. 3–10. Elsevier (2009)
7. Danos, V., Joinet, J.B., Schellinx, H.: The structure of exponentials: uncovering the dynamics of linear logic proofs. In: Mundici, D., Gottlob, G., Leitsch, A. (eds.) KGC 1993. LNCS, vol. 713, pp. 159–171. Springer, Heidelberg (1993)
8. Despeyroux, J., Chaudhuri, K.: A hybrid linear logic for constrained transition systems. In: Post-Proceedings of the 9th International Conference on Types for Proofs and Programs (TYPES 2013). Leibniz International Proceedings in Informatics, vol. 26, pp. 150–168. Schloss Dagstuhl-Leibniz-Zentrum fuer Informatik (2014)
9. Despeyroux, J., Olarte, C., Pimentel, E.: Hybrid and subexponential linear logics. In: Proceedings of the 11th workshop on Logical and Semantic Frameworks, with Applications (LSFA) (2016)
10. Fages, F., Soliman, S.: Formal cell biology in biocham. In: Bernardo, M., Degano, P., Zavattaro, G. (eds.) SFM 2008. LNCS, vol. 5016, pp. 54–80. Springer, Heidelberg (2008)
11. Gentzen, G.: Investigations into logical deductions, 1935. In: Szabo, M.E. (ed.) The Collected Papers of Gerhard Gentzen, pp. 68–131. North-Holland Publishing Co., Amsterdam (1969)
12. Girard, J.Y.: Linear logic. TCS **50**, 1–102 (1987)
13. de Maria, E., Despeyroux, J., Felty, A.P.: A logical framework for systems biology. In: Fages, F., Piazza, C. (eds.) FMMB 2014. LNCS, vol. 8738, pp. 136–155. Springer, Heidelberg (2014)
14. Miller, D., Pimentel, E.: A formal framework for specifying sequent calculus proof systems. Theoret. Comput. Sci. **474**, 98–116 (2013)
15. Nigam, V., Olarte, C., Pimentel, E.: A general proof system for modalities in concurrent constraint programming. In: D'Argenio, P.R., Melgratti, H. (eds.) CONCUR 2013 – Concurrency Theory. LNCS, vol. 8052, pp. 410–424. Springer, Heidelberg (2013)
16. Nigam, V., Pimentel, E., Reis, G.: Specifying proof systems in linear logic with subexponentials. Electron. Notes Theoret. Comput. Sci. **269**, 109–123 (2011)
17. Olarte, C., Chiarugi, D., Falaschi, M., Hermith, D.: A proof theoretic view of spatial and temporal dependencies in biochemical systems. Theoret. Comput. Sci. (TCS) **641**, 25–42 (2016)

18. Olarte, C., Pimentel, E., Nigam, V.: Subexponential concurrent constraint programming. TCS **606**, 98–120 (2015)
19. Phillips, A., Cardelli, L.: A correct abstract machine for the stochastic pi-calculus. In: BioConcur: Workshop on Concurrent Models in Molecular Biology. Electronic Notes in TCS (2004)
20. Reed, J.: Hybridizing a logical framework. In: International Workshop on Hybrid Logic (HyLo), Seattle, USA, August 2006

Regular Papers

Generalized Method of Moments for Stochastic Reaction Networks in Equilibrium

Michael Backenköhler[1], Luca Bortolussi[1,2], and Verena Wolf[1]([✉])

[1] Computer Science Department, Saarland University,
Saarbrücken, Germany
wolf@cs.uni-saarland.de
[2] Department of Mathematics and Geosciences,
University of Trieste, Trieste, Italy

Abstract. Calibrating parameters is a crucial problem within quantitative modeling approaches to reaction networks. Existing methods for stochastic models rely either on statistical sampling or can only be applied to small systems. Here we present an inference procedure for stochastic models in equilibrium that is based on a moment matching scheme with optimal weighting and that can be used with high-throughput data like the one collected by flow cytometry. Our method does not require an approximation of the underlying equilibrium probability distribution and, if reaction rate constants have to be learned, the optimal values can be computed by solving a linear system of equations. We evaluate the effectiveness of the proposed approach on three case studies.

1 Introduction

Stochastic models have proven to be a powerful tool for the analysis of biochemical reaction networks. Especially when chemical species are present in low copy numbers, a stochastic approach provides important insights on the randomness inherent to the system when compared to deterministic approaches. For the inference of parameters based on experimentally observed samples, more detailed descriptions given by stochastic models can substantially improve the quality of the estimation [18].

The arguably most popular stochastic modeling approach to chemical kinetics is based on a description in terms of continuous time Markov chains (CTMC) [10]. In this case, the exact time evolution of the entire probability distribution is given by the chemical master equation (CME). Although, this description is exact up to the numerical precision of the integration scheme, its solution is only feasible for simple systems with small molecular populations [22]. Therefore, the applicability of inference approaches based on a maximum likelihood estimation (MLE) is limited to this class of networks since they require an approximation of the probability distribution, i.e., a solution of the CME [2,3]. An alternative to ease the computational burden is to use stochastic simulation to estimate the likelihood function or to learn parameters in a Bayesian setting, e.g. by ABC methods [33]. However, the total number of simulations to be performed is huge, still resulting in a computationally intensive approach.

E. Bartocci et al. (Eds.): CMSB 2016, LNBI 9859, pp. 15–29, 2016.
DOI: 10.1007/978-3-319-45177-0_2

A computationally more feasible approach is to consider the statistical moments (such as the expected value or the variance) instead of the entire probability distribution. Moment-based analysis approaches rely on a derivation of a system of equations for the time-derivative of the moments [1,6,30]. Since the exact time evolution of the moments of order k may depend on moments of higher order, a closure method has to be applied to arrive at a finite system of equations. However, moment-based methods complicate the application of MLE since a reconstruction of the distribution is computationally expensive and may be inaccurate depending on the shape of the distribution [4].

In this paper we propose a parameter estimation approach that does not rely on MLE and distribution approximations, but on the generalized method of moments (GMM), which has been a widely used inference method in econometrics for over 30 years [12,14]. We consider the case in which experimentally observed samples are drawn when the process is in equilibrium. Population snapshot data of equilibrium processes are considered, for instance, if the (possibly multi-stable) steady-state expression in a gene regulatory network is investigated [7,17] or if the steady-state behavior of a mutant is compared the behavior of the wild type [19,27]. Modern high-throughput experimental techniques, like flow cytometry, deliver a large amount of measurements from a population of cells at steady state and thus give detailed information about the distribution of proteins and RNAs [13,16,25]. The idea of the GMM is to consider constraints of the form $\mathbb{E}[f(\boldsymbol{Y}_i, \boldsymbol{\theta}_0)] = 0$ where $\boldsymbol{Y}_i$ is a sample and $\boldsymbol{\theta}_0$ the parameter vector. We propose to choose f as the time derivatives of the statistical moments of the model, which can directly be derived from the CME. This follows from the fact that the time derivatives will become equal to zero when the process is in equilibrium. A major advantage, given the availability of steady state samples, is that, compared to time depended observations, no moment closure approximations are necessary. Instead exact equations for the steady state moments can be used. If the propensities are linear in the unknown parameters, as is the case for mass action kinetics, a closed linear form is possible. This results in an extremely fast inference procedure since no numerical optimization is needed. In case of propensities that are non-linear in the parameters numerical optimization is necessary. Still, no numerical integration of moment equations or probabilities is needed since the objective function corresponds to the right side of the steady state moment equations.

The moment equations may also contain moments of species whose quantity is hard to measure (e.g. the state of a promoter). Instead of treating these latent variables as unknown (probably non-linear) parameters, here we propose a clustering approach that estimates promoter states in a preprocessing step. Then, a closed linear solution is still possible, which again enables an accurate estimation in very short time.

We analyse the effectiveness of the GMM approach for the p53 oscillator model [9] and two variants of the genetic toggle switch [8,20]. Our results show that using moments of up to at least second order yields accurate estimates. The inclusion of higher order moments (higher than three) can lead to a further decrease of the estimators variances, but for the p53 model and only few samples the estimation becomes worse. Nevertheless, even for comparatively small sample sizes (100) the estimates are usually tightly distributed around the true parameter value when moments up to order two or three are considered.

The paper is organized as follows. We first provide some background on the model in Sect. 2 and present our inference approach based on GMM in Sect. 3. We discuss the inference results for the case studies in Sect. 4 and conclude the paper in Sect. 6.

2   Stochastic Chemical Kinetics

A stochastic model of a network of chemical reactions is usually specified by a set of n species, which are represented by a set of symbols $S_1, \ldots, S_n$. We are interested in the system state, i.e., the number of individuals of the species, and thus consider state space $\mathcal{S} \subseteq \mathbb{N}_{\geq 0}^n$. Furthermore, a set of J reactions is given describing the interactions between the different molecular populations. For $j \in \{1, \ldots, J\}$ reaction R_j is specified by its stoichiometry

$$R_j: \quad S_1 \nu_{j,1}^- + \cdots + S_n \nu_{j,n}^- \xrightarrow{c_j} S_1 \nu_{j,1}^+ + \cdots + S_n \nu_{j,n}^+, \tag{1}$$

where the vectors $\boldsymbol{\nu}_j^-$ and $\boldsymbol{\nu}_j^+ \in \mathbb{N}_{\geq 0}^n$ with entries $\nu_{j,i}^-$ and $\nu_{j,i}^+$ for $i \in \{1, \ldots, n\}$ specify how many molecules are consumed (produced) of each type, respectively. The vector $\boldsymbol{\nu}_j = \boldsymbol{\nu}_j^+ - \boldsymbol{\nu}_j^-$ is called the *change vector* of R_j. The propensity functions α_j are such that $\alpha_j : \mathcal{S} \times \Theta \to \mathbb{R}_{\geq 0}$, where Θ is the parameter space. If mass action kinetics are assumed, then α_j is the product of the rate constant c_j and the number of possible combinations of reactant molecules, i.e., $\alpha_j(\boldsymbol{x}, \boldsymbol{\theta}) = c_j \prod_{i=1}^n \binom{x_i}{\nu_{j,i}^-}$ for $\boldsymbol{x} \in \mathcal{S}$. Here, we do not restrict to mass action kinetics but only impose certain regularity conditions on the propensity functions, such as continuity and the existence of certain expected values. If a reaction does not follow mass action propensities, we give the propensity function separately from the stoichiometry (1).

Under the assumption of well-stirredness and thermal equilibrium such a system can be accurately described by a continuous-time Markov chain (CTMC) $\boldsymbol{X}(t) = (X_1(t), \ldots, X_n(t))$ over the state space $\mathcal{S}$ [10]. The time evolution of the probability distribution is given by the chemical master equation (CME):

$$\frac{d}{dt} P(\boldsymbol{X}(t) = \boldsymbol{x}) = \sum_{j=1}^J P(\boldsymbol{X}(t) = \boldsymbol{x} - \boldsymbol{\nu}_j) \, \alpha_j(\boldsymbol{x} - \boldsymbol{\nu}_j, \boldsymbol{\theta}) - \sum_{j=1}^J P(\boldsymbol{X}(t) = \boldsymbol{x}) \, \alpha_j(\boldsymbol{x}, \boldsymbol{\theta}) \tag{2}$$

Due to the largeness of the state space the integration of $\frac{d}{dt} P$ is computationally infeasible, especially if we have to integrate until convergence to determine the equilibrium distribution. Given (2) it is straight-forward to compute the time derivative of the expectation of some polynomial function $g : \mathcal{S} \to \mathbb{R}$ [6]:

$$\frac{d}{dt} \mathbb{E}[g(\boldsymbol{X})] = \sum_{j=1}^J \mathbb{E}[(g(\boldsymbol{X} + \boldsymbol{\nu}_j) - g(\boldsymbol{X})) \, \alpha_j(\boldsymbol{X}, \boldsymbol{\theta})], \tag{3}$$

where we omit the dependence of X on t. Here we are concerned with the *population moments* of the distribution, which are monomials $X^{\mathbf{m}}$ over X where we use the multi-index notation $x^{\mathbf{m}} = x_1^{m_1} \cdots x_n^{m_n}$ for the vectors $\mathbf{m} = (m_1, \ldots, m_n) \in \mathbb{N}_{\geq 0}^n$ and $x = (x_1, \ldots, x_n)$.[1] The order of a moment is given by the sum $m_1 + \cdots + m_n$. The first order moment of the i-th population, for example, is obtained from (3) by setting $g(x) = x_i$:

$$\frac{d}{dt}\mathbb{E}[X_i] = \sum_{j=1}^{J} \nu_{j,i}\mathbb{E}[\alpha_j(X, \boldsymbol{\theta})]. \tag{4}$$

In general, the equation of a moment of a certain order may depend on moments of higher order, except if α is constant or linear, i.e., of the form $c_j^\mathsf{T} x + b_j$ for some constant $c_j \in \mathbb{R}^n$ and $b_j \in \mathbb{R}$. Here, we do not aim at finding a finite system of ODEs to approximate the moments but we rather propose to use the exact moment equations when the system is in equilibrium. The *equilibrium probability* of a state x is defined as the limit of $P(X(t)=x)$ when $t \to \infty$, i.e.,

$$\pi(x) = \lim_{t \to \infty} P(X(t)=x). \tag{5}$$

and is uniquely defined for ergodic processes X. Since the equilibrium distribution is independent of time, the expected values in (3) are also time-independent when $t \to \infty$. Thus, we can use the right side of (3) to estimate propensity parameters given samples from the equilibrium distribution.

3   GMM Conditions at Equilibrium

We propose to use the moments of the equilibrium distribution as an input for a GMM inference, which is a very generic framework for parameter estimation [12,23]. It is most popular in econometrics, where often the exact distribution of a model is not known. In this case MLE cannot be used since it needs a sufficiently accurate description of the distribution for its optimality properties to hold. As opposed to this, the GMM is based on the construction and minimization of certain cost functions, called *moment conditions*, which relate the population and sample moments. A moment condition is given by a function whose expected value is zero for the true parameter value $\boldsymbol{\theta}_0$. Given independent samples $Y_1, \ldots, Y_N$ of the process X in equilibrium, a vector of moment conditions is given by

$$\mathbb{E}[\boldsymbol{f}(Y, \boldsymbol{\theta}_0)] = 0, \tag{6}$$

where we omit the index of the samples whenever they appear within the expectation operator since $Y_1, \ldots, Y_N$ are identically distribution according to the equilibrium distribution π. Moreover, let $\boldsymbol{f}$ be a vector of q different functions,

---

[1] The existence and convergence of moments is treated Gupta et al. [11]. It can be proved for the models in Sect. 4 with positive rate constants.

i.e., $f : (\mathcal{S} \times \Theta) \rightarrow \mathbb{R}^q$. The sample equivalent of (6) for the vector f of moment conditions is given by

$$f_N(\theta_0) = \frac{1}{N} \sum_{i=1}^{N} f(Y_i, \theta_0) = 0. \tag{7}$$

Depending on the number of such conditions q and the number of parameters to be estimated p, we distinguish for the estimated value the *non-identified case* $(q < p)$, the *exactly identified case* $(p = q)$ and the *over-identified case* $(q > p)$. In the exactly identified case, assuming (7) has a unique solution, we have Pearson's classical *method of moments* [28].

Since we are considering the system at equilibrium, the right-hand side of (3) must equal zero. In principle, it is possible to use any polynomial g meeting certain regularity conditions [12]. However, using population moments, i.e., monomials of Y is a natural choice that leads to the moment conditions

$$\frac{d}{dt}\mathbb{E}[Y^m] = \sum_{j=1}^{J} \mathbb{E}\left[\left((Y + \nu_j)^m - Y^m\right)\alpha_j(Y, \theta_0)\right] = 0 \tag{8}$$

for the estimation of θ. Therefore the moment condition vector f in (6) is determined by the functional form (8) of a selection of different vectors m, i.e., the entry in the vector f that corresponds to m is

$$f_m(Y, \theta) = \sum_{j=1}^{J} \left((Y + \nu_j)^m - Y^m\right)\alpha_j(Y, \theta).$$

Typically, we choose these vectors such that their entries correspond to the moments up to some fixed order. If, for example, we use first order moments only, the i-th entry of f is equal to $\sum_j \nu_{i,j}\alpha_j(Y, \theta)$ for $i = 1, 2, \ldots, n$ (see Eq. 4). For the moments of order two we extend f with entries according to the right side in (8) of the second order moments and so forth.

We may choose as many moment conditions as there are parameters to exactly identify the estimate. However, the inclusion of further information on the distribution may lead to a more accurate estimation. GMM provides a framework to deal with over-identified estimation problems. The estimator is given by

$$\hat{\theta}_N = \arg\min_{\theta \in \Theta} Q_N(\theta), \tag{9}$$

where $Q_N(\theta)$ is the objective function

$$Q_N(\theta) = f_N(\theta)^{\mathsf{T}} W f_N(\theta). \tag{10}$$

Here, W is some positive semi-definite matrix containing weights for each pair of moment conditions. Under certain regularity conditions [12], this estimator is asymptotically normal and consistent, i.e., the estimator converges in probability to θ_0. These regularity conditions mostly consist of the existence of expectations of $\boldsymbol{f}$ and $\partial \boldsymbol{f}/\partial\boldsymbol{\theta}$ and their continuity w.r.t. $\boldsymbol{\theta}$ on the parameter space Θ. Assuming convergence to equilibrium moments the validity of these conditions depends solely on the propensity functions. They hold for mass action and Hill's propensities, as they are smooth functions of the parameters. The parameter space itself is assumed to be bounded, which in practice can be done by either fixing a biologically relevant space or assuming a sufficiently large Θ [12]. A further necessary condition for normality is that $\boldsymbol{\theta}_0$ is a unique interior point of Θ such that $\mathbb{E}[\boldsymbol{f}(\boldsymbol{Y}, \boldsymbol{\theta}_0)] = 0$. However, if we have only samples from the steady state distribution this property may not hold if one tries to estimate all parameters at once. The reason is that often for a fixed steady-state distribution there is an infinite number of ergodic Markov chains having this steady-state distribution and the system is not fully identifiable.

Although the estimator's normality holds for all positive semi-definite weighting matrices, a good choice of W reduces the asymptotic variance of the estimator. It can be shown, that the asymptotically most efficient matrix W_0 is given by the inverse of $\lim_{N\to\infty} Var(\sqrt{N}\boldsymbol{f}_N(\boldsymbol{\theta}_0))$ [12,23]. In case of independent and identically distributed samples, W_0 can be estimated as follows [23]:

$$\hat{W}_N = \left(\frac{1}{N} \sum_{i=1}^{N} \boldsymbol{f}(\boldsymbol{Y}_i, \boldsymbol{\theta}_0) \boldsymbol{f}(\boldsymbol{Y}_i, \boldsymbol{\theta}_0)^{\mathsf{T}} \right)^{-1} . \tag{11}$$

Since this estimate depends on $\boldsymbol{\theta}_0$, which is unknown, GMM is usually applied in a iterative manner: A first estimate $\hat{\boldsymbol{\theta}}_1$ is computed using some positive-definite weight matrix, such as the identity matrix. The estimate $\hat{\boldsymbol{\theta}}_1$ is consistent, but likely asymptotically inefficient. This estimate is then used to approximate (11). The procedure of estimating $\boldsymbol{\theta}_0$ and computing $\hat{W}_N$ can be iteratively applied until some convergence criterion is met. Since W is constant at each iterative estimation, the solution to (9) can, under some restrictions on the propensities, be expressed as a linear system (cf. Sect. 3.1).

Beyond this iterative estimation scheme, the *continuously updating GMM* (CUGMM) [15] is a popular variant of the GMM estimator. Instead of recomputing the weight estimate between minimizations, the weight estimation (11) is substituted into the objective function (10). The resulting estimator is thus given by

$$\hat{\boldsymbol{\theta}}_{CU,N} = \arg\min_{\boldsymbol{\theta}\in\Theta} \boldsymbol{f}_N(\boldsymbol{\theta})^{\mathsf{T}} \left(\frac{1}{N} \sum_{i=1}^{N} \boldsymbol{f}(\boldsymbol{Y}_i, \boldsymbol{\theta}) \boldsymbol{f}(\boldsymbol{Y}_i, \boldsymbol{\theta})^{\mathsf{T}} \right)^{-1} \boldsymbol{f}_N(\boldsymbol{\theta}). \tag{12}$$

This estimator is often associated with improved finite sample properties and more reliable test statistics [23]. However, a closed form solution for linear propensities as described in Sect. 3.1 is not possible in a majority of cases. This necessitates numerical optimization to approximate (12).

3.1   Linear Propensities

In general, the minimization problem (9) can be solved using numerical optimization algorithms. However, depending on the rate functions, this may not be necessary, because a closed form solution, i.e., a linear system, can be obtained for many relevant cases, including mass action kinetics. This system results from the first order condition of the minimization $\partial Q_N(\hat{\theta}_N)/\partial\theta = 0$ which yields [12]

$$0 = \frac{\partial f_N(\hat{\theta}_N)}{\partial\theta}^{\mathsf{T}} W f_N(\hat{\theta}_N). \tag{13}$$

We now compute (13) under the condition that propensities are linear in θ and W is constant, as is the case in iterative GMM. To this end, let $\mathcal{R} \subseteq \{1,\ldots,J\}$ be the index set of functions α_j whose propensity is dependent on θ. Further let $\bar{f}$ be the part of f independent of θ such that the i-th entry equals

$$\bar{f}_{m_i}(Y) = \sum_{j\in\overline{\mathcal{R}}} \alpha_j(Y)\left((Y+\nu_j)^{m_i} - Y^{m_i}\right). \tag{14}$$

By computation of the matrix product (13) and splitting the moment condition based on (14), we get

$$\left(\frac{\partial f_N}{\partial\theta}^{\mathsf{T}} W f_N\right)_i = \underbrace{\sum_{h=1}^{p} \theta_h \sum_{\ell,k=1}^{q} \frac{\partial f_{N,m_k}}{\partial\theta_i} W_{k,\ell} \frac{\partial f_{N,m_\ell}}{\partial\theta_h}}_{(A\theta)_i} + \underbrace{\sum_{\ell,k=1}^{q} \frac{\partial f_{N,m_k}}{\partial\theta_i} W_{k,\ell} \bar{f}_{N,m_\ell}}_{-b_i}.$$

Note, that the sample derivatives $\partial f_N/\partial\theta_i$ are independent of θ. In vector notation this gives us the linear system $A\hat{\theta}_N = b$ as a solution to (13) where

$$A_{i,j} = \frac{\partial f_N}{\partial\theta_i}^{\mathsf{T}} W \frac{\partial f_N}{\partial\theta_j} \qquad b_i = -\frac{\partial f_N}{\partial\theta_i}^{\mathsf{T}} W \bar{f}_N. \tag{15}$$

Analogous to the general iterative scheme, we now solve (15) and use the estimate to in turn estimate W using (11). In the following discussion we will refer to this as the *closed form GMM* (CFGMM). One sees immediately that this method is far more efficient than numerically optimizing Q_N.

4   Case Studies

We evaluate the GMM estimation on three chemical reaction networks. Samples of the equilibrium distribution were generated by Gillespie's stochastic simulation algorithm (SSA) [10] and drawn by equidistant sampling after the initial warm-up period. For each case study 10^7 samples were generated and sample sets of different sizes were drawn at random from this large set. For each sample size considered, the estimation procedure was carried out on 100 random sample sets, in order to estimate the variance of the estimator.

4.1   P53 System

We first consider Model IV proposed in [9], that describes the interactions of the tumor suppressor p53. This system describes a negative feedback loop between p53 and the oncogene Mdm2, where pMdm2 is a Mdm2 precursor [9]. We chose the same parameter values as in [1], that is, $k_1 = 90$, $k_2 = 0.002$, $k_3 = 1.7$, $k_4 = 1.1$, $k_5 = 0.93$, $k_6 = 0.96$, $k_7 = 0.01$.

Model 1 (p53 System).

$$R_1: \qquad \varnothing \xrightarrow{k_1} \text{p53} \qquad\qquad R_2: \qquad \text{p53} \xrightarrow{k_2} \varnothing$$
$$R_3: \qquad \text{p53} \xrightarrow{k_4} \text{p53} + \text{pMdm2} \qquad R_4: \qquad \text{p53} \rightarrow \varnothing$$
$$R_5: \text{pMdm2} \xrightarrow{k_5} \text{Mdm2} \qquad\qquad R_6: \text{Mdm2} \xrightarrow{k_6} \varnothing$$

The degradation rate of p53 *is in part influenced by* Mdm2 *and is given by* $\alpha_4(\boldsymbol{x}, \boldsymbol{\theta}) = (k_3 x_{\text{p53}} x_{\text{Mdm2}})/(x_{\text{p53}} + k_7)$. *Terms of species with stoichiometric constant zero are omitted as well as stoichiometric constants equal to one.*

We estimated the four parameters k_3, k_4, k_5, and k_6 using the CFGMM as proposed in Sect. 3.1. Note that $\alpha_4(\cdot, \cdot)$ is linear in k_4. We fixed k_1 and k_2 to ensure identification as well as k_7 to avoid a time-consuming numerical optimization. The iterations were continued until either the parameter vector converged or the maximum number of four iterations was reached. The plot in Fig. 1 (left) shows that the best results were obtained already after the second step for moderate and large sample sizes, while for a small sample size of 100 further iterations were beneficial. It is important to note that for the first iteration, $\hat{W}_N$ is chosen as the identity matrix such that identical weights are assigned and mixed terms are not considered. Hence, the general idea of assigning appropriate weights gives significantly more accurate results compared to an estimation with identical weights.

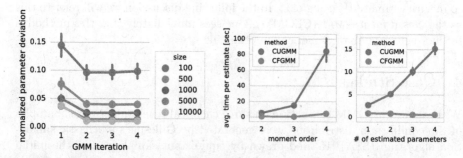

Fig. 1. p53 System: (left) The normalized parameter deviation $\|\hat{\boldsymbol{\theta}}_N - \boldsymbol{\theta}_0\| / \|\boldsymbol{\theta}_0\|$ over GMM iterations for different sample sizes. Moment conditions up to order two were used. (right) Comparison of the average running time for a single estimation, as a function of the number of parameters (maximal moment order three) and of the maximum order of moment conditions used (estimation for four parameters), for a sample size of 100.

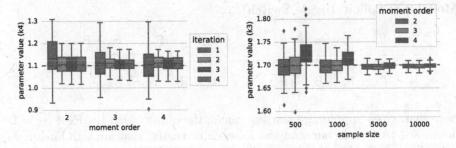

Fig. 2. p53 System: (left) Estimate of k_4 over GMM iterations (sample size 1000) (right) Estimates of parameter k_3 in relation to the sizes of the sample sets and the maximum order of used moment conditions. Results are presented as box plots (whiskers with a maximum of $1.5\,IQR$).

In Fig. 1 (right) we compare the running times of the CUGMM (using a numerical optimization scheme, the L-BFGS-B algorithm [36]) and of the iteration based method. The reported times are the average of 100 runs for a single estimate for different moment orders and different numbers of estimated parameters. As we can see, the iteration based method for linear propensities not only outperforms CUGMM, but also is essentially insensitive to including higher order moments and to increasing the number of estimated parameters. For CUGMM, an optimization is carried out since (12) is not linear in θ and this optimization becomes more costly when more moment conditions or parameters are considered. The advantage of CFGMM is that the Jacobian and $\bar{f}_N$ is only computed once for all iterations of a sample and no numerical optimization is needed.

In Fig. 2 we show the distribution of the estimate quality for different maximum moment orders against (left) different numbers of iterations for CFGMM and (right) for different sample sizes. The quality of the results is excellent for large sample sizes, while increasing the moment order beyond two does not result in significant improvements or may even (for small sample sizes) significantly decrease the quality (see Fig. 2 (right)). This bias may occur if the degree of overidentification $(q - p)$ is increased too much. It can be caused by the estimation of W and the dependence on the previous estimates and decreases proportional to N^{-1} [12,26]. In our evaluation estimators based on a maximal order of two and three showed the most reliable performance. Moreover, identical weights in the first step of the iteration lead to a very high variance of the corresponding estimator, as shown in Fig. 2 (left). In Fig. 2 (right) we also see that, when the number of samples is increased, the variance of the estimator becomes small.

4.2   Toggle Switch

The toggle switch is a widely known gene regulatory network [8,20] that models the production of two proteins A and B. Each protein can bind to the promoter of the opposite protein and thereby repress its production.

Model 2 (Explicit Toggle Switch). [20]

$$R_1: \qquad R_B \xrightarrow{\rho_B} R_B + A \qquad R_2: \qquad R_A \xrightarrow{\rho_A} R_A + B$$

$$R_3: \qquad A \xrightarrow{\delta_A} \varnothing \qquad\qquad R_4: \qquad B \xrightarrow{\delta_B} \varnothing$$

$$R_{5,6}: R_B + B \underset{\gamma_B}{\overset{\beta_B}{\rightleftharpoons}} \overline{R_B} \qquad\qquad R_{7,8}: R_A + A \underset{\gamma_A}{\overset{\beta_A}{\rightleftharpoons}} \overline{R_A}$$

Note that, given appropriate starting values, the conservation law $R_X + \overline{R_X} = 1$ holds ($X \in \{A, B\}$). In our study we focus on two cases, that are high binding-/unbinding rates and low binding-/unbinding rates with respect to the production and degradation of proteins.

Slow Binding Toggle Switch. In the case of low binding-/unbinding rates several attractor regions can arise that directly correspond to a given DNA state. Here, we use the parameters $\rho_X = 3$, $\delta_X = 0.5$, $\beta_X = 10^{-6}$, $\gamma_X = 3 \times 10^{-4}$, which are identical for $X = A$ and $X = B$. During the inference procedure, however, we did not make use of the information that the parameters are symmetric. For these parameters we get three distinct attractor regions corresponding to either one of the repressors being bound and both repressors being free[2].

Currently, our GMM-based approach requires all variables to be observed, which is in general unfeasible for the DNA state. One possible solution, when only proteins are observed, is to cluster the samples of the proteins using the k-Means algorithm (cf. Fig. 3 (left) for an example of a clustering of samples of the toggle switch). Then we can infer the state of the latent DNA state by assigning each cluster to a specific combination of DNA states and by looking at the cluster centroids, as illustrated in Fig. 3 (left). For low binding-/unbinding rates, the attractors are well separated and this approach is feasible, though more sophisticated approaches may be required when clusters overlap. After reconstruction of the state of the unobserved variables, we used the GMM estimation with the closed form solution for linear propensities. Results comparing different sample sizes are shown in Fig. 3 (right). The estimation quality is very good even in the case of only few samples, provided enough iterations are carried out. It is important to note that for these results, we excluded moment conditions corresponding to mixed moments involving the state of the gene as their moment conditions have very similar values. Including them leads to severe numerical instabilities (the matrix of the linear system for linear propensities becomes quasi-singular). However, ill-conditioned matrices are detected automatically when their determinant is calculated during the computation. Then, those entries responsible for the numerical instabilities can be excluded.

Fast Binding Toggle Switch. Often, it can be assumed that the repressor ($R_{A,B}$) binding and unbinding ($R_{5,6}$ and $R_{7,8}$) happens a lot faster than the

---

[2] The case of both repressors being bound, would result in samples around the origin, which can be neglected if there are no such samples.

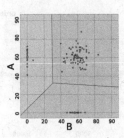 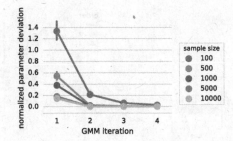

Fig. 3. Slow Switching Toggle Switch: (left) Clustering of a sample (size 100) using k-Means. (right) The normalized parameter deviation $\|\hat{\boldsymbol{\theta}}_N - \boldsymbol{\theta}_0\|/\|\boldsymbol{\theta}_0\|$ over GMM iterations for different sample sizes given the toggle switch with k-Means clustering. Moment conditions up to order 3 were used and 4 parameters were estimated.

protein production. Then, a *Michaelis-Menten* approximation is possible [20]. Therein the time derivative of the repressors is assumed to be zero. Applying this assumption to the mean-field equations of Model 2 yields the implicit toggle switch (Model 3). In this case, we no longer need the repressor state of each sample.

Model 3 (Implicit Toggle Switch).

$$R_1: \; \varnothing \to A \qquad R_2: \; \varnothing \to B \qquad R_3: \; A \xrightarrow{\delta_A} \varnothing \qquad R_4: \; B \xrightarrow{\delta_B} \varnothing$$

The rate function of reactions R_1 and R_2 resulting from the Michaelis-Menten approximation are

$$\alpha_1(\boldsymbol{x}, \boldsymbol{\theta}) = \frac{\rho_A}{1 + k_B x_B} \qquad \alpha_2(\boldsymbol{x}, \boldsymbol{\theta}) = \frac{\rho_B}{1 + k_A x_A}$$

where $\boldsymbol{\theta}$ is the vector of all parameters, $\boldsymbol{x} = (x_A, x_B)$, and $k_X = \frac{\beta_X}{\gamma_X}$ is the quotient of the binding and unbinding rate, $X \in \{A, B\}$.

The toggle switch exhibits bistability if the binding happens significantly faster than the unbinding, i.e., $k_A, k_B \gg 1$ [20]. However, the estimation of k_A and k_B is inherently difficult because switching between the attractors is a rare event.

In this case study, we simulated the explicit model using the symmetric constants $\beta_X = 100.0$, $\gamma_X = 50.0$, $\rho_X = 0.2$ and $\delta_X = 0.005$, assuming we could observe only the two proteins. Thus, we estimated the parameters k_X and δ_X of the implicit model and fixed ρ_X to ensure identification. Due to non-linear dependency of production rates on k_X, we cannot rely anymore on the method for linear propensities of Sect. 3.1, hence we resort to a numerical minimization routine, namely the L-BFGS-B algorithm [36], for the CUGMM scheme. The initial guess was chosen at random from $[0, 1]^p$. For detection of unsuccessful optimizations we used the J-Test statistic [14], which states that under the null hypothesis of a correctly specified model, $N Q_N(\hat{\boldsymbol{\theta}}_N)$ converges to the χ^2_{q-p} distribution. A confidence threshold of 90 % was fixed and the optimization was repeated for at

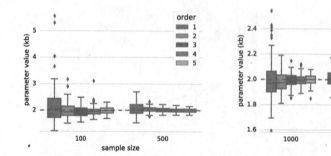

Fig. 4. Fast Binding Toggle Switch: Estimates of parameter k_B in relation to the sizes of the sample sets and the maximum order of used moment conditions. Only the parameter k_B was estimated. Results are presented as box plots (whiskers with a maximum of $1.5\,\mathrm{IQR}$).

most four times until the threshold was met. The use of numerical optimization increased the cost of a single estimate: For a sample size of 10,000 observations and order 2, the computation takes 1–2 of minutes.

In Fig. 4, we give statistics on the quality of estimates based on 100 runs of independently generated datasets. More specifically, we show how the quality of estimates varies with the maximum order of moments considered in the method and with sample size. For a fixed sample size, increasing the order from 1 to 2 improves considerably the quality of results. Use of higher order moments significantly reduces the variance of the estimator, in particular for the case of few samples.

5   Related Work

In the context of stochastic chemical kinetics, parameter inference methods are either based on Bayesian inference [5,32,34] or maximum likelihood estimation [2,3,29,31]. The advantage of the latter method is that the corresponding estimators are, in a sense, the most informative estimates of unknown parameters and have desirable mathematical properties such as unbiasedness, efficiency, and normality. On the other hand, the computational complexity of maximum likelihood estimation is high. If an analytic solution of the MLE is not possible, then, as a part of the non-linear optimization problem, the likelihood and its derivatives have to be calculated. Monte-Carlo simulation has been used to estimate the likelihood [31]. During the repeated random sampling it is difficult to explore those parts of the state space that are unlikely under the current rate parameters. Thus, especially if the rates are very different from the true parameters, many simulation runs are necessary to calculate an accurate approximation of the likelihood.

Therefore methods using computationally far more attractive moment expansion approximations have been proposed. Kügler [18] uses results of the moment closure approximations to apply an ad-hoc weighted least squares estimator. Milner et al. [24] construct a multi-variate normal distribution based on low order moments obtained from a moment closure approximation in order to apply MLE. Another approach based on moment closure and MLE relies on a normal distribution based on sample means and variances [35].

All of the aforementioned moment-based inference methods are, in contrast to the scenario discussed in this paper, based on samples of the transient distribution before equilibrium is reached. Therefore they have to rely on moment closure approximations, which is not necessary in our approach based on the equilibrium distribution. Recently, the performance of GMM estimators has been studied for transient (non-equilibrium) data [21] together with a (hybrid) moment closure approach.

6 Conclusion

Parameter inference methods for stochastic models of reaction networks require huge computational resources. The proposed approach based on the generalized method of moments is based on an adjustment of the statistical moments of the model in equilibrium and therefore does not require the computation of likelihoods. This makes the approach appealing for complex networks where stochastic effects play an important role, since no statistical sampling or numerical integration of master or moment equations is necessary. The proposed approach gives accurate results in seconds when the parameters are linear because a closed form of the solution is available. For non-linear parameters, a global optimization problem must be solved and therefore the inference takes longer but is still fast compared to other approaches based on the numerical computation of likelihoods.

Our results show that the GMM estimator yields accurate results, where its variance decreases when moments of higher order are considered. We found that when moments of order higher than three are included, the results become slightly worse in case of the p53 system while for the toggle switch quality improved (variance decreased). A general strategy could be to start with as many cost functions as unknown parameters and increase the maximal order until appropriate statistical tests suggest that higher orders do not lead to an improvement.

Currently, a major drawback of the method is that all species must be observed in order to apply it. For populations of at most one individual, the proposed clustering approach circumvents the problem that such species can usually not be observed. In general, however, the clustering may not always be possible and there may be other species that can not be observed. To deal with such cases, we plan to develop an extension of the method that treats the moments of such species as (additional) unknown parameters. Moreover, we will investigate how measurement errors could be taken into account within the GMM framework.

References

1. Ale, A., Kirk, P., Stumpf, M.: A general moment expansion method for stochastic kinetic models. J. Chem. Phys. **138**(17), 174101 (2013)
2. Andreychenko, A., Mikeev, L., Spieler, D., Wolf, V.: Parameter identification for markov models of biochemical reactions. In: Gopalakrishnan, G., Qadeer, S. (eds.) CAV 2011. LNCS, vol. 6806, pp. 83–98. Springer, Heidelberg (2011)
3. Andreychenko, A., Mikeev, L., Spieler, D., Wolf, V.: Approximate maximum likelihood estimation for stochastic chemicalkinetics. EURASIP J. Bioinf. Syst. Biol. **9**, 1–14 (2012)
4. Andreychenko, A., Mikeev, L., Wolf, V.: Model reconstruction for moment-based stochastic chemical kinetics. ACM Trans. Model. Comput. Simul. (TOMACS) **25**(2), 12 (2015)
5. Boys, R., Wilkinson, D., Kirkwood, T.: Bayesian inference for a discretely observed stochastic kinetic model. Stat. Comput. **18**, 125–135 (2008)
6. Engblom, S.: Computing the moments of high dimensional solutions of the master equation. Appl. Math. Comput. **180**(2), 498–515 (2006)
7. Fournier, T., Gabriel, J.-P., Mazza, C., Pasquier, J., Galbete, J.L., Mermod, N.: Steady-state expression of self-regulated genes. Bioinformatics **23**(23), 3185–3192 (2007)
8. Gardner, T.S., Cantor, C.R., Collins, J.J.: Construction of a genetic toggle switch in escherichia coli. Nature **403**(6767), 339–342 (2000)
9. Geva-Zatorsky, N., Rosenfeld, N., et al.: Oscillations and variability in the p53 system. Mol. Syst. Biol. 2(1) (2006)
10. Gillespie, D.T.: Exact stochastic simulation of coupled chemical reactions. J. Phys. Chem. **81**(25), 2340–2361 (1977)
11. Gupta, A., Briat, C., Khammash, M.: A scalable computational framework for establishing long-term behavior of stochastic reaction networks. PLoS Comput. Biol. **10**(6), e1003669 (2014)
12. Hall, A.R.: Generalized Method of Moments. Oxford University Press, New York (2005)
13. Hanley, M.B., Lomas, W., Mittar, D., Maino, V., Park, E.: Detection of low abundance RNA molecules in individual cells by flow cytometry. PloS one **8**(2), e57002 (2013)
14. Hansen, L.P.: Large sample properties of generalized method of moments estimators. Econometrica **50**(4), 1029–1054 (1982)
15. Hansen, L.P., Heaton, J., Yaron, A.: Finite-sample properties of some alternative GMM estimators. J. Bus. Econ. Stat. **14**(3), 262–280 (1996)
16. Hasenauer, J., Waldherr, S., Doszczak, M., Radde, N., Scheurich, P., Allgöwer, F.: Identification of models of heterogeneous cell populations from population snapshot data. BMC Bioinf. **12**(1), 1 (2011)
17. Isaacs, F.J., Hasty, J., Cantor, C.R., Collins, J.J.: Prediction and measurement of an autoregulatory genetic module. PNAS **100**(13), 7714–7719 (2003)
18. Kügler, P.: Moment fitting for parameter inference in repeatedly and partially observed stochastic biological models. PloS one **7**(8), e43001 (2012)
19. Lee, Y.J., Holzapfel, K.L., Zhu, J., Jameson, S.C., Hogquist, K.A.: Steady-state production of il-4 modulates immunity in mouse strains and is determined by lineage diversity of INKT cells. Nat. Immunol. **14**(11), 1146–1154 (2013)
20. Lipshtat, A., Loinger, A., Balaban, N.Q., Biham, O.: Genetic toggle switch without cooperative binding. Phys. Rev. Lett. **96**(18), 188101 (2006)

21. Lück, A., Wolf, V.: Generalized method of moments for estimating parameters of stochastic reaction networks. ArXiv e-prints, May 2016
22. Mateescu, M., Wolf, V., Didier, F., Henzinger, T.A.: Fast adaptive uniformisation of the chemical master equation. IET Syst. Biol. $\mathbf{4}$(6), 441–452 (2010)
23. Mátyás, L.: Generalized Method of Moments Estimation, vol. 5. Cambridge University Press, New York (1999)
24. Milner, P., Gillespie, C.S., Wilkinson, D.J.: Moment closure based parameter inference of stochastic kinetic models. Stat. Comput. $\mathbf{23}$(2), 287–295 (2013)
25. Munsky, B., Fox, Z., Neuert, G.: Integrating single-molecule experiments and discrete stochastic models to understand heterogeneous gene transcription dynamics. Methods $\mathbf{85}$, 12–21 (2015)
26. Newey, W.K., Smith, R.J.: Higher order properties of gmm and generalized empirical likelihood estimators. Econometrica $\mathbf{72}$(1), 219–255 (2004)
27. Nishihara, M., Ogura, H., Ueda, N., Tsuruoka, M., Kitabayashi, C., et al.: IL-6-gp130-STAT3 in T cells directs the development of IL-17+ Th with a minimum effect on that of Treg in the steady state. Int. Immunol. $\mathbf{19}$(6), 695–702 (2007)
28. Pearson, K.: Contributions to the mathematical theory of evolution. Philos. Trans. R. Soc. Lond. A $\mathbf{185}$, 71–110 (1894)
29. Reinker, S., Altman, R.M., Timmer, J.: Parameter estimation in stochastic biochemical reactions. IEEE Proc. Syst. Biol $\mathbf{153}$, 168–178 (2006)
30. Singh, A., Hespanha, J.P.: Lognormal moment closures for biochemical reactions. In: 2006 45th IEEE Conference on Decision and Control, pp. 2063–2068. IEEE (2006)
31. Tian, T., Xu, S., Gao, J., Burrage, K.: Simulated maximum likelihood method for estimating kinetic rates in gene expression. Bioinformatics $\mathbf{23}$, 84–91 (2007)
32. Toni, T., Welch, D., Strelkowa, N., Ipsen, A., Stumpf, M.: Approximate Bayesian computation scheme for parameter inference and model selection in dynamical systems. J. R. Soc. Interface $\mathbf{6}$(31), 187–202 (2009)
33. Toni, T., Welch, D., Strelkowa, N., Ipsen, A., Stumpf, M.P.H.: Approximate Bayesian computation scheme for parameter inference and model selection in dynamical systems. J. R. Soc. Interface $\mathbf{6}$(31), 187–202 (2009)
34. Wilkinson, D.J.: Stochastic Modelling for Systems Biology. C & H, Sesser (2006)
35. Zechner, C., Ruess, J., Krenn, P., Pelet, S., Peter, M., Lygeros, J., Koeppl, H.: Moment-based inference predicts bimodality in transient gene expression. PNAS $\mathbf{109}$(21), 8340–8345 (2012)
36. Zhu, C., Byrd, R.H., Lu, P., Nocedal, J.: Algorithm 778: L-BFGS-B: fortran subroutines for large-scale bound-constrained optimization. ACM Trans. Math. Softw. (TOMS) $\mathbf{23}$(4), 550–560 (1997)

Inference of Delayed Biological Regulatory Networks from Time Series Data

Emna Ben Abdallah[1]([✉]), Tony Ribeiro[1], Morgan Magnin[1,2], Olivier Roux[1], and Katsumi Inoue[3]

[1] Institut de Recherche en Communications et Cybernétique de Nantes, LUNAM Université, École Centrale de Nantes, IRCCyN UMR CNRS 6597, 1 rue de la Noë, 44321 Nantes, France
{emna.ben-abdallah,olivier.roux}@irccyn.ec-nantes.fr
[2] National Institute of Informatics, 2-1-2, Hitotsubashi, Chiyoda-ku, Tokyo 101-8430, Japan
[3] Department of Computer Science, Tokyo Institute of Technology, 2-12-1 Oookayama, Meguro-ku, Tokyo 152-8552, Japan

Abstract. The modeling of Biological Regulatory Networks (BRNs) relies on background knowledge, deriving either from literature and/or the analysis of biological observations. But with the development of high-throughput data, there is a growing need for methods that automatically generate admissible models. Our research aim is to provide a logical approach to infer BRNs based on given time series data and known influences among genes. In this paper, we propose a new methodology for models expressed through a timed extension of the Automata Networks [22] (well suited for biological systems). The main purpose is to have a resulting network as consistent as possible with the observed datasets. The originality of our work consists in the integration of quantitative time delays directly in our learning approach. We show the benefits of such automatic approach on dynamical biological models, the DREAM4 datasets, a popular reverse-engineering challenge, in order to discuss the precision and the computational performances of our algorithm.

Keywords: Inference model · Dynamic modeling · Delayed biological regulatory networks · Automata network · Time series data

1 Introduction

With both the spread of numerical tools in every part of daily life and the development of NGS methods (New Generation Sequencing methods), like DNA microarrays in biology, a large amount of time series data is now produced [6,8,18]. This means that the produced data from the experiments led on a biological system grows drastically. The newly produced data - as long as the associated noise does not raise an issue with regard to the precision and relevance of the corresponding information - can give us some new insights on the behavior of a system. This justifies the urge to design efficient methods for inference.

© Springer International Publishing AG 2016
E. Bartocci et al. (Eds.): CMSB 2016, LNBI 9859, pp. 30–48, 2016.
DOI: 10.1007/978-3-319-45177-0_3

Reverse engineering of gene regulatory networks from expression data have been handled by various approaches [14, 16, 29, 34]. Most of them are only static. However, other researchers are rather focusing on incorporating temporal aspects in inference algorithms. The relevance of these various algorithms have been recently assessed in [15]. The authors of [17] tackled the inference problem of time-delayed gene regulatory networks through Bayesian networks. As this is a complex problem, in [33], the authors propose a Time-Window-Extension Technique based on time series segmentation in different successive phases. These approaches take gene expression data into account as input and generate the associated regulations. But the discrete approaches that simplify this problem by abstractions, need to determine the relevant thresholds of each gene to define its active and inactive state. Various approches have been designed to tackle the discretization problem. We can cite for example [33], in which the authors have proposed an alternative methodology that considers not a concentration level, but the way the concentration changes (in other words: the derivative of the function giving the concentration w.r.t time) in the presence/absence of one regulator. On the other hand, the major problem for modeling lies on the quality of the expression data. Indeed, noisy data may be the main origin of the errors in the inference process. Thus, the pre-processing of the biological data is crucial pertinence of the inferred relations between components. In this work, the input data is considered to be pre-processed and the result is reliable discretized time series data.

In this paper, we aim to provide a logical approach to tackle the learning of qualitative models of biological dynamic systems, like gene regulatory networks. In our context, we assume the set of interacting components as fixed and we consider the learning of the interactions between those components. As in [3], in which the authors targeted the completion of stationary Boolean networks, we suppose that the topology of the network is given, providing us the influences among each gene as background knowledge. From time series data of the evolution of the system, given its topology, we learn the dynamics of the system. The main originality of our work is that we address this problem in a timed setting, with quantitative delays potentially occurring between the moment an interaction activated and the moment its effect is visible.

During the past decade, there has been a growing interest for the hybrid modeling of gene regulatory networks with delays. These hybrid approaches consider various modeling frameworks. In [19], the authors hybridize Petri Nets: the advantage of hybrid with regard to discrete modeling lies in the possibility of capturing biological factors, e.g., the delay for the transcription of RNA polymerase. The merits of other hybrid formalisms in biology have been studied, for instance timed automata [28], hybrid automata [2] and boolean representation [21]. Finally, in [7], the authors investigate a direct extension of the discrete René Thomas' modeling approach by introducing quantitative delays. These delays represent the compulsory time for a gene to turn from a discrete qualitative level to the next (or previous) one. They exhibit the advantage of such a framework for the analysis of mucus production in the bacterium Pseudomonas aeruginosa. The approach we propose in this paper inherits from this idea that some models need to capture these timing features.

2   Background

The definition and semantics of automata networks is presented in Sect. 2.1. The enrichment of the automata networks with delays and the corresponding new semantics is presented in Sect. 3.

2.1   Automata Network

Definition 1 introduces the Automata Network (AN) [22–24] as a model of finite-state machines having transitions between their *local state* conditioned by the state of other automata in the network. A *local state* of an automaton is noted by a_i, where a is the automaton identifier, and i is the expression level within a. At any time, each automaton has exactly one *active local state*, and the set of *active local states* is called the *global state* of the network.

The concurrent interactions between automata are defined by a set of *local transitions*. Each *local transition* has this form $\tau = a_i \overset{\ell}{\rightarrow} a_j$, with a_i, a_j being local states of an automaton a called respectively *origin* and *destination* of t and ℓ is a (possibly empty) set of local states of automata other than a (with at most one local state per automaton).

Notation: Given a finite state A, $\wp(A)$ is the power set of A. Given a network N, state(N, t) is the state of N at a time step $t \in \mathbb{N}$.

Definition 1 (Automata Network). *An* Automata Network *is a triple* $(\Sigma, \mathcal{S}, \mathcal{T})$ *where:*

- *$\Sigma = \{a, b, \dots\}$ is the finite set of automata identifiers;*
- *For each $a \in \Sigma$, $\mathcal{S}(a) = \{a_i, \dots, a_j\}$, is the finite set of local states of automaton a; $\mathcal{S} = \prod_{a \in \Sigma} \mathcal{S}(a)$ is the finite set of global states;*
  $\mathbf{LS} = \cup_{a \in \Sigma} \mathcal{S}(a)$ denotes the set of all the local states.
- *$\mathcal{T} = \{a \mapsto \mathcal{T}_a \mid a \in \Sigma\}$, where $\forall a \in \Sigma, \mathcal{T}_a \subset \mathcal{S}(a) \times \wp(\mathbf{LS} \setminus \mathcal{S}(a)) \times \mathcal{S}(a)$ with $(a_i, \ell, a_j) \in \mathcal{T}_a \Rightarrow a_i \neq a_j$, is the mapping from automata to their finite set of local transitions.*

Example 1. The Fig. 1 represents an Automata Network, $\mathcal{AN} = (\Sigma, \mathcal{S}, \mathcal{T})$ with 4 automata ($\Sigma = \{a, b, c, d\}$) such that $\mathcal{S}(a) = \{a_0, a_1\}$, $\mathcal{S}(b) = \{b_0, b_1\}$, $\mathcal{S}(c) = \{c_0, c_1, c_2\}$, $\mathcal{S}(d) = \{d_0, d_1, d_2\}$ and 5 local transitions,
$\mathcal{T} = \{\ b_0 \overset{\{a_1\}}{\longrightarrow} b_1,\ a_1 \overset{\{b_1, d_2\}}{\longrightarrow} a_0,\ c_2 \overset{\{a_1\}}{\longrightarrow} c_1,\ d_2 \overset{\{a_0\}}{\longrightarrow} d_1,\ b_1 \overset{\{a_1, c_2\}}{\longrightarrow} b_0, \}.$

A *global state* of a given AN consists in a set of all active *local states* of each automaton in the network. The active *local state* of a given automaton $a \in \Sigma$ in a state $\zeta \in \mathcal{S}$ is noted $\zeta[a]$. For any given *local state* a_i we also note, $a_i \in \zeta$ if and only if $\zeta[a] = a_i$. For each automaton, it cannot have more than one active *local state* at one *global state*.

Definition 2 (Playable Local Transition). *Let* $\mathcal{AN} = (\Sigma, \mathcal{S}, \mathcal{T})$ *be an Automata Network and $\zeta \in \mathcal{S}$, with $\zeta = $ state$(\mathcal{AN}, t)$. We note P_t the set of playable local transitions in $\mathcal{AN}$ at time step t by:*
$P_t = \{\ a_i \overset{\ell}{\rightarrow} a_j \in \mathcal{T} \mid \ell \subseteq \zeta \wedge a_i \in \zeta \text{ with state}(\mathcal{AN}, t) = \zeta\}.$

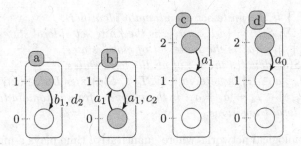

Fig. 1. Example of Automata Network with 4 automata: a, b, c and d presented by labeled boxes and their local states are presented by circles (for instance a is either at level 0 or 1). A local transition is a labeled directed edge between two local states within the same automaton: its label stands for the set of necessary conditions local states of the automata to play the transition. The grayed circles stand for the global state: $\langle a_1, b_0, c_2, d_2 \rangle$.

The dynamics of the AN is performed thanks to the *global transitions*. Indeed, the transition from one state ζ_1 to its successor ζ_2 is satisfied by a set of the playable local transitions (Definition 2) at ζ_1.

Definition 3 (Global Transitions). *Let* $\mathcal{AN} = (\Sigma, \mathcal{S}, \mathcal{T})$ *be an Automata Network and* $\zeta_1, \zeta_2 \in \mathcal{S}$, *with* $\zeta_1 = state(\mathcal{AN}, t)$ *and* $\zeta_2 = state(\mathcal{AN}, t+1)$. *Let* P_t *be the set of playable local transitions at* t. *We note* G_t *the power set of global transitions at* t:

$$G_t := \wp(P_t)$$

In the semantics that we are based on, parallel application of local transitions in different automata is permitted but it is not enforced. Thus the set of *global transitions* is a power set of all the playable *local transitions* (also empty set).

3 Timed Automata Networks

In some dynamics it is crucial to have information about the *delays* between two events (two states of a $\mathcal{AN}$). The discrete transition, described above, cannot exhibit this information: we just process chronological information, that the state ζ_2 will be after ζ_1 in the next step but it is not possible to know chronometry, i.e., how much time this transition takes to occur and whether it blocks some transitions during this time. In fact some local transitions could not be played any more because of concurrency about shared resources (necessary components to play the transition) between them. We thus need to restrain the general dynamics to capture more realistic behavior w.r.t biology. So we propose in this section to add the *delays* in the *local transitions* attributes and give the associated semantics that we based on to infer biological networks.

Definition 4 (Timed Automata Network (T-AN)). Timed Automata Network *is a triple* $(\Sigma, \mathcal{S}, \mathcal{T})$ *where:*

- $\Sigma = \{a, b, \dots\}$ is the finite set of automata identifiers;
- For each $a \in \Sigma$, $\mathcal{S}(a) = \{a_i, \dots, a_j\}$, is the finite set of local states of automaton a; $\mathcal{S} = \prod_{a \in \Sigma} \mathcal{S}(a)$ is the finite set of global states;
  $\boldsymbol{LS} = \cup_{a \in \Sigma} \mathcal{S}(a)$ denotes the set of all the local states.
- $\mathcal{T} = \{a \mapsto \mathcal{T}_a \mid a \in \Sigma\}$, where $\forall a \in \Sigma, \mathcal{T}_a \subset \mathcal{S}(a) \times \wp(\boldsymbol{LS} \setminus \mathcal{S}(a)) \times \mathcal{S}(a) \times \mathbb{N}$
  with $(a_i, \ell, a_j, \delta) \in \mathcal{T}_a \Rightarrow a_i \neq a_j$, is the mapping from automata to their finite set of timed local transitions.

To model biological networks where quantitative time plays a major role, we will use T-AN (Timed Automata Network). This formalism enriches $\mathcal{AN}$ with timed local transitions: $a_i \xrightarrow[\delta]{\ell} a_j$. In the latter, δ is called a *delay* and represents the time needed for the transition to be performed. When modeling a regulation phenomenon, this allows to capture the delay between the activation order of the production of the protein and its effective synthesis. and the synthesis of the product.

We note $a_i \xrightarrow[\delta]{\ell} a_j \in \mathcal{T} \Leftrightarrow (a_i, \ell, a_j, \delta) \in \mathcal{T}(a)$ and $a_i \xrightarrow{\ell} a_j \in \mathcal{T} \Leftrightarrow \exists \delta \in \mathbb{N}, a_i \xrightarrow[\delta]{\ell} a_j \in \mathcal{T}$. Given $\tau = a_i \rightarrow a_j \in \mathcal{T}$, $orig(\tau) = a_i$, $dest(\tau) = a_j$. Definition 2 also applies to timed local transitions.

Considering delays in the evolution of timed automata networks creates concurrency between the timed local transitions. This concurrency is mainly justified by the shared resources between local transitions. Indeed, transitions that have the same origins and/or destinations could not be fired synchronously. Besides, during the delay of the execution of a transition τ_1, it is possible that another transition τ_2 could be activated. Then we need to take care of the following possible conflicts between resources: transition τ_2 may change the local states of the automata participating in τ_1. We make the following assumptions, that is similar to the one adapted in [12]: we consider τ_2 needs to be blocked until the current transition τ_1 finishes. Nevertheless, we allow the resources of τ_1 to participate to other transitions. In addition we do not forbid the process involved in $orig(\tau_1)$ to participate to other transition τ_2 if and only if that the remaining $delay(\tau_1)$ is greater than $delay(\tau_2)$ (see Definition 5). Those considerations lead to the followings definitions.

Definition 5 (Blocked Timed Local Transition). Let $AN = (\Sigma, \mathcal{S}, \mathcal{T})$ be a T-AN and $t \in \mathbb{N}$. Let P be a set of pairs $\mathcal{T} \times \mathbb{N}$. The set of blocked timed local transitions of $\mathcal{AN}$ by P at t is defined as follows:

$B(\mathcal{AN}, P, t) := \{a_i \xrightarrow[\delta]{\ell} a_j \in \mathcal{T} \mid \exists (b_k \xrightarrow[\delta']{\ell'} b_l, t') \in P$ such that $(a = b) \vee (a_i \in \ell' \wedge \delta' > t' - (t + \delta)) \vee (b_k \in \ell \wedge \delta' < t' - (t + \delta))\}$

In Definition 5, if P is the set of currently ongoing timed local transition, it allows us to prevent the execution of transitions that would alternate the resources currently being used or that would rely on resources that will be modified before the end of those transitions. Let t_1 be a transition such that $\tau_1 = a_i \xrightarrow[\delta]{\ell} a_j$ is fired at time step t. So $t + \delta$ is the ending time of τ_1 and $(t' - (t + \delta))$ is the interval of time between the ending of the transition $\tau_2 = b_k \xrightarrow{\ell}_{\delta'} b_l$ and the beginning of transition τ_1 with $t' > t$. According to the Definition 5, τ_2 is blocked if a_i (resp. b_k) is a necessary resource for τ_2 (resp. τ_1) and the τ_1 (resp. τ_2) finishes before τ_2 (resp. τ_1): $\delta' > t' - (t + \delta)$ (resp. $\delta' < t' - (t + \delta)$) i.e. a_i (resp. b_k) is not available to participate in the transition τ_2 (resp. τ_1) during δ' (δ).

Definition 6 (Fireable Timed Local Transition). *Let $\mathcal{AN} = (\Sigma, \mathcal{S}, \mathcal{T})$ be a T-AN, $\zeta \in \mathcal{S}$ the state of $\mathcal{AN}$ at $t \in \mathbb{N}$. Let P be a set of pairs $\mathcal{T} \times \mathbb{N}$ and $B(\mathcal{AN}, P, t)$ be the set of blocked timed local transitions of $\mathcal{AN}$ by P at t. The set of fireable local transitions of $\mathcal{AN}$ in ζ w.r.t. P at t is defined as follows:*

$$F(\mathcal{AN}, \zeta, P, t) := \{a_i \xrightarrow{\ell} a_j \in \mathcal{T} \setminus B(\mathcal{AN}, P, t)) \mid \ell \subseteq \zeta, a_i \in \zeta\}$$

Definition 6 extends the notion of playable transition by considering concurencies with the currently ongoing transition of P.

Definition 7 (Set of Fireable Sets of Timed Local Transition). *Let $\mathcal{AN} = (\Sigma, \mathcal{S}, \mathcal{T})$ be a T-AN, $\zeta \in \mathcal{S}$ the state of $\mathcal{AN}$ at $t \in \mathbb{N}$. Let P be a set of pairs $\mathcal{T} \times \mathbb{N}$ and $F(\mathcal{AN}, \zeta, P, t)$ the set of fireable local transitions of $\mathcal{AN}$ in ζ w.r.t. P at t. The set of firable sets of timed local transition of $\mathcal{AN}$ in ζ w.r.t. P at t is defined as:*

$$SFS(\mathcal{AN}, \zeta, P, t) := \{FS \subseteq F(\mathcal{AN}, \zeta, P, t) \mid$$

$$(\forall \tau = (b_k \xrightarrow{\ell}_{\delta} b_l) \in FS, \nexists (b_k \xrightarrow{\ell'}_{\delta'} b_{l'}) \in FS, b_l \neq b_{l'}, \tau \notin B(\mathcal{AN}, FS \setminus \{\tau\}, t)\}$$

Definition 7 prevents the execution of two transitions that would affect the same automaton.

Definition 8 (Active Timed Local Transitions). *Let $\mathcal{AN} = (\Sigma, \mathcal{S}, \mathcal{T})$ be a T-AN, $\zeta \in \mathcal{S}$ the state of $\mathcal{AN}$ at $t \in \mathbb{N}$. Let $SFS(\mathcal{AN}, \zeta, P, t)$ be the set of firable sets of timed local transition. The set of active timed local transitions of $\mathcal{AN}$ at t is:*

$$A(\mathcal{AN}, t) := \begin{cases} \{(\tau \in FS, t) \mid FS \in SFS(\mathcal{AN}, \zeta, \emptyset, t)\} & if\ t = 0 \\ \{(\tau \in FS, t) \mid FS \in SFS(\mathcal{AN}, \zeta, A(\mathcal{AN}, t-1), t)\} \\ \cup \{(b_k \xrightarrow{\ell'}_{\delta'} b_l, t') \in A(\mathcal{AN}, t-1) \mid t - t' < \delta\} & if\ t > 0 \end{cases}$$

Definition 8 provides us the evolution of the possible set of ongoing actions. Supposing that in the initial state of a trajectory (at $t = 0$) no transition is blocked and all playable timed transitions are fireable. Then, when $t > 0$, at each time step it should be verified that a playable timed transition is also fireable, in other words that it is not blocked by the active timed local transitions fired in previous steps. Furthermore the timed local transitions fired at the same time should not block each other.

Delays of local transitions can now be represented in an Automata Network thanks to *timed local transitions*. Note that if all delays of local transitions are set to 0 it is equivalent to an AN without delays (original AN). The way these new local transitions should be used is described as follows.

At any time, each automaton has one and only one *local state*, forming the *global state* of the network. Choosing arbitrary ordering between automata identifiers, the set of *global states* of the network is referred to as S with $S = \prod_{a \in \Sigma} S(a)$. Given a global state $\zeta \in S$, $\zeta(a)$ is the local state of automaton a in ζ, i.e., the a-th coordinate of ζ. We write also $a_i \in \zeta \Leftrightarrow \zeta(a) = a_i$; and for any $ls \in \mathbf{LS}$, $ls \subset \zeta \Leftrightarrow \forall a_i \in \zeta, \zeta(a) = a_i$. In this paper, we allow, but do not force, applying in parallel transitions in different automata such in Definition 3 but adding delays in the local transitions and considering concurrency between transitions require further study of the semantics of the model (Definition 9).

Definition 9 (Semantics of Timed Automata Network). *Let* $\mathcal{AN} = (\Sigma, S, T)$ *be a T-AN and* $t \in \mathbb{N}$. *The set of timed local transition fired at* t *is:*
$$FS := \{(a_i \xrightarrow{\ell}_{\delta} a_j) \mid ((a_i \xrightarrow{\ell}_{\delta} a_j), t) \in A(\mathcal{AN}, t)\} \text{ then}$$

$$(a_i \xrightarrow{\ell}_{\delta} a_j) \in FS \implies \zeta(a) = a_j \, with \, \zeta = state(\mathcal{AN}, t + \delta).$$

The state of $\mathcal{AN}$ *at* $t + 1$ *is denoted* $\zeta_{t+1} = state(\mathcal{AN}, t + 1)$ *and defined according to the set of timed local transitions that finished at* $t + 1$:

$$F_{t+1} := \{(b_k \xrightarrow{\ell'}_{\delta'} b_l) \mid ((b_k \xrightarrow{\ell'}_{\delta'} b_l), t') \in A(\mathcal{AN}, t), t + 1 - t' = \delta\}$$

then $\forall c \in \Sigma$, *such that* $\nexists (c_k \xrightarrow{\ell''}_{\delta''} c_l) \in F_{t+1} \implies \zeta_{t+1}(c) = \zeta_t(c) \, with \, \zeta_t = state(\mathcal{AN}, t) \, and \, \zeta_{t+1} = state(\mathcal{AN}, t + 1).$

We note that at any time step t such that $\zeta = state(\mathcal{AN}, t)$ and P the set of ongoing transitions, we have: $FS \in F(\mathcal{AN}, \zeta, P, t) \in T \setminus B(\mathcal{AN}, P, t)$.

Where synchronous biological regulatory networks have been studied, little has been done on the asynchronous counterpart [31], although there is evidence that most living systems are governed by synchronous and asynchronous updating. According to Harvey and Bossomaier [13], asynchronous systems are biologically more plausible for many phenomena than their synchronous counterpart and observed global synchronous behavior in nature usually simply arises from the local asynchronous behavior. In this paper, we defend these assumptions and we consider an asynchronous behavior for each automata in one hand and a synchronous behavior in the global network.

The assumptions in the synchronous model that all components could change at the same time and take an equivalent amount of time in changing their expression levels, is biologically unrealistic. But there is seldom enough informations to be able to discern the precise order and duration of state transitions. The timed extension of Automata Network we propose in this paper allows both asynchronous and synchronous behavior by proposing a non-deterministic application of the timed local transitions. Figure 2 shows a trajectory of a Timed Automata Network when we choose to apply timed local transition in a synchronous maner.

We presented above the semantics of the T-AN that we are based on to modeling BRNs from experimental data. Even if it already exists a few hybrid formalisms like time Petri Nets, hybrid automata, etc., we propose this extension of the AN framework for several reasons. First, AN is a general framework that, although it was mainly used for biological networks [9,23], allows to represent any kind of dynamical models, and converters to several other representations are available. Indeed, a T-AN is a subclass of time Petri nets [10]. Finally, the particular form of the timed local transition in a AN model allows to easily represent them in ANSWER SET PROGRAMMING (ASP), with one fact per timed local transition, as described in this work [1]. Later we propose a new approach to resolve the generation problem of T-AN models from time series data.

Taking the following timed automata network as an example, we generate a possible trajectory of the network starting from a known initial state.

Example 2. Let $\mathcal{AN} = (\Sigma, \mathcal{S}, \mathcal{T}))$ be a timed automata extended with delays from Example 1. Such that $\mathcal{T} = \{\tau_1 = b_0 \xrightarrow[2]{\{a_1\}} b_1, \tau_2 = a_1 \xrightarrow[3]{\{b_1,d_2\}} a_0, \tau_3 = c_2 \xrightarrow[5]{\{a_1\}} c_1, \tau_4 = d_2 \xrightarrow[2]{\{a_0\}} d_1, \tau_5 = b_1 \xrightarrow[2]{\{a_1,c_2\}} b_0, \}$.

T	0	1	2	3	4	5	6	7
$B(\mathcal{AN}, P, t)$	∅	$\{\tau_1, \tau_2, \tau_3\}$	$\{\tau_2, \tau_3\}$	$\{\tau_2, \tau_3\}$	$\{\tau_2, \tau_3, \tau_5\}$	∅	$\{\tau_4\}$	∅
$F(\mathcal{AN}, \varsigma, P, t)$	$\{\tau_1, \tau_3\}$	∅	$\{\tau_2, \tau_5\}$	∅	∅	$\{\tau_4\}$	∅	∅
$SFS(\mathcal{AN}, \varsigma, P, t)$	$\{\emptyset, \{\tau_1\}, \{\tau_3\}, \{\tau_1, \tau_3\}\}$	$\{\emptyset\}$	$\{\emptyset, \{\tau_2\}, \{\tau_5\}\}$	$\{\emptyset\}$	$\{\emptyset\}$	$\{\emptyset, \{\tau_4\}\}$	$\{\emptyset\}$	$\{\emptyset\}$
FS	$\{\tau_1, \tau_3\}$	∅	$\{\tau_2\}$	∅	∅	$\{\tau_4\}$	∅	∅
$A(\mathcal{AN}, t)$	$\{(\tau_1, 0), (\tau_3, 0)\}$	$\{(\tau_1, 0), (\tau_3, 0)\}$	$\{(\tau_3, 0), (\tau_2, 2)\}$	$\{(\tau_3, 0), (\tau_2, 2)\}$	$\{(\tau_3, 0), (\tau_2, 2), (\tau_5, 3)\}$	$\{(\tau_4, \ )\}$	$\{(\tau_4, 5)\}$	∅
state($\mathcal{AN}, T$)	$< a_1, b_0, c_2, d_2 >$	$< a_1, b_0, c_2, d_2 >$	$< a_1, b_1, c_2, d_2 >$	$< a_1, b_1, c_2, d_2 >$	$< a_1, b_1, c_2, d_2 >$	$< a_0, b_0, c_1, d_2 >$	$< a_0, b_0, c_1, d_2 >$	$< a_0, b_0, c_1, d_1 >$

Fig. 2. Example of a trajectory of the timed automata network of Example 2 starting from an initial state $<a_1, b_0, c_2, d_2>$ (at $t = 0$) to a stable state $< a_0, b_1, c_1, d_1 >$ (at $t = 10$). With $P = A(AN, t - 1)$ as in Definition 8.

4   Learning Timed Automata Networks

This algorithm takes as input a model expressed as a Timed Automata Network, which the set of local transitions is empty, and time series data capturing the dynamics of the studied system. Given the influences between the components (or assuming all possible influences if no background knowledge is available), this algorithm generates the *timed local transitions* that could result in the the same changes of the model than the ones observed through the observation data.

4.1   Algorithm

In this section we propose an algorithm to build Timed Automata Networks from time series data. We assume that the latter data observations are provided as a *chronogram* of size T: the value of each variable is given for each time point t, $0 \leq t \leq T$, through a time interval discretization (see Definition 10 below).

Definition 10 (Chronogram). *A chronogram is a discretization of the time series data for each component of a biological regulatory network. It is presented by the following function Γ,*

$$\Gamma : [0, T] \subset \mathbb{N}^+ \longrightarrow \{0, ..., n\}$$

$$t \longmapsto i$$

with T is the maximum time point regarding the time series data called the size of the chronogram and n is the maximum level of discretization.

We note Γ_a a chronogram of the time series data for a component a.

Algorithm 1, MoT-AN (Modeling Timed Automata networks) shows the pseudo code of our implemented algorithm. It will generate all possible *timed local transitions* that can realize each observed change. Because of the delays and the non-determinism of the semantics, it is not possible to decide whether a timed local transition is absolutely correct or not. But we can output the minimal sets of time local transitions necessary to realize all the changes.

Theorem 1 (Completeness). *Let $\mathcal{AN} = (\Sigma, \mathcal{S}, \mathcal{T})$ be a Timed Automata Network, Γ be a chronogram of the components of $\mathcal{AN}$, $i \in \mathbb{N}$ and $R \in \mathcal{T}$ be the set of timed local transitions that realized the chronogram Γ such that $(a_i, l, a_j, \delta) \in R \implies |l| \leq i$. Let χ be the regulation influences of all $a \in \Sigma$. Let $\mathcal{AN}' = (\Sigma, \mathcal{S}, \emptyset)$ be a Timed Automata Network. Given $\mathcal{AN}'$, Γ, χ and i as input, Algorithm 1 is complete: it will output a set of Timed Automata Networks ϕ, such that $\exists \mathcal{AN}'' = (\Sigma, \mathcal{S}, \varphi') \in \phi$ with $R \subseteq \varphi'$.*
Proof is given in appendix.

Theorem 2 (Complexity). *Let $\mathcal{AN} = (\Sigma, \mathcal{S}, \mathcal{T})$ be a Timed Automata Network, $|\Sigma|$ be the number of automata of $\mathcal{AN}$ and η be the total number of local states of an automaton of $\mathcal{AN}$. Let Γ be a chronogram of the components of $\mathcal{AN}$ over τ units of time, such that c is the number of changes of Γ. The memory use of Algorithm 1 belongs to $O(\tau \cdot i^{|\Sigma|+1} \cdot 2^{\tau \cdot i^{|\Sigma|+1}})$ that is bounded by $O(\tau \cdot |\Sigma|^{T \cdot |\Sigma|^{|\Sigma|+1}})$. The complexity of learning $\mathcal{AN}$ by generating timed local transitions from the observations of Γ with Algorithm 1 belongs to $O(c \cdot i^{|\Sigma|+1} + 2^{2 \cdot \tau \cdot i^{|\Sigma|+1}} + c \cdot 2^{\tau \cdot i^{|\Sigma|+1}})$, that is bounded by $O(\tau \cdot 2^{3 \cdot \tau \cdot |\Sigma|^{|\Sigma|+1}})$.*
Proof is given in appendix.

4.2 Case Study

In this section we show how this method generates a T-AN model consistent with the set of biological regulatory time series data. First, the method uses discretized observations as an input (i.e. *chronogram*), thus it is necessary to treat first the time series data with another method in order to discretize it.

Our method may be summarized as follows:

- Detect biological components changes;
- Compute the *candidate timed local transitions* responsible for the network changes;
- Generate minimal subset of *candidate timed local transitions* that can realize all changes.

We apply the Algorithm 1 on learning the timed local transition $\tau \in \mathcal{T}$ of a simple example of a network $\mathcal{AN} = (\Sigma, \mathcal{S}, \mathcal{T})$ with 3 components ($|\Sigma| = 3$) whose chronogram is detailed in Fig. 3:

The first change occurs at $t_{min} = t_1 = 2$, denoted by change(2). It is the gene z whose value changes from 0 to 1, thus the timed local transition that has realized this change has this form $z_0 \xrightarrow[\delta]{\ell} z_1$, where ℓ can be any combination of the values of the regulators at $t_1 - 1$ of z.

Let $\chi = \{b \to z, a \to z, a \to a\}$ be the set of regulation influences among the components of the network. According to χ the set of genes having influence on z is $\chi_z = \{a, b\}$. It means that $\ell = \{a_?, b_?\}$ or $\ell = \{a_?\}$ or $\ell = \{b_?\}$. The expression

---

Algorithm 1. MoT-AN: Modeling Timed Automata Networks

INPUT:
- Timed Automata Network $\mathcal{AN} = (\Sigma, \mathcal{S}, \mathcal{T})$ with $\mathcal{T} = \emptyset$;
- a chronogram $\Gamma = \bigcup_{a \in \Sigma} \Gamma_a$;
- the regulation influences $\chi = \bigcup_{a \in \Sigma} \chi_a$ and
- a maximal in-degree $i \in \mathbf{N}^*$

OUTPUT: ϕ a set of Timed Automata Networks that realize the time series data.

- Let $\varphi := \emptyset$
- Step 1: According to the chronogram Γ, for each time step where a component a changes its value from a_i to a_j, with $a_i, a_j \in \mathcal{S}(a)$:
  - Let $\delta(b) < t$ be the last time step where b has changed with $b \in \chi_a$
  - For each $l' \in \wp(\chi_a), |l'| \leq i$ generates all timed local transitions:

$$\tau := (a_k, l, a_l, t - \delta)$$

  such that $\delta = \delta(b'), b' \in l', \nexists b'' \in l', \delta(b'') > \delta(b')$ and $l = \{b_i \in \zeta(\delta) \mid b \in l'\}$
  - Add all timed local transition τ in φ
- Step 2: Generate ϕ the set of all Timed Automata Networks $\mathcal{AN}' = (\Sigma, \mathcal{S}, \varphi')$ with $\varphi' \subseteq \varphi$ a set of timed local transitions that can realize Γ such that φ' is minimal:

$$\forall \mathcal{AN}' = (\Sigma, \mathcal{S}, \varphi') \in \phi, \nexists \varphi'' \subseteq \varphi, \varphi'' \subset \varphi', \text{such that } \varphi' \text{can realize } \Gamma$$

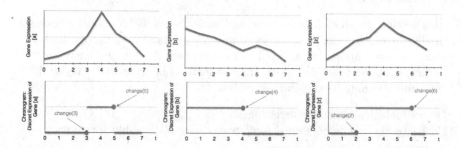

Fig. 3. Examples of the discretization of continous time series data into bi-valued *chronograms*. Abscissa (resp. ordinate) represents time (resp. gene expression levels). In this example, the expression level is discretized according to a threshold fixed to the half of the maximum gene expression value. `change(t)` indicates that the expression level of a biological component, here a gene, changes its value at a time point `t`.

level of the genes of χ_z when the researched candidate timed local transition (τ_i) is ongoing, i.e. during the partial steady state between two successive changes (t_i and t_{i-1}). This level is computed from the chronograms as follows:

   - $a \in \chi_z$: $[a]_t = 0 \ \forall t \in [0, 2]$ - $b \in \chi_z$: $[b]_t = 1 \ \forall t \in [0, 2]$

Thus $\ell=\{a_0, b_1\}$ or $\ell=\{a_0\}$ or $\ell=\{b_1\}$ and the set of *candidate timed local transitions* is: $\mathcal{T}_{change(2)} = \{\tau_1 = z_0 \xrightarrow[\delta_1]{a_0} z_1, \tau_2 = z_0 \xrightarrow[\delta_2]{b_1} z_1, \tau_3 = z_0 \xrightarrow[\delta_3]{a_0 \wedge b_1} z_1\}$. Since it is the first change, the delay of each timed local transition is the same: $\delta_1 = \delta_3 = \delta_3 = 2$.

The second change occurs at $t_2 = 3$ and denoted by $change(3)$. Here it is the gene a whose state changes from a_0 to a_1, thus the timed local transition that realize this change has this form $\tau = a_0 \xrightarrow[\delta]{\ell} a_1$ where ℓ can be any combination of the regulators value at t_1 of z. According to χ the genes influencing a are $\chi_a = \{a\}$. It means that $\ell = \{a_?\}$ and the expression level of a between t_1 and t_2 is a_0. So $\ell = \{a_0\}$. Thus there is only one *candidate timed local transition*: $\mathcal{T}_{change(3)} = \{\tau = a_0 \xrightarrow[1]{\emptyset} a_1\}$.

The third change occurs at $t_3 = 4$, $change(4)$. Here it is the gene b whose value changes from b_1 to b_0, thus the timed local transition that realize this change is of this form, $\tau = b_1 \xrightarrow[\delta]{\ell} b_0$ where ℓ can be any combination of the regulators value at $t_3 - 1$ of b. According to χ there is no gene that can influence b, thus no timed local transition can realize this change.

The fourth change occurs at $t_4 = 5$, $change(5)$. Here it is a whose expression decreases and changes from a_1 to a_0, thus the candidate timed local transition that could realize this change has this form, $\tau = a_1 \xrightarrow[\delta]{\ell} a_0$ where ℓ can be any combination of the regulators value at $t_4 - 1$ of a. According to χ the set of genes having influences on a is $\chi_a = \{a\}$. Again $\ell = \{a_?\}$ and since the expression level of a since its last change is a_1. we have $A = \{a_1\}$ and there is only one candidate timed local transition: $\mathcal{T}_{change(5)} = \{\tau = a_1 \xrightarrow[1]{\emptyset} a_0\}$.

The fifth change occurs at $t_5 = 6$, $change(6)$. Here it is z whose value changes from z_1 to z_0, thus the time local transition that has realized this change has the form of: $\tau = z_1 \xrightarrow{\ell}{\delta} z_0$ where ℓ can be any combination of the regulators value at $t_3 - 1$ of b. Since $\chi_z = \{a, b\}$, it means that $\ell = \{a_?, b_?\}$ or $\ell = \{a_?\}$ or $\ell = \{b_?\}$ The expression level of a and b at $t_5 - 1$ is respectively a_0 and b_0. Thus $\ell = \{a_0, b_1\}$ or $\ell = \{a_0\}$ or $\ell = \{b_0\}$. The candidate timed local states are:
$$\mathcal{T}_{change(6)} = \{\tau_1 = z_1 \xrightarrow[\delta_1]{a_0} z_0, \tau_2 = z_1 \xrightarrow[\delta_2]{b_0} z_0, \tau_3 = z_1 \xrightarrow[\delta_3]{a_0 \wedge b_0} z_0\}.$$

The last change of a is at $t_4 = 5$ and the last change of b is at $t_3 = 4$. Thus $\delta_1 = t_5 - t_4 = 1$, $\delta_2 = t_5 - t_3 = 2$, $\delta_3 = t_5 - max(t_4, t_3) = 1$.

After processing all changes, the set of *timed local transitions* that could realize the chronograms are:
$$\mathcal{T}_{change(2)} = \{\tau_1 = z_0 \xrightarrow{a_0}{2} z_1, \tau_2 = z_0 \xrightarrow{b_1}{2} z_1, \tau_3 = z_0 \xrightarrow{a_0 \wedge b_1}{2} z_1\}$$
$$\mathcal{T}_{change(3)} = \{\tau_4 = a_0 \xrightarrow{\emptyset}{1} a_1\}, \mathcal{T}_{change(5)} = \{\tau_5 = a_1 \xrightarrow{\emptyset}{1} a_0\}$$
$$\mathcal{T}_{change(6)} = \{\tau_6 = z_1 \xrightarrow{a_0}{1} z_0, \tau_7 = z_1 \xrightarrow{b_0}{2} z_0, \tau_8 = z_1 \xrightarrow{a_0 \wedge b_0}{1} z_0\}.$$

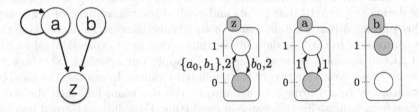

Fig. 4. Left: influence graph modeling of the case study example (Fig. 3). Right, one of the Timed Automata Networks generated by the Algorithm 1. The labels of each local transition stands for the local states of the automata which make the transition playable and its delay (time needed for the transition to be performed).

All *timed local transitions* learned are consistent with all observed time series data and the regulation influences given as input. The used method ensures completeness, we have the full set of timed local transitions that can explain the observations. By generating all minimal subsets of this set of timed local transitions, one of those subset will be the set who realized the observations (Fig. 4).

5   Evaluation

In this section, we provide two evaluations of Algorithm 1. We evaluate the capacity of our algorithm[1] to construct models for prediction and the impact of the quantity of observations on run time. Here we process chronograms obtained from time series data of the DREAM4 challenge [25].

5.1   DREAM4

In this section, we assess the efficiency of our algorithm through case studies coming from the DREAM4 challenge. DREAM challenges are annual reverse engineering challenges that provide biological case studies. In this section, we focus on the datasets coming from DREAM4. It provides data for systems of different size (10 genes on one hand, 100 genes on the other hand), allowing us to assess the scalability of our approach. The input data that we tackle here consists of the following: 5 different systems each composed of 100 genes, all coming from E. coli and yeast networks. For every such system, the available data are the following: (i) 10 time series data with 21 time points and 1000 is the duration of each time series; (ii) steady state for wild type; (iii) steady states after knocking out each gene; (iv) steady states after knocking down each gene (i.e. forcing its transcription rate at 50 %); (v) steady states after some random multifactorial perturbations. We processed all the data. Here, we focus on the management of time series data.

Each time series includes different perturbations that are maintained all time along during the first 10 time points and applied to at most 30 % of the genes. In this setting, a perturbation means a significant increase or decrease of the gene expression. In the raw data of the time series, gene expression values are given as real numbers between 0 and 1. To apply our approach, we chose to discretize those data into two to six qualitative values. Increasing the number of qualitatives values from 2 to 4 improves the precision, but then the score decrease from 5, must likely because of overfitting: the relations learned become too precised and can't be applied on something else than the training data. The best score we obtain were with 4 qualitatives values and are reported in Fig. 5. Each gene is discretized in an independent manner, with respect to the following procedure: we compute the average value of the gene expression among all data of a time series, then the values between the average and the maximal/minimal value are divided into as many levels. Discretizing the data according to the average value of expression is expected to reduce the impact of perturbation on the discretization and thus on the model learned.

The DREAM4 challenge offers two different problems, which consist in predicting (i) the structure of the gene interactions (in terms of an unsigned directed graph); (ii) attractors in some given conditions. Our method is not designed to tackle the first issue, indeed we need to know those influences. But the models

---

[1] All programs, described in this article, for Timed Automata Network generation are implemented in ASP and are available online at: http://www.irccyn.ec-nantes.fr/~benabdal/modeling-biological-regulatory-networks.zip.

we learn can be applied to predict trajectories and thus attractors. Here we use the influences graphs expected in the first problem as background knowledge (given in appendix) to tackle the attractor prediction part of the challenge.

5.2 Results

For this evaluation, we are given an initial state and 5 different dual gene knock-outs conditions. The goal is to predict the attractor in which the system will fall from the initial state for each dual knockout. Here, we just choose the first model that our algorithm output and use the biggest set of fireable timed local transition at each time step to produce a trajectory until a cycle is detected. The first state of this cycle is reverse discretized and proposed as the predicted state. In the challenge, the quality of the prediction is evaluated by computing the mean square error (MSE) between the predicted state and the expected one. As shown in Fig. 5, the precision we achieved in those experiments is quite good considering the results of the competitors of the DREAM4 challenge [1]. Their results range between 0.010 and 0.075 for the same evaluation settings, which we are comparable to (0.033 to 0.086) giving us encouraging results. Regarding run time, learning and predicting the trajectories of the benchmarks of 10 genes took less than 30 s and the same experiements for the benchmarks of 100 genes took about 3 h and 20 min on one processor Intel Core2 Duo (P8400, 2.26 GHz).

Benchmark	Number of genes	MSE		Benchmark	Number of genes	MSE
insilico_size10_1	10	0.086		insilico_size100_1	100	0.052
insilico_size10_2	10	0.080		insilico_size100_2	100	0.042
insilico_size10_3	10	0.076		insilico_size100_3	100	0.033
insilico_size10_4	10	0.039		insilico_size100_4	100	0.033
insilico_size10_5	10	0.076		insilico_size100_5	100	0.052

Fig. 5. Evaluation of our method on learning and prediction of the evolution of gene regulatory network benchmarks from the DREAM4 challenge through the Mean Square Error (MSE): 10 variables benchmarks (left) and 100 variables benchmarks (right).

To achieve this score, we had to perform several tests by varying the discretization precision and the complexity of the dynamics learned. Those tests also allows us to assess the scalability of our approach in practice. Figure 6 shows the impact of both timed local transition indegree and discretization level on run time.

In the results obtained from the experimentation of our algorithm on the time series data of the DREAM4 we can see the exponential influence on the run time of the indegree per local transition considered as well as the level of discretization chosen for all the 5 different networks. But it also shows that in practice our approach can tackle big network, here 100 genes.

5.3 Discussion

We propose a new method MoT-AN (Algorithm 1) to automatically infer models that could explain the dynamic evolution of the biological system. We illustrated the merits of this method by applying it on a large real biological system

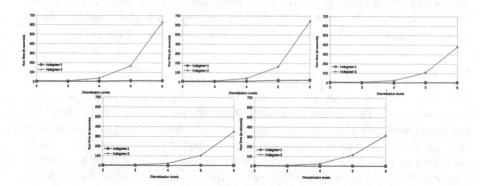

Fig. 6. Evolution of run time on processing different models inferred from time series data of DREAM4 (100 variables benchmarks), varying indegree of local timed transitions and discretization levels. These tests were performed on a processor Intel Core i7 (4700, 3 GHz) with 16 GB of RAM.

(DREAM4 challenge). As a result we obtain in few seconds models that are proved to be relevant (this relevance is qualified in terms of mean square error with gold standard network) This algorithm is implemented using Answer Set Programming [4,5], thus provides the exhaustive enumeration of all models.

The main limit of the approach presented in this paper is the fact that topology of the network is considered as granted. As discussed in the introduction of the paper, there is a wide range of algorithms designed to address this issue. Furthermore, such interaction graphs could be deduced from the available reliable databases of biological networks. Some examples of data bases for human regulatory knowledge are: Pathways Interaction Database [20], Human Integrated Pathway DB [32] and Causal Biological Network Database [30].

Various inference approaches [11,20,26] from time series data based on prior knowledge about component interactions have been proposed. But they share a common limit: they focus on static characterization of the interactions and they do not allow to infer dynamic behaviors where delays are involved. The merits of our contribution lie in the fact that we overcome such limits, and we infer delays in a qualitative dynamic modeling of the network.

6   Conclusion and Perspectives

In this paper, we propose an approach takes a background knowledge under the form of regulation graph and time series data as an input. The contribution of our method lies in the fact that it identifies the set of interactions between biological components by (1) concertizing the signs (negative or positive) (2) providing thresholds and associating the quantitative time delays. As a result, we have a set of Timed Automata networks that explain the biological network evolution. The Algorithm 1 is implemented in ASP. We illustrated the applicability and limits of the proposed method through benchmarks from DREAM4. This opens the way to promising applications in the cooperation between biologists and computer scientists. Further works now consist in discussing the kind of information one can get on Timed Automata Network by analyzing the associated untimed model. We also plan to improve our implementation to make it robust against noisy and scarse data as like in the DREAM8: Heritage-DREAM breast cancer network inference challenge.

A   Appendixes

A.1   Proof of Theorem 1

Theorem 3 (Completeness). *Let $\mathcal{AN} = (\Sigma, \mathcal{S}, \mathcal{T})$ be a Timed Automata Network, Γ be a chronogram of the components of $\mathcal{AN}$, $i \in \mathbb{N}$ and $R \in \mathcal{T}$ be the set of timed local transitions that realized the chronogram Γ such that $(a_i, l, a_j, \delta) \in R \implies |l| \leq i$. Let χ be the regulation influences of all $a \in \Sigma$. Let $\mathcal{AN}' = (\Sigma, \mathcal{S}, \emptyset)$ be a Timed Automata Network. Given $\mathcal{AN}'$, Γ, χ and i as input, Algorithm 1 is complete: it will output a set of Timed Automata Network ϕ, such that $\exists \mathcal{AN}'' = (\Sigma, \mathcal{S}, \varphi') \in \phi$ with $R \subseteq \varphi'$.*

Proof. Let us suppose that the algorithm is not complete, then there is a timed local transition $h \in R$ that realized Γ and $h \notin \varphi'$. In Algorithm 1, after step 1, φ contains all timed local transitions that can realize each change of the chronogram Γ. Here there is no timed local transition $h \in R$ that realizes Γ which is not generated by the algorithm, so $h \in \varphi$. Then it implies that at step 2, $\forall \varphi', h \notin \varphi'$. But since h realizes one of the change of Γ and h is generated at step 1, then it will be present in one of the minimal subset of timed local transitions. Such that h will be in one of the networks outputted by the algorithm. □

A.2   Proof of Theorem 2

Theorem 4 (Complexity). *Let $\mathcal{AN} = (\Sigma, \mathcal{S}, \mathcal{T})$ be a Timed Automata Network, $|\Sigma|$ be the number of automaton of $\mathcal{AN}$ and η be the total number of local state of a automaton of $\mathcal{AN}$. Let Γ be a chronogram of the components of $\mathcal{AN}$ over τ units of time, such that c is the number of changes of Γ. The memory use of Algorithm 1 belongs to $O(\tau \cdot i^{|\Sigma|+1} \cdot 2^{\tau \cdot i^{|\Sigma|+1}})$ that is bounded by $O(\tau \cdot |\Sigma|^{T \cdot |\Sigma|^{|\Sigma|+1}})$. The complexity of learning $\mathcal{AN}$ by generating timed local transitions from the observations of Γ with Algorithm 1 belongs to $O(c \cdot i^{|\Sigma|+1} + 2^{2 \cdot \tau \cdot i^{|\Sigma|+1}} + c \cdot 2^{\tau \cdot i^{|\Sigma|+1}})$, that is bounded by $O(\tau \cdot 2^{3 \cdot \tau \cdot |\Sigma|^{|\Sigma|+1}})$.*

Proof. Let i be the maximal indegree of a timed local transition in $\mathcal{AN}$, $0 \leq i \leq |\Sigma|$. Let p be an automaton local state of $\mathcal{AN}$ then $|\Sigma|$ is maximal the number of automaton that can influence p. There is $i^{|\Sigma|}$ possible combinations of those regulators that can influences p at the same time forming a timed local transition. There is at most τ possible delays, so that there are $\tau \cdot |\Sigma| \cdot i^{|\Sigma|}$ possibles *timed local transitions, thus in Algorithm 1 at step 1, the memory is bounded by* $O(\tau \cdot i^{|\Sigma|+1})$, *which belongs to* $O(\tau \cdot |\Sigma|^{|\Sigma|+1})$ *since* $0 \leq i \leq |\Sigma|$. *Generating all minimal subsets of timed local transitions* φ *of* $\mathcal{AN}$ *that can realize* Γ *can require to generate at most* $2^{\tau \cdot |\Sigma| \cdot i^{|\Sigma|+1}}$ *set of rules. Thus, the memory of our algorithm belongs to* $O(\tau \cdot i^{|\Sigma|+1} \cdot 2^{\tau \cdot i^{|\Sigma|+1}})$ *and is bounded by* $O(\tau \cdot |\Sigma|^{|\Sigma|+1} \cdot 2^{\tau \cdot |\Sigma|^{|\Sigma|+1}})$.

The complexity of this algorithm belongs to $O(c \cdot i^{|\Sigma|} + 1)$. *Since* $0 \leq i \leq |\Sigma|$ *and* $0 \leq c \leq \tau$ *the complexity of Algorithm 1 is bounded by* $O(\tau \cdot |\Sigma|^{|\Sigma|+1}))$.

Generating all minimal subsets of timed local transitions φ *of* $\mathcal{AN}'$ *that realize* Γ *can require to generate at most* $2^{\tau \cdot i^{|\Sigma|+1}}$ *set of timed local transitions. Each set has to be compared with the others to keep only the minimal ones, which costs* $O(2^{2 \cdot \tau \cdot i^{|\Sigma|+1}})$. *Furthermore, each set of timed local transitions has to realize each change of* Γ, *it requires to check* c *changes and it costs* $O(c \cdot 2^{\tau \cdot i^{|\Sigma|+1}})$. *Finally, the total complexity of learning* $\mathcal{AN}$ *by generating timed local transitions from the observations of* Γ *belongs to* $O(c \cdot i^{|\Sigma|+1} + 2^{2 \cdot \tau \cdot i^{|\Sigma|+1}} + c \cdot 2^{\tau \cdot i^{|\Sigma|+1}})$. *that is bounded by* $O(3\tau \cdot 2^{2 \cdot \tau \cdot |\Sigma|^{|\Sigma|+1}})$.

□

A.3   DREAM4: Influence Network

The Fig. 7 presents the regulatory graph that we are based on to identify the signs (negative or positive), the thresholds and the quantitative time delays of the learned transitions.

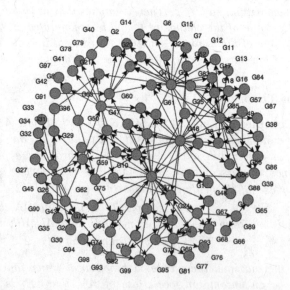

Fig. 7. The influence network of the DREAM4 challenge model (100 genes) given by GeneNetWeaver (GNW) data generator [27]. Each node is a gene and each edge is an influence from the source to the target gene.

References

1. Ben Abdallah, E., Folschette, M., Roux, O., Magnin, M.: Exhaustive analysis of dynamical properties of biological regulatory networks with answer set programming. In: IEEE International Conference on Bioinformatics and Biomedicine (BIBM), pp. 281–285. IEEE (2015)
2. Ahmad, J., Bernot, G., Comet, J.-P., Lime, D., Roux, O.: Hybrid modelling and dynamical analysis of gene regulatory networks with delays. ComPlexUs 3(4), 231–251 (2006)
3. Akutsu, T., Tamura, T., Horimoto, K.: Completing networks using observed data. In: Gavaldà, R., Lugosi, G., Zeugmann, T., Zilles, S. (eds.) ALT 2009. LNCS, vol. 5809, pp. 126–140. Springer, Heidelberg (2009)
4. Anwar, S., Baral, C., Inoue, K.: Encoding higher level extensions of petri nets in answer set programming. In: Cabalar, P., Son, T.C. (eds.) LPNMR 2013. LNCS, vol. 8148, pp. 116–121. Springer, Heidelberg (2013)
5. Baral, C.: Knowledge representation, reasoning and declarative problem solving. Cambridge University Press, New York (2003)
6. Callebaut, W.: Scientific perspectivism: a philosopher of sciences response to the challenge of big data biology. Stud. Hist. Philos. Sci. Part C. Stud. Hist. Philos. Biol. Biomed. Sci. 43(1), 69–80 (2012)
7. Comet, J.-P., Fromentin, J., Bernot, G., Roux, O.: A formal model for gene regulatory networks with time delays. In: Chan, J.H., Ong, Y.-S., Cho, S.-B. (eds.) CSBio 2010. CCIS, vol. 115, pp. 1–13. Springer, Heidelberg (2010)
8. Fan, J., Han, F., Liu, H.: Challenges of big data analysis. Nat. Sci. Rev. 1(2), 293–314 (2014)
9. Folschette, M., Paulevé, L., Inoue, K., Magnin, M., Roux, O.: Identification of biological regulatory networks from process hitting models. Theoret. Comput. Sci. 568, 49–71 (2015)
10. Freedman, P.: Time, petri nets, and robotics. IEEE Trans. Robot. Autom. 7(4), 417–433 (1991)
11. Gallet, E., Manceny, M., Le Gall, P., Ballarini, P.: An LTL model checking approach for biological parameter inference. In: Merz, S., Pang, J. (eds.) ICFEM 2014. LNCS, vol. 8829, pp. 155–170. Springer, Heidelberg (2014)
12. Goldstein, Y.A.B., Bockmayr, A.: A lattice-theoretic framework for metabolic pathway analysis. In: Gupta, A., Henzinger, T.A. (eds.) CMSB 2013. LNCS, vol. 8130, pp. 178–191. Springer, Heidelberg (2013)
13. Harvey, I., Bossomaier, T.: Time out of joint: attractors in asynchronous random boolean networks. In: Proceedings of the Fourth European Conference on Artificial Life, pp. 67–75. MIT Press, Cambridge (1997)
14. Kim, S.Y., Imoto, S., Miyano, S.: Inferring gene networks from time series microarray data using dynamic bayesian networks. Briefings Bioinf. 4(3), 228–235 (2003)
15. Koh, C., Fang-Xiang, W., Selvaraj, G., Kusalik, A.J.: Using a state-space model and location analysis to infer time-delayed regulatory networks. EURASIP J. Bioinf. Syst. Biol. 2009(1), 1 (2009)
16. Koksal, A.S., Yewen, P., Srivastava, S., Bodik, R., Fisher, J., Piterman, N.: Synthesis of biological models from mutation experiments. ACM SIGPLAN Not. 48, 469–482 (2013). ACM
17. Liu, T.-F., Sung, W.-K., Mittal, A.: Learning multi-time delay gene network using bayesian network framework. In: 16th IEEE International Conference on Tools with Artificial Intelligence, ICTAI 2004, pp. 640–645. IEEE (2004)

18. Marx, V.: Biology: the big challenges of big data. Nature **498**(7453), 255–260 (2013)

19. Matsuno, H., doi, A., Nagasaki, M., Miyano, S.: Hybrid petri net representation of gene regulatory network. In: Pacific Symposium on Biocomputing, vol. 5, p. 87. World Scientific Press, Singapore (2000)

20. Ostrowski, M., Paulevé, L., Schaub, T., Siegel, A., Guziolowski, C.: Boolean network identification from multiplex time series data. In: Roux, O., Bourdon, J. (eds.) CMSB 2015. LNCS, vol. 9308, pp. 170–181. Springer, Heidelberg (2015)

21. Paoletti, N., Yordanov, B., Hamadi, Y., Wintersteiger, C.M., Kugler, H.: Analyzing and synthesizing genomic logic functions. In: Biere, A., Bloem, R. (eds.) CAV 2014. LNCS, vol. 8559, pp. 343–357. Springer, Heidelberg (2014)

22. Paulevé, L.: Goal-oriented reduction of automata networks. In: CMSB 2016–14th Conference on Computational Methods for Systems Biology (2016)

23. Paulevé, L., Chancellor, C., Folschette, M., Magnin, M., Roux, O.: Logical Modeling of Biological Systems, chapter Analyzing Large Network Dynamics with Process Hitting, pp. 125–166. Wiley, Hoboken (2014)

24. Paulevé, L., Magnin, M., Roux, O.: Refining dynamics of gene regulatory networks in a stochastic π-calculus framework. In: Priami, C., Back, R.-J., Petre, I., de Vink, E. (eds.) Transactions on Computational Systems Biology XIII. LNCS, vol. 6575, pp. 171–191. Springer, Heidelberg (2011)

25. Prill, R.J., Saez-Rodriguez, J., Alexopoulos, L.G., Sorger, P.K., Stolovitzky, G.: Crowdsourcing network inference: the dream predictive signaling network challenge. Sci. Signal. **4**(189), mr7 (2011)

26. Saez-Rodriguez, J., Alexopoulos, L.G., Epperlein, J., Samaga, R., Lauffenburger, D.A., Klamt, S., Sorger, P.K.: Discrete logic modelling as a means to link protein signalling networks with functional analysis of mammalian signal transduction. Mol. Syst. Biol. **5**(1), 331 (2009)

27. Schaffter, T., Marbach, D., Floreano, D.: Genenetweaver: in silico benchmark generation and performance profiling of network inference methods. Bioinformatics **27**(16), 2263–2270 (2011)

28. Siebert, H., Bockmayr, A.: Temporal constraints in the logical analysis of regulatory networks. Theoret. Comput. Sci. **391**(3), 258–275 (2008)

29. Sima, C., Hua, J., Jung, S.: Inference of gene regulatory networks using time-series data: a survey. Curr. Genomics **10**(6), 416–429 (2009)

30. Talikka, M., Boue, S., Schlage, W.K.: Causal biological network database: a comprehensive platform of causal biological network models focused on the pulmonary and vascular systems. Comput. Syst. Toxicol. **2015**, 65–93 (2015)

31. Thomas, R.: Regulatory networks seen as asynchronous automata: a logical description. J. Theoret. Biol. **153**(1), 1–23 (1991)

32. Namhee, Y., Seo, J., Rho, K., Jang, Y., Park, J., Kim, W.K., Lee, S.: Hipathdb: a human-integrated pathway database with facile visualization. Nucleic Acids Res. **40**(D1), D797–D802 (2012)

33. Zhang, Z.-Y., Horimoto, K., Liu, Z.: Time series segmentation for gene regulatory process with time-window-extension (2008)

34. Zhao, W., Serpedin, E., Dougherty, E.R.: Inferring gene regulatory networks from time series data using the minimum description length principle. Bioinformatics **22**(17), 2129–2135 (2006)

Matching Models Across Abstraction Levels with Gaussian Processes

Giulio Caravagna[1](✉), Luca Bortolussi[2,3,4], and Guido Sanguinetti[1]

[1] School of Informatics, University of Edinburgh, Edinburgh, UK
giulio.caravagna@ed.ac.uk
[2] DMG, University of Trieste, Trieste, Italy
[3] ISTI-CNR, Pisa, Italy
[4] MOSI, Department of Informatics, Saarland University, Saarbücken, Germany

Abstract. Biological systems are often modelled at different levels of abstraction depending on the particular aims/resources of a study. Such different models often provide qualitatively concordant predictions over specific parametrisations, but it is generally unclear whether model predictions are quantitatively in agreement, and whether such agreement holds for different parametrisations. Here we present a generally applicable statistical machine learning methodology to automatically reconcile the predictions of different models across abstraction levels. Our approach is based on defining a correction map, a random function which modifies the output of a model in order to match the statistics of the output of a different model of the same system. We use two biological examples to give a proof-of-principle demonstration of the methodology, and discuss its advantages and potential further applications.

Keywords: Computational abstraction · Emulation · Gaussian Processes · Heteroschedasticity

1 Introduction

Computational modelling in the sciences is founded on the notion of abstraction, the process of identifying and representing mathematically the salient features and interactions of a real system. Abstraction is a human led and interdisciplinary activity: many factors influence the decision of which features/interactions are eventually represented in the abstracted model, including the specialist interests of the scientists formulating the model, as well as computational constraints on the level of detail which can feasibly be implemented on the available hardware. Such factors inevitably vary between different research groups and at different times, leading to a proliferation of different models representing the same underlying phenomena.

GC and GS gratefully acknowledge support from the European Research Council under grant MLCS306999. LB acknowledges partial support from the EU project QUANTICOL, 600708, and by FRA-UniTS. We thank Dimitris Milios for useful discussions and for providing us with the MATLAB for heteroscedastic regression.

E. Bartocci et al. (Eds.): CMSB 2016, LNBI 9859, pp. 49–66, 2016.
DOI: 10.1007/978-3-319-45177-0_4

Systems biology is a prominent field where models with different level of abstraction coexist. As an example, consider the process of gene expression, whereby genetic material stored in the DNA is transcribed into messenger RNA and eventually translated into protein. At the highest level of abstraction, which is frequently employed when studying high-throughput data, the process may be considered as a black box, and one may only be interested in characterising the statistical structures underlying observed levels of gene expression in different conditions [8]. Alternatively, one may want to mechanistically represent the dynamics of some of the steps involved in the process. The choice of which steps to include is again largely driven by extrinsic factors: examples in the literature range from highly detailed models where the synthesis of mRNA/protein is modelled through the binding and elongation of polymerase/ribosomes, to models where protein production is modelled as a single reaction with first order kinetics [1,10].

Representing the same physical process by multiple models at different levels of abstraction immediately engenders the question of how different model outputs can be reconciled. As the models all represent the same underlying physical system, it can be plausibly assumed that such models will agree at least qualitatively for suitable parametrisations. In general, however, models may not agree quantitatively, and their discrepancy may be a function of the parameters. Understanding and quantifying such discrepancies would often be very valuable: first of all, it can shed light on how simplifications within models affect predictions, and secondly it may open the opportunity to construct computationally smaller surrogates of complex models. Such surrogates can be precious when modelling requires intrinsically computationally intensive tasks like inference, as they have less parameters.

In this paper, we approach the problem of reconciling models from a statistical machine learning angle. We start by sampling a sparse subset of the parameter space over which we evaluate the models' outputs (generally by simulation). These evaluations are used as a *training set* to learn a *correction map* via a nonparametric regression approach based on Gaussian Processes. We show that our approach yields a consistent stochastic equivalence between models, which provably reconciles the predictions of the two models up to the second moment. We demonstrate the approach on two biological examples, showing that it can lead to non-trivial insights into the structure of the models, and provide an efficient way to simulate a complex model via a simpler model. Correction maps could be used to reduce the complexity of some common problems that we briefly discuss in the paper, e.g., model selection, synthesis and parameter estimation.

The rest of the paper is organised as follows: we start by giving a high level description of the problem and how we attack it. This is followed by a formal definition and a thorough illustration of the proposed solution, discussing its desirable theoretical properties. We then demonstrate the approach on two proof of principle examples, showing the potential for practical application of the method. We conclude the paper by discussing the relationship of our approach to existing ideas in model reduction and statistical emulation, as well as some possible extensions of our results.

2   Problem Definition

High level description of the approach. In this paper we consider the problem of performing analyses which require exhaustive sampling of a model's outputs, M, the dynamics of which are expensive to compute. We are not interested in the origin of such a complexity, and *we assume* to be given an *abstraction/surrogate* of this model, m, which is reasonably less challenging to analyze[1]. For this reason, *we want to investigate a general methodology to use* m *as a reliable proxy to get statistics over* M. Possible applications of this framework, which we discuss in Sect. 4, regard *model selection and synthesis, inference,* and *process equivalence/control.*

In general, we will assume both models to be stochastic processes, e.g. continuous time Markov Chains (CTMCs). Furthermore, we assume that the highly detailed model M and the reduced model m share some parameters θ_m and some observables/state variables X, but the large model will generally have many more state variables and parameters. In general we can compute some statistics of the shared state variables X (e.g. mean), and that such computation will be considerably more expensive using the detailed model M.

As both models are devised as abstractions of the same physical system, it is not unreasonable to assume that the expected values of the shared variables will be similar for some parametrisations of the models. However, it is in general not the case that the *distribution* of the shared variables implied by the two models will be the same, and, as we vary the shared parameters θ_m, even the expected values may show non-negligible discrepancies. Our aim is to develop a generally applicable machine learning approach to correct the output of the reduced model, in order to match the distribution of the larger model. This has strong practical motivations: one of the primary uses of models in general is to test hypotheses statistically against observed data, and it is therefore essential to capture as accurately as possible the implied variability on observable variables.

The strategy we will follow is simple: we start by sampling values of the shared parameters θ_m, and compute the first two statistics of the observed variables as implied by both the large and reduced models (by simulation). In general, one may expect the variance implied by the larger model to be larger, as a more detailed model will imply more stochastic steps. We can therefore correct the first two statistics (mean and variance) of the reduced model output by adding a random function of the shared parameters θ_m, which can be learned by rephrasing it as a regression task. We will work with heteroschedastic Gaussian Processes [5].

---

[1] M could be complex to analyze either because of its structure, e.g., it might have many variables, or its numerical hurdles, e.g., the degree of non-linearity or parameters stiffness. For similar reasons, we do not care whether m is has been derived by means of independent domain-knowledge or automatic techniques.

2.1   Formal Problem Statement

We assume to be given two models of the same underlying physical system:

– a highly detailed model M, with state variables $\mathcal{Y}$ and parameters $\boldsymbol{\theta}_M$.
– a reduced model m, with state variables $\mathcal{X}$ and parameters $\boldsymbol{\theta}_m$.

We will have $|\mathcal{Y}| \gg |\mathcal{X}|$ and $|\boldsymbol{\theta}_M| \gg |\boldsymbol{\theta}_m|$.

Assumptions. In general, the problem we want to tackle draws immediate connection to that of using m as a *statistical emulation* of M. However, to exploit solutions from *regression analysis* and *machine learning*, we make further assumptions and discuss their limitations thorough the paper.

1. (*Observables*) we assume that it exists a non-empty set of state variables (or, more generally, observables) X, common to both models, that is sufficient to compute our statistics of interest.
2. (*Parameters*) we assume that model M is fully parametrized by $\boldsymbol{\theta}_M$, a vector of real-valued parameters that breaks down as $\boldsymbol{\theta}_M = [\boldsymbol{\theta}_m \quad \boldsymbol{\theta}_f]^\top$, with $\boldsymbol{\theta}_m$ shared between models m and M. Here, we assume that m is fully parametrized[2] by $\boldsymbol{\theta}_m$, which ranges in a domain set Θ. We term $\boldsymbol{\theta}_f$ free parameters in M, given $\boldsymbol{\theta}_m$. We further assume to have a probability density $p(\boldsymbol{\theta}_f)$ for the free parameters, and a probability density $p(\boldsymbol{\theta}_m)$ for the shared parameters, encoding our knowledge about them.
3. (*Sampling*) we assume that, for every parametrization, each model's dynamics is computable, i.e. it can be simulated.

In this work, we will consider correction maps conditioned to a reference statistic of interest, in the following sense.

Definition 1 (Statistic). *A statistic η is* any observable that we can compute from one, or from an ensemble of simulations of a model. *We denote its* estimator *computed from model* q *with parameters* x *as* $q_{\hat{\eta}}(x)$, *and its* true value $q_\eta(x)$.

Valid examples of such statistics are, e.g., a single value or the expectation of a variable in X, the satisfiability of a temporal logical formula depending on variables X that could be model-checked, etc. The richer the estimator, the higher number of samples are required for the estimator to converge to the true statistics. We will make use of estimators that are *consistent* in the following sense: $q_{\hat{\eta}}(x) \rightarrow q_\eta(x)$ as the number of samples goes to infinity.

Definition 2 (Correction map). *We define an ϵ-correction map $\mathcal{M}_\eta : \Theta \rightarrow \mathbb{R}^w$ for a statistics η to be a function that for any $\boldsymbol{\theta}_m \in \Theta$, satisfies*

$$\hat{\mathsf{M}}_\eta(\boldsymbol{\theta}_m) \triangleq \mathsf{m}_\eta(\boldsymbol{\theta}_m) + \mathcal{M}_\eta(\boldsymbol{\theta}_m) \ and \ \int_\Theta \| \ \hat{\mathsf{M}}_\eta(\boldsymbol{\theta}_m) - \mathbb{E}_{\boldsymbol{\theta}_f}[\mathsf{M}_\eta(\boldsymbol{\theta}_m)] \ \|_2 \ p(\boldsymbol{\theta}_m)d\boldsymbol{\theta}_m < \epsilon \quad (1)$$

---

[2] In principle, even m might have a set of free variables, with respect to M. However, as we have full control over that model, we could assume a parametrization of such variables and all what follows would be equivalent.

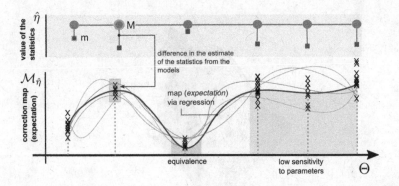

Fig. 1. Correction maps as regression problems. From sampled estimates of a statistics $\hat{\eta}$ computed over domain Θ by models M and m, we can define a correction $\hat{M}_{\hat{\eta}}(x) \triangleq m_{\hat{\eta}}(x) + \mathcal{M}_{\hat{\eta}}(x)$ from their difference. According to the variables/parameters involved, we collect multiple values of such correction; see also Fig. 2. Then, a regression over such values leads to a correction model for M's prediction, from m's ones; this differs from standard emulation as we retain the mechanistic description of the system in m. A map is modeled as a random function via Gaussian Processes (GPs) with heteroschedastic variance; GPs induce a distribution over functions (colored lines), and the map will be the expectation (red line). Maps allow to detect regions of Θ where the predictions of the models are in agreement, $\mathcal{M}_{\hat{\eta}} \to 0$, or roughly constant (i.e., at low sensitivity). (Color figure online)

where $\epsilon > 0$ *is the precision, and* $\mathbb{E}_{\boldsymbol{\theta}_f}[M_{\eta}(\boldsymbol{\theta}_m)] = \int M_{\boldsymbol{\eta}}(\boldsymbol{\theta}_m; \boldsymbol{\theta}_f) p(\boldsymbol{\theta}_f) d\boldsymbol{\theta}_f$ *is the* expectation of the statistics *computed from* M, *with respect to its free parameters* $\boldsymbol{\theta}_f$. $\hat{M}$ *is our* corrected prediction *of* M.

Thus, $\mathcal{M}_{\hat{\eta}}$ can correct the outcomes of η assessed over m, $m_{\eta}(\boldsymbol{\theta}_m)$, to match (with a given precision) those that we would have by computing the statistics over M. Notice that the corrected outcome has no more dependence from the free parameters, as this is a correction in expectation with respect to $\boldsymbol{\theta}_f$.

Notice that the correction is a w-dimensional vector, and hence $\| \cdot \|_2$ is used as distance metric, and that the term ϵ allows for tolerance in the correction's precision. It is easy to define the optimal, zero-error, correction map:

$$\mathcal{M}_{\eta}{}^{\star}(\boldsymbol{\theta}_m) \triangleq \mathbb{E}_{\boldsymbol{\theta}_f}[M_{\eta}(\boldsymbol{\theta}_m)] - m_{\eta}(\boldsymbol{\theta}_m). \qquad (2)$$

However, the correction function $\mathcal{M}_{\eta}{}^{\star}(\boldsymbol{\theta}_m)$ is impossible to compute exactly, as we cannot compute neither M_{η} nor its marginalisation over $\boldsymbol{\theta}_f$. Hence, we will learn an approximation of $\mathcal{M}_{\eta}{}^{\star}(\boldsymbol{\theta}_m)$ trying to keep its error low so to satisfy Definition 2. We turn this problem into a *regression task*, as graphically explained in Figs. 1 and 2, and exploit Gaussian Processes.

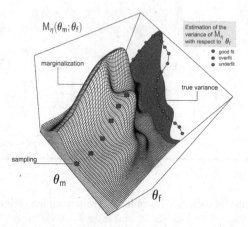

Fig. 2. For any pair of θ_m and θ_f we can compute the statistics for M, $M_\eta(\theta_m; \theta_f)$. Since we do regression over θ_m, we model the relation between such values and θ_f as the variance of a random variable, indexed by θ_m, whose samples are the values, as a function of θ_f. Marginalization is the exponential strategy that estimates such variance correctly; all downsampling strategies possibly over or under fitting. Accounting for the relation between this variance and θ_m can be achieved by heteroschedastic regression.

3 Learning the Correction Map

In this section we will present our machine learning strategy in more detail. We consider the case of a scalar statistics, as w-dimensional ones can be treated by solving w independent learning problems.

3.1 Marginalising θ_f

In order to evaluate (2), we need to be able to compute or approximate the value $\mathbb{E}_{\theta_f}[M_\eta(\theta_m)]$ with respect to the free parameters of M, for a any given θ_m. As this integral cannot be treated analytically or numerically, due to the large dimensionality involved (the cost is exponential in $|\theta_f|$), we will resort to statistical approximations. Before discussing them, let us comment on the distribution $p(\theta_f)$, which is an input for our method. In particular, this distribution encodes our knowledge on the more plausible values of the free parameters. In case we have no information, we can choose an uniform distribution. On the other side of the spectrum, we may know the true value θ_f^* of θ_f, and choose a delta Dirac distribution, which will dramatically simplify the evaluation of the integral. In this case, we can evaluate (2) as

$$\mathcal{M}_\eta(\theta_m) \triangleq M_\eta(\theta_m; \theta_f^*) - m_\eta(\theta_m), \tag{3}$$

Moreover, the approximation, $\int M_\eta(\theta_m; \theta_f)p(\theta_f)d\theta_f \approx M_\eta(\theta_m; \theta_f^*)$ is appropriate when the distribution $p(\theta_f)$ is tightly concentrated around its mode θ_f^*.

In general, however, when $p(\boldsymbol{\theta}_\mathsf{f})$ does not have this special form, we can resort to downsampling $\mathbb{E}_{\boldsymbol{\theta}_\mathsf{f}}[\mathsf{M}_\eta(\boldsymbol{\theta}_\mathsf{m})]$, by generating k samples $\boldsymbol{\theta}_\mathsf{f}^{(1)}, \ldots, \boldsymbol{\theta}_\mathsf{f}^{(k)}$ and approximating $\mathbb{E}_{\boldsymbol{\theta}_\mathsf{f}}[\mathsf{M}_\eta(\boldsymbol{\theta}_\mathsf{m})] \approx \frac{1}{k} \sum_j \mathsf{M}_\eta(\boldsymbol{\theta}_\mathsf{m}; \boldsymbol{\theta}_\mathsf{f}^{(j)})$. In the following, however, we will not necessarily aggregate the values $\mathsf{M}_\eta(\boldsymbol{\theta}_\mathsf{m}; \boldsymbol{\theta}_\mathsf{f}^{(j)})$, and treat them individually to better account for the variance in the observed predictions.

3.2 Gaussian Processes

We will solve the learning problem resorting to Gaussian Process (GP) regression [12]. GPs are random functions, i.e. probability distributions over a function space, in our case functions $f : \Theta \to \mathbb{R}$, with the property that any finite dimensional projection $f(\boldsymbol{\theta}_1), \ldots, f(\boldsymbol{\theta}_k)$ is a multidimensional Gaussian random variable. It follows that GP are defined by a mean function $\mu(\boldsymbol{\theta})$, returning the mean at any point in Θ, and by a covariance or kernel function $k(\boldsymbol{\theta}_1, \boldsymbol{\theta}_2)$, for giving the covariance between any pair of points in Θ. GP can be used to solve regression tasks in a Bayesian setting. The idea is as follows: we put a GP prior on the space of functions $\{f \mid f : \Theta \to \mathbb{R}\}$, typically by assuming constant zero mean and fixing a kernel function[3], and then consider the posterior distribution given a set of observations $Y = \{y^{(i)}\}_I$, $i \in I$, at input points $X = \{\boldsymbol{\theta}_\mathsf{m}^{(i)}\}$. If we assume that $y^{(i)} = f(\boldsymbol{\theta}_\mathsf{m}^{(i)}) + \epsilon_i$, with ϵ_i a zero mean Gaussian noise term with variance σ^2, then we obtain that the posterior distribution given the observed data is still a GP, with mean and kernel functions that can be obtained analytically, see [12] for further details. GP regression is essentially the same if the observation noise σ^2 is different at different input points, i.e. $\sigma^2 = \sigma(\boldsymbol{\theta}_\mathsf{m}^{(i)})^2$, in which case we talk about heteroschedastic regression.

3.3 The Regression Task

Let $\boldsymbol{\theta}_\mathsf{m}^{(i)}$ for some index set $i \in I$ be the *input space*, and $\{\langle \boldsymbol{\theta}_\mathsf{m}^{(i)}, y^{(i)} \rangle\}_I$ our *training points*. In case we use Eq. (3) to evaluate the correction map, each $y^{(i)}$ is a scalar value, and a standard regression schema based on Gaussian Processes can be used. In that case we assume samples of the correction map y to be observations from a random variable centered at a value given by the *latent function*

$$y^{(i)} \sim \mathcal{N}(\mathcal{M}_\eta(\boldsymbol{\theta}_\mathsf{m}^{(i)}), \sigma^2). \tag{4}$$

In this standard Gaussian Processes regression the noise model in the observations is assumed to be a constant σ^2 for all sampled points.

In the more general case we work with downsampling solutions that exploit k samples for the free variable, $\boldsymbol{\theta}_\mathsf{f}^{(1)}, \ldots, \boldsymbol{\theta}_\mathsf{f}^{(k)}$. In that case, we have k correction values per training point, $\left\{\langle \boldsymbol{\theta}_\mathsf{m}^{(i)}, [y^{(i,1)} \cdots y^{(i,k)}]^\top \rangle\right\}_I$, that we can use in a straightforward way to reduce to the above schema, or to estimate the variance of M conditioned to its free variables. In these cases, the training set is

---

[3] In this work, we use the classic Gaussian kernel fixing hyperparameters by maximising the type-II likelihood; see [12].

$\{\langle \boldsymbol{\theta}_{\mathsf{m}}^{(i)}, \overline{y}^{(i)} \rangle\}_I$, namely we do regression on the point-wise expectation of the correction (i.e., $\overline{y}^{(i)} = \frac{1}{k}\sum_{j=1}^{k} y^{(i,j)}$).

Estimator 1 (Empirical $\overline{\sigma}$-estimator). *Set $\overline{\sigma}$ to the empirical estimate of the variance* across all correction values *to exploit the schema in Eq. (4) with $\sigma^2 \triangleq \overline{\sigma}$.*

Besides the simple first case, it is more interesting to account for a *model of the variance in the observations of the predictions from a model*; we discuss how this could be done in two ways.

Estimator 2 (Point-wise σ-estimator). *Let $\sigma(\cdot)$ the empirical estimator of the variance of the correction values, per training-point*

$$\sigma(\boldsymbol{\theta}_{\mathsf{m}}^{(i)}) \triangleq \mathsf{Var}[y^{(i,1)}, \dots, y^{(k,1)}], \tag{5}$$

then define a model of the variance as a point-wise function of the regression parameter, that is

$$y^{(i)} \sim \mathcal{N}\left(\mathcal{M}_\eta(\boldsymbol{\theta}_{\mathsf{m}}^{(i)}), \sigma(\boldsymbol{\theta}_{\mathsf{m}}^{(i)})^2\right). \tag{6}$$

In this case, the variance that we expect in each prediction of the latent function is estimated from the data, thus leading to a form of *heteroscedastic* regression.

We can estimate with higher precision a model of the variation in the variance across the input space; to do that we perform *regression of the variance change*, and then inform the outer regression task of that prediction.

Estimator 3 (Nested σ-estimator). *Consider the same estimator of the variance as above, and collect the variance estimates as $\{\langle \boldsymbol{\theta}_{\mathsf{m}}^{(i)}, w^{(i)} \rangle\}_I$. Learn a latent function model of the true variance $\sigma_\star(\cdot)$, which we assume to have a fixed variance $\sigma_\star^2$*

$$w^{(i)} \sim \mathcal{N}\left(\sigma_\star(\boldsymbol{\theta}_{\mathsf{m}}^{(i)}), \sigma_\star^2\right) \qquad\qquad y^{(i)} \sim \mathcal{N}\left(\mathcal{M}_\eta(\boldsymbol{\theta}_{\mathsf{m}}^{(i)}), \sigma_\star(\boldsymbol{\theta}_{\mathsf{m}}^{(i)})^2\right). \tag{7}$$

This is also a form of heteroschedastic regression, but nesting two GP regressions to account in a finer way for the variance's changes.

3.4 Properties of the Correction Map

The correction map that we learn, for all variance schemes, is a statistically sound estimator of $\mathbb{E}_{\boldsymbol{\theta}_{\mathsf{f}}}[\mathsf{M}_\eta(\boldsymbol{\theta}_{\mathsf{m}})]$, in the sense of being consistent.

Theorem 1 (Correctness). *Let $\mathsf{m}_{\hat{\eta}}(\boldsymbol{\theta}_{\mathsf{m}})$ and $\mathbb{E}_{\boldsymbol{\theta}_{\mathsf{f}}}[\mathsf{M}_{\hat{\eta}}(\boldsymbol{\theta}_{\mathsf{m}})]$ be consistent estimators of $\mathsf{m}_\eta(\boldsymbol{\theta}_{\mathsf{m}})$ and $\mathbb{E}_{\boldsymbol{\theta}_{\mathsf{f}}}[\mathsf{M}_\eta(\boldsymbol{\theta}_{\mathsf{m}})]$, then $\hat{\mathsf{M}}_{\hat{\eta}}(\boldsymbol{\theta}_{\mathsf{m}}) \triangleq \mathsf{m}_{\hat{\eta}}(\boldsymbol{\theta}_{\mathsf{m}}) + \mathcal{M}_{\hat{\eta}}(\boldsymbol{\theta}_{\mathsf{m}})$ is a consistent estimator of $\mathbb{E}_{\boldsymbol{\theta}_{\mathsf{f}}}[\mathsf{M}_\eta(\boldsymbol{\theta}_{\mathsf{m}})]$, for any estimation strategy of $\mathcal{M}_{\hat{\eta}}$.*

The result follows because $\mathcal{M}_{\hat{\eta}}$ converges to $\mathcal{M}_\eta$ due to properties of GPs [12], and because of the consistency of $\mathsf{m}_{\hat{\eta}}$ and $\mathbb{E}_{\boldsymbol{\theta}_{\mathsf{f}}}[\mathsf{M}_{\hat{\eta}}(\boldsymbol{\theta}_{\mathsf{m}})]$. The proof is sketched in Appendix A.2.

The correction map $\mathcal{M}_{\hat{\eta}}(\theta_m)$ is estimated from samples of the system, hence it is an approximation of the exact map defined by Eq. (2). Thus, it is a correction map in the sense of Definition 2. However, being the result of a GP regression, $\mathcal{M}_{\hat{\eta}}(\theta_m)$ is in fact a random function. Therefore, in measuring the error according to Eq. (1), we need to consider the average value of the random function $\mathbb{E}[\mathcal{M}_{\hat{\eta}}(\theta_m)]$. The variance $\mathsf{Var}[\mathcal{M}_{\hat{\eta}}(\theta_m)]$, instead, provides a way of computing the error ϵ, but in a statistical sense with confidence α: ϵ can be estimated by averaging over Θ (with respect to $p(\theta_m)$) the half-width of the region containing $\alpha\%$ of the probability mass of $\mathcal{M}_{\hat{\eta}}(\theta_m)$.

The cost of all our approaches inherently depends on how many samples we pick from Θ, the way parameters in θ_f are accounted for and the number of parameters in θ_m. The sampling cost in general grows exponentially with $|\theta_m|$, and each Gaussian regression is cubic in the number of sampled values. Notice that, asymptotically, the cost of the empirical and nested σ-estimators is the same, as the two regressions are executed in series.

4   Applications

We discuss now several potential applications of our framework. The advantages of using our approach are mostly computational: the reduced model is simpler to analyze, yet it retains a mechanistic interpretation, compared to statistical emulation.

Model Building. Many common problems in the area of dynamical systems can be tackled by resorting to correction maps.

Problem 1 (Model selection). *Let* M *be a model, and* m_1, ..., m_k *a set of candidate models for correction, each one having a correction map* $\mathcal{M}_{\eta,1}$, ..., $\mathcal{M}_{\eta,k}$. *The smallest-correction model* m^* *for a statistic* η *is* $m^* \triangleq$ $\arg\min_{m_i} \int \mathcal{M}_{\eta,i}(\theta)d\theta$.

This criterion is certainly alternative to structural Bayesian approaches [3], which can be used to select the structurally smaller model *within* an equivalence class (see below). Also, allows to frame a model synthesis problem.

Problem 2 (Model synthesis). *For a model* M *with parameters* θ_M *and for a statistic* η: (*i*) *partition* θ_M *into sets* θ_m *and* θ_f, (*ii*) *generate a finite set of plausible reduced models with parameters* θ_m *and* (*iii*) *select the best one, according to the above model selection criterion.*

In this case, the partition might aim at identifying in θ_M the model's parameters which contribute the most to the variance for the statistics η. Opportunities for *control* are also plausible if one can reduce the problem of controlling M to "controlling and correcting" a reduced model m. This should be easier as m is structurally smaller than M, and so is lower the complexity of solving a *controller synthesis problem*.

Parameter Estimation. Another application of our framework is in Bayesian parameter estimation of the parameters $\boldsymbol{\theta}_m$ of the big model M, given observations D of variables X, using the small model and the corrected statistics to build approximations of the likelihood function $p(D \mid \boldsymbol{\theta}_m)$. For instance, this can be done by correcting the mean of variables X, and using the variance of the correction map as a proxy of the variance X in M after marginalisation of $\boldsymbol{\theta}_f$. We can then plug the distribution of X in a Bayesian inference scheme, and compute an approximate posterior over $\boldsymbol{\theta}_m$.

Model Equivalence. Correction maps can also be used to compare processes, via weak forms of *bisimilarity* with respect to the observations and a statistics.

Definition 3 (Model equivalence). *Models* M_1 *and* M_2, *for a statistic* η *and parameter sets* $\boldsymbol{\theta}_m$ *and* $\boldsymbol{\theta}_f$, *are* η-*bisimilar conditioned to* m, $M_1 \equiv_m^\eta M_2$, *if and only if for all* $\boldsymbol{\theta}_m \in \Theta$, *it holds* $\mathcal{M}_{\eta,1}(\boldsymbol{\theta}_m) = \mathcal{M}_{\eta,2}(\boldsymbol{\theta}_m)$. *A class of equivalence of models with respect to* m *and* η *is the set of all such bisimilar models.*

This notion of bisimilarity resembles conditional dependence, as we are saying that two models are equivalent if an observer corrects both the same way. In this case, m is a *universal corrector* of $\equiv_m^\eta$, as it can correct for all the models in the class. The class of models that are equivalent to a model M is $\{M^* \mid M^* \equiv_M^\eta M\}$ – i.e., the class of models with correction zero; notice that in this case $\boldsymbol{\theta}_f = \emptyset$. The previous definition can be relaxed by asking that $|\mathcal{M}_{\eta,1}(\boldsymbol{\theta}_m) - \mathcal{M}_{\eta,2}(\boldsymbol{\theta}_m)| \leq \epsilon$, for all $\boldsymbol{\theta}_m \in \Theta$.

Remark 1. Criterion $\equiv_m^\eta$ is a weaker form of *probabilistic bisimilarity*, namely if $M_1 \equiv M_2$ are bisimilar, then $M_1 \equiv_m^\eta M_2$ for some m and any statistics of interest. For instance, for CTMCs, this follows because $\equiv$ implies that M_1 and M_2 have the same infinitesimal generators for any parameter $\boldsymbol{\theta}_m$ and $\boldsymbol{\theta}_f$, hence the outcomes of M_1 and M_2 are indistinguishable, and so are their statistics.

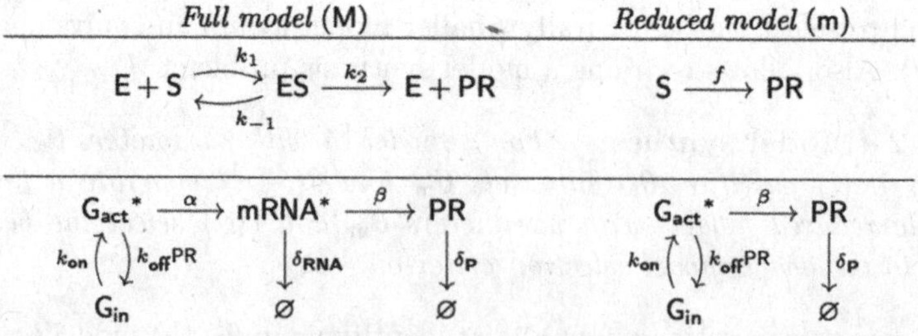

Fig. 3. Example models tested in this paper. Top panel: the Henri-Michaelis-Menten model, where m is derived when $\mathbf{C} : [E]_0 + [ES]_0 \ll [S]_0 + K_{MM}$. Bottom panel, a protein translation network where m when $\mathbf{C} : \beta \gg \alpha$.

5 Examples

We investigate two examples to better illustrate our method.

5.1 Model Reduction via QSSA

Consider the irreversible canonical enzyme reaction with its exact representation (here, model M), for enzyme E, complex ES , substrate S and product PR (Fig. 3, top left panel). When the concentration of the intermediate complex does not change on the time-scale of product formation, product is produced at rate $f \triangleq V_M S/(K_M + S)$ where $V_M = k_2([E]_0 + [ES]_0)$ and $K_{MM} = (k_{-1} + k_2)/k_1$. This is the Henri-Michaelis-Menten kinetics and is technically derived by a quasi-steady-state assumption (QSSA), i.e., $\dot{ES} = \dot{E} = 0$, that is in place when $\mathbf{C}$: $[E]_0 + [ES]_0 \ll [S]_0 + K_{MM}$, where $[x]_0$ is the initial amount of species x. m is thus the QSSA reduced model (Fig. 3, top right panel).

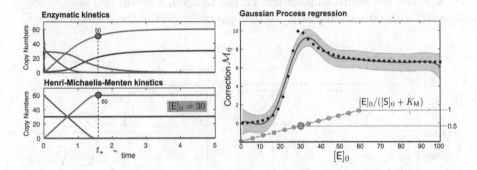

Fig. 4. Correction of the product formation at the transient time $t_* = 1.5$, for a mean field model of irreversible canonical enzyme reaction and its simplified Henri-Michaelis-Menten form. Here $k_1 = 2$, $k_{-1} = 1$ and $k_2 = 1.5$, $[S]_0 = 60$ and $[P]_0 = 0$. Regression over $[E]_0$ is done with 40 training points from $(0, 100]$, and the correction in Eq. (3) as M's free variables are part of the Michaelis-Menten constant.

We interpret these two models as two systems of *ordinary differential equations*, see Appendix A.1, and learn a correction for the following statistics

$$\eta : \mathbb{E}[PR(t_*)], \text{ with } PR(t_*) \text{ the number of products at time } t_* \qquad (8)$$

For non-degenerate parameter values both models predict the same equilibrium state, where a total transformation of substrate into product has happened, $\mathbb{E}[PR(t)] \to [S]_0$ for large t. Thus, we are not interested in correcting the dynamics of m for long-run times, but rather in the transient (i.e., for small t_*).

Also, as the QSSA hypothesis does not hold for certain initial conditions, we set $\boldsymbol{\theta}_m = \{[E]_0\}$ as the regression variable, and set $[S]_0 = 60$ and $[P]_0 = 0$. The other parameters are $k_1 = 2$, $k_{-1} = 1$ and $k_2 = 1.5$ with unit *(mole* *per second)*$^{-1}$. In terms of regression, we pick 40 samples of the initial enzyme amount from $(0, 100]$, and set $t_* = 1.5$ as it is a time in the transient (manual tuning). This particular example is rather simple, as the free parameters of M are part of the Michaelis-Menten constant and fixed, so we use the simpler correction of Eq. (3). Also, knowing when the QSSA holds gives us an interval where we expect the correction to shrink towards zero. The map is shown in Fig. 4, which depicts the expected concordance among the correction map and validity of the QSSA.

5.2   Model Reduction via Time-Scale Separation

Consider a gene switching among active and inactive states of mRNA transcription to be ruled by a *telegraph process* with rates k_{off}/k_{on}. A reaction model of such gene G, protein PR, messenger mRNA with transcription/translation rates α/β as in Fig. 3, bottom left panel.

Here the gene switches among active and inactive states, with rates k_{on} and k_{off}, and PR feedbacks positively on inactivation. Proteins and mRNAs are degraded with rates δ_P and δ_{RNA}. In the uppermost part of the diagram species

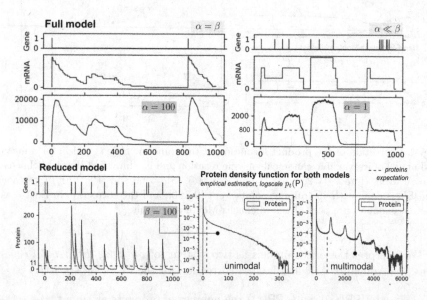

Fig. 5. Comparison between the dynamics of the full and the reduced models from Sect. 5.2, with $k_{off} = k_{on} = \delta_P = 10^{-2}$, $\delta_{RNA} = 10$, $[G_{act}]_0 = 1$. Values for transcription (α) and translation (β) are reported in the figure. The reduced model predicts spiked dynamics, leading to a unimodal distribution of proteins. The larger model, instead, can either predict protein buffers, when there is no time scale separation ($\alpha = \beta$), or multiple equilibria, leading to a multimodal distribution of protein counts. Observe that the expectation on the number of proteins in the long run spans over different order of magnitudes, according to the relation between α and β.

marked with a ∗ symbol are not consumed by a reaction, i.e., mRNA transcription is $G_{act} \rightarrow G_{act} + mRNA$. This model can be easily encoded in a Markov process, as discussed in A.1.

We can derive an approximation to M where the transcription step is omitted. This is valid when $\mathbf{C} : \beta \gg \alpha$ (*time-scales separation*), namely for every copy of mRNA a protein is immediately translated.

Correction of protein dynamics. We build a correction map with $\boldsymbol{\theta}_m = \{\beta\}$. In this case the telegraph process is common to both models, but α and δ_{RNA} are free variables of M; here we assume to have a prior delta distribution over the values of mRNA's degradation, so we set $\boldsymbol{\theta}_f = \{\alpha\}$. For some values of the transcription rate condition $\mathbf{C}$ does not hold; in this case it is appropriate to account for α's contribution to the variance in the statistics that we want to correct, when we do a regression over β. Note that also β is part of $\mathbf{C}$.

Model M leads to stochastic bursts in PR's expression when the baseline gene switching is slower than the characteristic times of translation/transcription. Here we set $k_{off} = k_{on} = 10^{-2}$, and assume mRNA's lifespan to be longer than protein's ones (also in light of condition $\mathbf{C}$), so $\delta_{RNA} = 10\delta_P = 10^{-2}$. We simulate both models with one active gene, $[G_{act}]_0 = 1$; example dynamics are shown in Fig. 5, for $\beta = 100$ and $\alpha \in \{1, 100\}$. We observe that, when $\mathbf{C}$ does not hold ($\alpha = \beta$) the protein bursts increases of one order of magnitude, and the long-run probability density function for the proteins, $p_t(PR)$, becomes multimodal.

We define two statistics. One measures the *first moment* of the random variable that models the number of proteins in the long run; the other is a metric interval temporal logic formula [2], expressing the probability of a protein burst within the first 100 time units of simulation.

$$\eta_1 : \mathbb{E}[PR(t_*)], \text{ with } PR(t_*) \text{ the number of proteins at time } t_* \gg 0 \qquad (9)$$

$$\eta_2 : \mathbb{E}[p(\varphi)], \text{ with } \varphi \triangleq \mathbf{F}_{[0,100]}PR(t) > 200 \qquad (10)$$

The former is evaluated by a unique long-run simulation of the model, as its dynamics are ergodic. For the latter we estimate the *satisfaction probability of the formula* via multiple ensembles, as in a parametric *statistical model checking problem* [4].

For the regression task we sample 50 values of β, in the range $[0.1, 100]$. For α, instead, we sample 50 random values in the same interval, for each value of β; notice that with high probability we pick values where $\mathbf{C}$ does not hold, so we might expect high correction factors. Data generated and the regression results are shown in Fig. 6, for both fixed-variance regression, empirical σ-estimator in Eq. (4) and with the σ-estimator, Eq. (6). Because variance spans over many orders of magnitude, regression is performed in the logarithmic space. Results highlight a general difference between the posterior variance between the two estimators.

For the second statistics, data is generated from an initial condition where one gene is inactive, $[G_{in}]_0 = 1$. Notice that the expected time for the gene to trigger its activation is $1/k_{on} = 100$ (the time upper-bound of the formula),

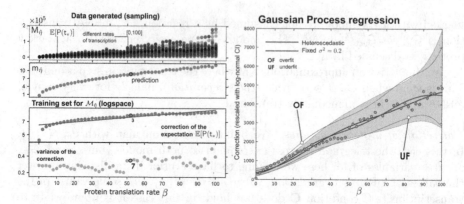

Fig. 6. For the regression task we sample 50 values of β from $[0.1, 100]$. For each value of β, we sample 50 random values of α in the same interval. For each pair (α, β) we estimate the statistics for both models (green, M; red, m), and obtain 50 correction values indexed by each β (green). Correction values (blue and pink) are the expectation and variance of the difference between M's and m's predictions. We transform them logarithmically before doing regression; observe that, on the linear scale, the correction is of the order of 10^3 with variance 10^7 (midpoint value $\beta \approx 50$). Gaussian Process regression (right panel) is performed with a constant $\sigma^2 = 0.2$, Eq. (4) and with the σ-estimator, Eq. (6). Values are re-scaled linearly, and 95 % log-normal confidence intervals are shown; regression highlights that the posterior variances are similar, but the fixed-variance schema tends to underfit or overfit the heteroscedastic variance (*assumed it to be closer to the true one*). (Color figure online)

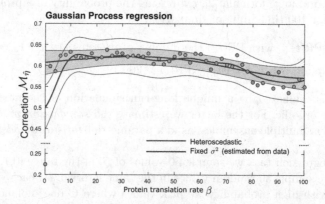

Fig. 7. Correction map for the expected satisfaction probability of the linear logic formula η_2 in Eq. (9). Comparison between the point-wise σ-estimator and the empirical $\bar{\sigma}$-estimator. Sampled data are shown in Appendix Fig. 8.

so for some parametrization there will be non-negligible probability of having no protein spike above threshold 200. The formula satisfaction probability is evaluated with 100 independent model runs, and data generated are shown in Appendix Fig. 8. Regression results are shown in Fig. 7, where the point-wise

σ-estimator and the empirical $\overline{\sigma}$-estimator are compared, highlighting the more robustness of the former with respect to the sampled bottom-left outlier.

6   Conclusions

Abstraction represents a fundamental tool in the armoury of the computational modeller. It is ubiquitously used in biology as an effective mean to reduce complexity, however systematic analysis tools to quantify discrepancies introduced by abstraction are lacking. Prominent examples, beyond the ones already mentioned, include models with delays, generally introduce to approximately lump multiple biochemical steps [6], or physiological models of organ function which adopt multi-scale abstractions to model phenomena at different organisational levels. These include some of the most famous success stories of systems biology, such as the heart model of [11], which also constitutes perhaps the most notable example of a physical systems which has been modelled multiple times at different levels of abstraction. Employing our techniques to clarify the quantitative relationship between models of cardiac electrophysiology would be a natural and very interesting next step.

Our approach has deep roots in the statistical literature. In this paper we have focussed on the scenario where we try to reconcile the predictions of two separate models, however the complex model was simply used as a sample generator black box. As such, nothing would change if instead of a complex model we used a real system which can be interrogated as we vary some control parameters. Our approach would then reduce to fitting the simple model with a structured, parameter dependent error term. This is closely related with the use of Gaussian Processes in the geostatistics literature [7], where simple (generally linear) models are used to explain spatially distributed data, with the residual variance being accounted for by a spatially varying random function. Another important connection with the classical statistics literature is with the notion of *emulation* [9]. Emulation constructs a statistical model of the output of a complex computational model by interpolating sparse model results with a Gaussian Process. Our approach can be viewed as a *partial emulation*, whereby we are interested in retaining mechanistic detail for some aspects of the process, and emulate statistically the residual variability. In this light, our work represents a novel approach to bridge the divide between the model-reduction techniques of formal computational modelling and the statistical approximations to complex models.

A   Appendix

All the code that replicate these analysis is available at the corresponding author's webpage, and hosted on Github (repository GP-correction-maps).

A.1 Further Details on the Examples

The two models from Sect. 5.1 correspond to these systems of differential equations

$$M := \begin{cases} \dot{E} = -k_1 E \cdot S + k_{-1} ES + k_2 ES \\ \dot{S} = -k_1 E \cdot S + k_{-1} ES \\ \dot{ES} = k_1 E \cdot S - k_{-1} ES - k_2 ES \\ \dot{PR} = +k_2 ES \end{cases} \qquad m := \begin{cases} \dot{E} = 0 \\ \dot{S} = -V_M S / (K_M + S) \\ \dot{ES} = 0 \\ \dot{PR} = +V_M S / (K_M + S) \end{cases}$$

which we solved in MATLAB with the ode45 routine with all parameters (Initial-Step, MaxStep, RelTol and AbsTol) set to 0.01.

Concerning the Protein Translation Network (PTN) in Sect. 5.2, the set of reactions and their propensity functions that we can use to derive a Continuous Time Markov Chain model of the network are the following. Here x denotes a generic state of the system and, for instance, x_{mRNA} the number of mRNA copies in x.

event	reaction	propensity
activation	$G_{in} \rightarrow G_{act}$	$a_1(x) = k_{on} \cdot x_{G_{in}}$
inactivation	$G_{act} \rightarrow G_{in}$	$a_2(x) = k_{off} \cdot x_{PR}$
transcription	$G_{act} \rightarrow G_{act} + mRNA$	$a_3(x) = \alpha \cdot x_{G_{act}}$
mRNA decay	$mRNA \rightarrow \varnothing$	$a_4(x) = \delta_{RNA} \cdot x_{mRNA}$
translation	$mRNA \rightarrow mRNA + PR$	$a_5(x) = \beta \cdot x_{mRNA}$
PR decay	$PR \rightarrow \varnothing$	$a_4(x) = \delta_P \cdot x_{PR}$

The reduced PTN model is a special of this reactions set where transcription and mRNA decay are omitted. In this case we used StochPy to simulate the models and generate the input data per regression – see http://stochpy.sourceforge.net/; data sampling exploits python parallelism to reduce execution times.

For regression, we used the Gaussian Processes for Machine Learning toolbox for fixed-variance regression, see http://www.gaussianprocess.org/gpml/code/matlab/doc/ and a custom implementation of the other forms of regression.

A.2 Proofs

Proof of Theorem 1

Proof. Both the empiricals and nested estimator rely on an *unbiased* estimator of the mean/variance, which means that if $k \rightarrow \infty$, i.e., we sample all possible values for the free variables, we would have a true model of $\overline{y}$ σ. This means that, for each sampled value from Θ, even the simplest $\overline{\sigma}$-estimator would be equivalent, in expectation, to the marginalization of the free variables. This is enough, combined with properties of Gaussian Processes regression (i.e., the convergence to the true model with infinite training points), to state that the overall approach leads to an unbiased estimator of the correction map.

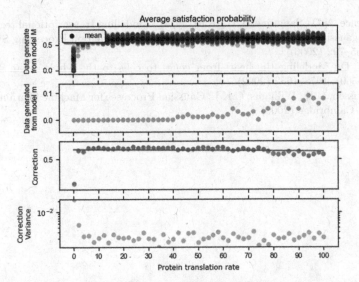

Fig. 8. Data generated to compute the satisfaction probability of the linear logic formula η_2 in Eq. (9). For each model 100 independent simulations are used to estimate the expectation of the probability. The regression input space is the same used to compute η_1, but the models are simulated with just one inactive gene in the initial state. The heteroscedastic variance in the regression is computed as the variance of the correction of the expected satisfaction probability (point-wise σ-estimator); the fixed-variance regression is computed by estimating the variance from the data (empirical $\bar{\sigma}$-estimator).

References

1. Aitken, S., Alexander, R.D., Beggs, J.D.: A rule-based kinetic model of rna polymerase ii c-terminal domain phosphorylation. J Roy. Soc. Interface **10**(86), 20130438 (2013)
2. Alur, R., Feder, T., Henzinger, T.A.: The benefits of relaxing punctuality. J. ACM **43**(1), 116–146 (1996)
3. Barber, D.: Bayesian Reasoning and Machine Learning. Cambridge University Press, Cambridge (2012)
4. Bortolussi, L., Milios, D., Sanguinetti, G.: Smoothed model checking for uncertain continuous-time markov chains. Inf. Comput. **247**, 235–253 (2016)
5. Bortolussi, L., Sanguinetti, G.: Learning and designing stochastic processes from logical constraints. In: Joshi, K., Siegle, M., Stoelinga, M., D'Argenio, P.R. (eds.) QEST 2013. LNCS, vol. 8054, pp. 89–105. Springer, Heidelberg (2013)
6. Caravagna, G.: Formal modeling and simulation of biological systems with delays. Ph.D. thesis, University of Pisa (2011)
7. Cressie, N., Wikle, C.K.: Statistics for Spatio-Temporal Data. Wiley, New York (2015)
8. Hoyle, D.C., Rattray, M., Jupp, R., Brass, A.: Making sense of microarray data distributions. Bioinformatics **18**(4), 576–584 (2002)
9. Kennedy, M.C., O'Hagan, A.: Bayesian calibration of computer models. J. Roy. Stat. Soc.: Ser. B (Stat. Methodol.) **63**(3), 425–464 (2001)

10. Lawrence, N.D., Sanguinetti, G., Rattray, M.: Modelling transcriptional regulation using gaussian processes. In: Advances in Neural Information Processing Systems, pp. 785–792 (2006)
11. Noble, D.: Modeling the heart-from genes to cells to the whole organ. Science **295**(5560), 1678–1682 (2002)
12. Rasmussen, C.E., Williams, C.K.I.: Gaussian Processes for Machine Learning. MIT Press, Cambridge (2006)

Target Controllability of Linear Networks

Eugen Czeizler[✉], Cristian Gratie, Wu Kai Chiu, Krishna Kanhaiya,
and Ion Petre

Computational Biomodeling Laboratory, Department of Computer Science and Turku
Centre for Computer Science, Åbo Akademi University, Turku, Finland
eugen.czeizler@abo.fi

Abstract. Computational analysis of the structure of intra-cellular
molecular interaction networks can suggest novel therapeutic approaches
for systemic diseases like cancer. Recent research in the area of network
science has shown that network control theory can be a powerful tool
in the understanding and manipulation of such bio-medical networks. In
2011, Liu et al. developed a polynomial time optimization algorithm for
computing the size of the minimal set of nodes controlling a given linear
network. In 2014, Gao et al. generalized the problem for target structural
control, where the objective is to optimize the size of the minimal set
of nodes controlling a given target within a linear network. The working
hypothesis in this case is that partial control might be "cheaper" (in
the size of the controlling set) than the full control of a network. The
authors developed a Greedy algorithm searching for the minimal solution
of the structural target control problem, however, no suggestions were
given over the actual complexity of the optimization problem. In here we
prove that the structural target controllability problem is NP-hard when
looking to minimize the number of driven nodes within the network, i.e.,
the first set of nodes which need to be directly controlled in order to
structurally control the target. We also show that the Greedy algorithm
provided by Gao et al. in 2014 might in some special cases fail to provide a valid solution, and a subsequent validation step is required. Also,
we improve their search algorithm using several heuristics, obtaining in
the end up to a 10-fold decrease in running time and also a significant
decrease of the size of the minimal solution found by the algorithms.

1 Introduction

The intrinsic robustness of living systems against perturbations is a key factor
that explains why many single-target drugs have been found to provide poor
efficacy or lead to significant side effects [5]. The efficacy of multi-target therapies can be understood from a robustness of disease-networks point of view to
deal with single node perturbations, due to inherent diversity and redundancy
of compensatory signaling pathways that result in highly resilient and resistant
network architecture with modular and interconnected topology [5]. Rather than
trying to design selective ligands that target individual receptors only, network

© Springer International Publishing AG 2016
E. Bartocci et al. (Eds.): CMSB 2016, LNBI 9859, pp. 67–81, 2016.
DOI: 10.1007/978-3-319-45177-0_5

polypharmacology aims to modify multiple cellular targets to tackle the compensatory mechanisms and robustness of disease-associated cellular systems, as well as to control unwanted off-target side effects that often limit the clinical utility of many conventional drug treatments [1,3,5]. However, the exponentially increasing number of potential drug target combinations makes the pure experimental approach quickly unfeasible, and translates into a need for design principles to determine the most promising target combinations to effectively control complex disease systems, without causing drastic toxicity or other side-effects.

Network biology, with the help of mathematical modeling, has revolutionized the human diseasome research and paved the way towards the development of new therapeutic approaches and personalized medicine. Recent work on network controllability has shown that full controllability and reprogramming of intercellular networks can be achieved by a minimum number of control targets [10]. However, the computer-based experimental tests of Liu et al. [10] suggest that the approach is totally unfeasible in practice, as achieving full control over gene regulatory networks requires roughly 80 % of the nodes (i.e., on the order of 800 – 1000 nodes) to be directly controlled by an external controller.

Although diseased cells may harbor hundreds of genomic alterations in various biological pathways [8,17], only a subset of these alterations are driving the disease initiation and progression. These genes form together the sets of (disease specific) essential genes, see [2]. Due to the new CRISPR gene editing technology, researchers can now pinpoint the sets of essential genes, for a very large class of illnesses [11,16], including many types of cancers [18].

In this research we concentrate over the target structural controllability problem, where the aim is to select a minimal set of driver/driven nodes which can control a given target within a linear network. That is, for every initial configuration of the system and any desired final configuration of the target nodes, there exists a finite sequence of input functions for the driver nodes such that the target nodes can be driven to the desired final configuration, in finite time.

The target controllability problem for linear networks is a particular case of output controllability [14] and a generalization of the full controllability problem, which requires the control over the entire system. In 2011 Liu et al. [10] have provided a polynomial time algorithm (in the size of the network) computing the optimal solution for the full structural controllability problem. Few years latter, Gao et al. [4] developed a Greedy algorithm searching for the minimal solution of the structural target controllability problem. However, the overall complexity of the target control optimization problem was not tackled.

In this study we prove that the structural target controllability problem is NP-hard when looking to minimize the number of driven nodes within the network. The driven nodes of a network are those to be directly controlled from an outside agent in order to structurally control the given target. We also show that the Greedy algorithm provided by Gao et al. [4] might sometimes fail to provide a valid solution (i.e., a driver/driven set of nodes actually controlling the target), and thus a subsequent validation step is required. Also, we improve their search algorithm using several heuristics, obtaining up to a 10-fold decrease in the average running time and a significant decrease in the size of the average minimal solution found by the algorithms, especially in the case of proportionally small targets, i.e., less than 15 % of the total number of nodes.

2   Background and Definitions

A *linear, time invariant dynamical system* (in short LTIS)is a system of the form

$$\frac{dx(t)}{dt} = Ax(t) \tag{1}$$

where $x(t) = (x_1(t), ..., x_n(t))^T$ is the n-dimensional vector describing the system's state at time t, and $A \in R^{n \times n}$ is the time-invariant state transition matrix, describing how each of these states are influencing the dynamics of the system. The elements in x are called the variables of the system; we abuse notation and denote with X the set of these variables. If the system is influenced by a size-m external input controller $u(t) = (u_1(t), ..., u_m(t))^T$, then system (1) becomes:

$$\frac{dx(t)}{dt} = Ax(t) + Bu(t) \tag{2}$$

where $B \in R^{n \times m}$ is the time-invariant input matrix describing how each of the n variables are affected by the m inputs. In the additional case when at each time step t the system is also exporting a set of k output values, $y(t) = (y_1(t), ..., y_k(t))^T$ depending on the current state $x(t)$, the system (1) becomes:

$$\begin{aligned} \frac{dx(t)}{dt} &= Ax(t) \\ y(t) &= Cx(t) \end{aligned} \tag{3}$$

where $C \in R^{k \times n}$ is the output matrix describing how each of the k outputs are influenced by the n variables of the system at time t. For example, in the particular case when the desired output is represented just by the numerical values of a k subset $T \subseteq X$ of the total n variables, such as a target set, the output matrix C_T is a 0–1 matrix, with $C_T(i, j) = 1$ iff $i = j$ and $i, j \in T$, i.e., C_T is the the identity matrix restricted to the subset T. For ease of notation, such m-input, k-output linear, time invariant, dynamical systems are denoted as (A, B, C) with $A \in R^{n \times n}$, $B \in R^{n \times m}$, and $C \in R^{k \times n}$.

Given a target set T, a linear time-invariant dynamical system (A, B, C_T) is said to be *target controllable* if there exists a time-dependent input vector $u(t) = (u_1(t), ..., u_m(t))^T$ that can drive the state of the target variables to any desired numerical setup in finite time. It is known, see e.g. [4], that a system (A, B, C_T) is target controllable if and only if

$$\text{rank}[C_T B, C_T AB, , C_T A^2 B, ..., C_T A^{n-1} B] = |T| \tag{4}$$

In the particular case when the target is the entire n variable set X, we can see that the above condition is reduced to the well known Kalman condition for full controllability [6], i.e., an LTIS (A, B) is controllable if and only if

$$\text{rank}[B, AB, A^2 B, ..., A^{n-1} B] = n \tag{5}$$

A big step forward in the search for algorithms looking for efficient solutions for the (target) controllability problem has been achieved by translating the problem to graphs. A first step in this direction is implemented by detaching from particular numerical setups of a linear system, and focussing on the intrinsic wiring of the system's variables. We say that a linear time-invariant dynamical system (A, B, C_T) is *structurally target controllable* (with respect to a given size-k target set T) if there exists a time-dependent input vector $u(t) = (u_1(t), ..., u_m(t))^T$ and a numerical setup for the non-zero values within the matrices A and B, that can drive the state of the target nodes to any desired numerical setup in finite time. According to Eq. (4) above, a system (A, B, C_T) is structurally target controllable if and only if there exist values for the non-zero entries in A and B such that

$$\text{rank}[C_T B, C_T AB, , C_T A^2 B, ..., C_T A^{n-1} B] = k \tag{6}$$

The case of full structural controllability is obtained from the above when $T = X$ and $C_T = I_n$. It is known, see e.g. [9,15], that if a system is structurally (target) controllable, then it is (target) controllable in almost all numerical setups of the non-zero values within the state transition matrix A.

Linear systems can be represented in terms of directed weighted graphs. The n variables of the systems are the nodes of the graphs, while directed edges correspond to non-zero values in the state transition matrix. That is, there exists a directed edge between variables x_i an x_j with weight v if and only if $A(x_j, x_i) = v \neq 0$. Similarly, the size-m controller vector u corresponds to m input nodes, $u_1, ...u_m$, called driver nodes, while the input matrix B determines the edges between the driver nodes and the network. That is, there exists a directed edge between u_i an x_j with weight w if and only if $B(x_j, u_i) = w \neq 0$. The nodes x_j such that there exists i with $B(x_j, u_i) \neq 0$ are called the driven nodes of the network; these are the first nodes in the network which are directly manipulated in order to drive the entire system to the desired state.

Instead of (target) structural controllability we can now talk of the equivalent network controllability problem, where the variables and the targets are now nodes in the directed network graph. It is known, see e.g. [9], that the structural controllability problem has a counterpart formulation in terms of network graphs. The n variable system (A) is structurally controllable from the m-input/driver controller u and control matrix B if and only if we can select a set of n directed paths from the input/driver nodes (i.e., as starting points) to each of the network nodes (i.e., as ending points), such that no two paths would intersect at the same distance d from their end points. In case of the target controllability problem for a given target set T, with $|T| = m$, the above condition must hold for a path family containing m paths, connecting all the targets to the driver nodes.

From the point of view of bio-medical disease network analysis and control, it is sometimes more advantageous to consider the set of driven nodes instead of that of the driver nodes. To a rough understanding, the set of driver nodes is describing the complexity of an outside controller, assuming this controller can interact/influence with equal impact several of the network nodes; such an interaction could be seen for example as the influence of a drug over the expression of some particular genes. Meanwhile, the set of driven nodes provide the exact collection of network nodes, i.e., genes, that will be used in order to ultimately control the entire set of target nodes. In particular, if we require that each driver node is interacting with at most one network node, i.e., the control matrix B has at most one non-zero entry for every column, then there is a one-to-one correspondence between driver and driven nodes. From now on, within this study we will concentrate over minimizing the set of driven nodes for the control of a given target within a directed network.

Definition 1. *We say that for an LTIS (A, B, C_T) the input controller is 1-bounded if and only if matrix B contains only one non-zero value on every column, i.e., each of the m inputs $u_j(t)$ control exactly one of the variables $x_i(t)$, and that variable is independent of the choice for the time point t.*

3 Driven Target Control Is NP-hard

In this section we prove that the problem of minimizing the number of driven nodes for a given LTIS (A, B, C_T) and a target set T is NP-hard. If moreover the system (A, B, C_T) is 1-bounded, the problem is equivalent to minimizing its number of driver nodes. We are providing this result by proving that the corresponding decision problem, i.e., whether there exists a size-k 1-bounded controller B which can structurally control the target T, is itself NP-hard. This will be done via a reduction to 3SAT.

We recall that in a directed graph, we say that a node X_i is an *ancestor* of a node X_j if there exists a directed path (possibly empty) from X_i to X_j.

Theorem 1. *The 1-Bounded Target Control Optimization Problem is NP-hard, as the following associated decision problem is itself NP-hard. Given a network graph $G = (V, E)$, a target subset $T \subseteq V$, and a number $n \leq |V|$, is there a size-n 1-bounded control scheme for the target T, i.e., a set of n driver nodes, each interacting with exactly one node from the graph, such that we obtain the full control of the target nodes T? In matrix representation, is there a matrix B of size $|V| \times n$, with exactly one non-zero entry per each column, such that rank $[C_T B, C_T AB, , C_T A^2 B, ..., C_T A^{n-1} B] = |T|$?*

Proof (Sketch): We are proving the NP-hardness result via a reduction from the 3SAT problem. Let P be an arbitrary 3SAT boolean formula instance, containing n boolean variables $x_1, ..., x_n$ and m clauses $Cl_1, ..., Cl_m$. We are going to construct a graph $G = (V, E)$ with $|V| = 3m(m+1)/2 + 3n + m$ nodes and select a target subset C with $|C| = m + n$ such that the formula P is satisfiable if and only if the cardinality of a minimal control set for C is n.

The graph G, presented also in Fig. 1, can be described as follows. It consists of five types of nodes: *valuation-nodes, clause-nodes, tautology-nodes, and path-nodes*. The valuation set of nodes contains $2n$ nodes, $X_j^T, X_j^F, 1 \le j \le n$, one for each possible truth assignment of a variable x_j. The clause set of nodes contains m nodes, $CL_1, ..., CL_m$, one for each of the formula's clauses. The tautology set of nodes contains n nodes, $TA_1, ..., TA_n$, each corresponding to a variable-specific tautological clause $(x_j \vee \neg x_j)$. Finally, the path-level set of nodes contains $3m(m+1)/2$ nodes, that is, for each of the clauses Cl_i, with $1 \le i \le m$, we have $3i$ nodes $Pa_j^{(1;i)}, Pa_j^{(2;i)},$ and $Pa_j^{(3;i)}$, with $1 \le j \le i$.

$$P = (x_1 \vee x_2 \vee x_3) \wedge (\neg x_1 \vee x_2 \vee x_n) \wedge \cdots \wedge (x_3 \vee x_7 \vee x_n)$$

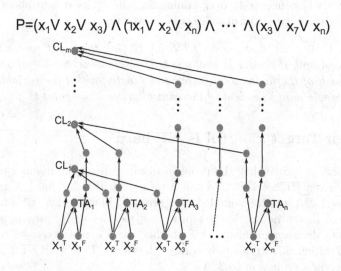

Fig. 1. The graph associated to a boolean formula P. For reducing the complexity of the notations, the path nodes $Pa_j^{k;i}$ are not labeled on the figure.

The directed edges of G can be easily described as follows. In the formula P, every clause Cl_i has exactly three valuations of the variables $x_1, ..., x_n$ which may validate Cl_i; let these be $x_{i_1}^{v_1}, x_{i_2}^{v_2},$ and $x_{i_3}^{v_3}$. Using the 3 disjoint sets of vertices $Pa_j^{(1;i)}, Pa_j^{(2;i)},$ and $Pa_j^{(3;i)}$, with $1 \le j \le i$, we connect the nodes $X_{i_1}^{v_1}, X_{i_2}^{v_2},$ and $X_{i_3}^{v_3}$ to CL_i, using 3 disjoint directed paths (each) of length i. The direction of these edges are from the variable-type nodes towards the clause-type nodes.

In addition to the above edges, for any of the tautology-level nodes TA_j, $1 \le j \le n$, we have two directed edges (X_j^T, TA_j) and (X_j^F, TA_j), representing the two valuations which would validate the corresponding clause.

We fix the set $T = \{TA_j \mid 1 \le j \le n\} \cup \{CL_i \mid 1 \le i \le m\}$ containing n tautology nodes and m clause nodes as our target set.

We prove that the formula P is satisfiable if and only if the target T can be controlled from exactly n control nodes. Moreover, at most one of the valuation nodes associated to a boolean variable can be connected to a driver node.

The above result can be generalized for the case when each of the driver nodes is connected to the network by at most k edges, for any given constant k. That is, each driver node is controlling at most k driven nodes; we call such a system k-bounded.

Definition 2. *We say that for a linear time invariant dynamical system* (A, B, C_T) *the input controller is k-bounded if and only if the matrix B contains at most k non-zero values on every column, i.e., each of the m inputs $u_j(t)$ control at most k of the variables in X.*

Theorem: The k-Bounded Target Control Optimization Problem is NP-hard. Namely, given a linear time invariant dynamical system (A, B, C_T) with a k-bounded input controller and target control set T, finding the minimal set of driven nodes controlling the target T is an NP-hard optimization problem.

Due to space limitations we omit the proof here.

4   Approximation Algorithms for Target Control

We have demonstrated in the previous section that trying to provide the optimal solution for the Target Control problem is computationally hard. An alternative choice is to develop approximation algorithms, trying to get close to the optimal solution in a time-efficient manner.

A first Greedy algorithm for the Target Control problem has been described by Gao et al. [4]. The authors approach the problem from a different perspective, that of generating a linking in an associated network, called the dynamic graph; the method has its roots in earlier studies of Poljak and Murota [13,14].

In the following we present first the approach of Poljak and Murota [14] which connects the target control problem to the linking graph structure. Then, we proceed to presenting the Gao et al. approximation algorithm for target controllability, show its connection to the linking graph approach, and analyze the algorithm's shortcomings. Finally, we introduce three new heuristic improvements of the optimization algorithm, and analyze their performance.

Let (A, B, C_T) be an LTIS over n variables, m inputs, and l targets (i.e., $|T| = l$), and let $G = (V, E)$ be the associated network graph. The dynamical graph $\overline{G}$ is a time-disjoint representation of the network graph, where each state (from $t = 1$ to $t = n$) and each input variable (from $t = 0$ to $t = n - 1$) is viewed as a distinct node at different time points, whereas the target states are associated only with the time-point $t = n+1$. Formally, it is defined as the graph $\overline{G} = (\overline{V}, \overline{E})$, with the set of nodes $\overline{V} = \overline{V_A} \cup \overline{V_B} \cup \overline{V_C}$, where

- $\overline{V_A} = \{v_{i,t} \mid i = 1..n,\ t = 1..n\}$,
- $\overline{V_B} = \{v_{n+j,t} \mid j = 1..m,\ t = 0..n - 1\}$, and
- $\overline{V_C} = \{v_{n+m+k} \mid k = 1..l\}$.

Note that the nodes in $\overline{V_C}$ are in one-to-one correspondence with the nodes V_C, as well as with the target T. The graph $\overline{G}$ has the following set of edges $\overline{E}$:

- $\{(v_{j,t}v_{i,t+1}) \mid$ for all i and j such that $A_{i,j} \neq 0,\ t = 1..n\} \cup$
- $\{(v_{n+j,t}v_{i,t+1}) \mid$ for all i and j such that $B_{i,j} \neq 0,\ t = 0..n-1\} \cup$
- $\{(v_{j,n}v_{n+m+i}) \mid$ for all i and j such that $C_{i,j} \neq 0\}$.

A collection $L = (p_1, p_2, ..., p_k)$ of k edge disjoint paths in the dynamical graph $\overline{G}$ is called a linking of size k. If $S, T \subseteq \overline{V}$ are the sets of initial and terminal nodes of the path L, then we say that L is an (S, T)-linking.

It has been shown in [14] that if (A, B) is an LTIS with m driver nodes (i.e., the number of columns of B is m) and T is a size-l target set which is controllable from these driver nodes, then there must exist an $(\overline{V_B}, \overline{V_C})$-linking of size l. It has been a question for many years whether the converse of the above result also holds. Namely, if for an LTIS (A, B, C_T) there exists an $(\overline{V_B}, \overline{V_C})$-linking of size l, then does it imply that the size-m driver set associated to B is controlling the target T, i.e., $\text{rank}[C_T B, C_T AB, , C_T A^2 B, ..., C_T A^{n-1}B] = l$? Although the answer to this question was proved in [14] to be negative, it became clear that any counter-example for this claim must obey some very strict design conditions regarding the controlling path from the driver nodes to the target.[1] Thus, in practice, finding a collection of nodes V_B such that there exists a $(\overline{V_B}, \overline{V_C})$-linking of size l provides a good candidate for the set of driver nodes controlling the target V_C.

The above approach has been employed by Gao et al. [4] which introduced a Greedy algorithm for the target control problem. Namely, given an LTIS A and a target T, their algorithm searches for a small set V_B for which there exists a $(\overline{V_B}, \overline{V_C})$-linking. In turns, such a set V_B would have a very high probability for defining a set of driver nodes for the target T. However, after applying this algorithm, one has to perform a validation step which verifies whether the selected set of driver/driven nodes selected by the algorithm are indeed controlling the target. This can be done by checking that the rank of the controllability matrix $[C_T B, C_T AB, , C_T A^2 B, ..., C_T A^{n-1}B]$ is indeed equal to $|T|$.

In the following we describe the Gao et al. algorithm [4] and we introduce three new heuristically improved variants of it. The comparative analysis of all these algorithms is performed in Sect. 5

4.1   The Basic Target Control Algorithm (TarCo)

Let A be an LTIS over n variables and let $G = (V_A, E_A)$ be the directed graph associated to it. Let $T \subseteq V_A$ be a set of target variables/nodes. The following algorithm outputs a set of driven nodes D which has a one-to-one correspondence to the searched set V_B for which there exists a $(\overline{V_B}, \overline{V_C})$-linking.

---

[1] An intuitive description of those systems for which a linking is not translated to a valid controlling path is when there exist two targets t_1 and t_2 such that for every path from a driver note d to t_1 there exists another path from d to t_2 using the exact same collection of edges (as a multiset).

Step 1: Let $i = 0$, $C^i = T$, and $D = D^i = \emptyset$.

Step 2: Define a bipartite graph G_{bi} with nodes $L \cup R$, where $L = V_A$, $R = C^i$, and any node appearing both in V_A and in C^i is treated distinctly in L and R. For $l \in L$ and $r \in R$ there exists an edge (l, r) in G_{bi} iff $(l, r) \in E_A$ is an edge in the initial directed graph G.

Step 3: Find a maximum matching (M_L, M_R) in G_{bi}, $M_L \subseteq L$ and $M_R \subseteq R$, and let $C^{i+1} = M_L$ be the set of the left sided matched nodes and let $D_i = R \backslash M_R$ be the set of right sided un-matched nodes. Let $D = D \cup D_i$.

Step 4: We consider C^{i+1} as the new set of target nodes. If $C^{i+1} = \emptyset$ then we complete the algorithm and output D. If not, we proceed to Step 5.

Step 5: If $i < n$ then $i = i+1$ and proceed to Step 2 with the updated target C^i and driven set D. Else, proceed to Step 6.

Step 6: Output D as the set of driven nodes.

Note: The previous algorithm is focussed on minimizing the set of generic driver nodes, and not the set of 1-bounded driver nodes (i.e., driven nodes) focussed on this research. In particular, the algorithm might not output the complete set of driven nodes, but rather a subset of it which is in one-to-one correspondence with the set of generic driver nodes. Indeed, if the algorithm ends in Step 6, then it implies that the target set C^n is non-empty. Since the total number of nodes in G is n, it implies that all the remaining nodes in C^n can be partitioned into a number of cycles. Since the 1-bounded condition for driver nodes is not imposed, all the nodes in these cycles, including the ones in C^n, can be controlled from any driver nodes.

In order to modify the $TarCo$ algorithm for finding a suitable set of driven nodes, instead of driver nodes, we implemented an update/optimization step.

4.2   The Optimized Target Control Algorithm (OpTarCo)

In the $TarCo$ algorithm, once a node x is selected for being a driven node, i.e., added to D in Step 3, we do not check whether until that stage the node x appeared before in some previous control path. If so, now that we know that node x is selected for being a driven node, we can prune that control path after reaching node x. This leads to the following modified algorithm:

As before, let A be an LTIS over n variables and let $G = (V_A, E_A)$ be the directed graph associated to it. Let $T \subseteq V_A$ be the set of target variables/nodes.

Step 1 (Similar to $TarCo$): Let $i = 0$, $C^i = T$, and $D = D^i = \emptyset$.

Step 2 (Similar to $TarCo$): Define a bipartite graph G_{bi} with nodes $L \cup R$, where $L = V_A$, $R = C^i$, and any node appearing both in V_A and in C^i is treated distinctly in L and R. For $l \in L$ and $r \in R$ there exists an edge (l, r) in G_{bi} iff $(l, r) \in E_A$ is an edge in the initial directed graph G.

Step 3.1: Find a maximum matching (M_L, M_R) in G_{bi}, $M_L \subseteq L$ and $M_R \subseteq R$, and let $C^{i+1} = M_L$ be the set of the left sided matched nodes and $D_i = R \backslash M_R$ be the set of right sided un-matched nodes.

Step 3.2: For each $x \in D_i \setminus D$, do:

Step 3.2.1: If node x appears in any previously computed C^j, $j < i$, then remove the entire control path from that occurrence (in C^j) onward, and update all the sets C^k, D^k with $j \leq k \leq i+1$ accordingly. Then update D as $D = \bigcup_{p=0,...,i} D^p$.

End For (from Step 3.2)

Step 4: We consider $D = D \cup D^i$ as the new set of driven nodes, and $C^{i+1} \setminus D$ as the new set of targets. If $C^{i+1} = \emptyset$ then we complete the algorithm and output D. If not, we proceed to Step 5.

Step 5 (Similar to $TarCo$): If $i < n$ then $i = i+1$ and proceed to Step 2 with the updated target C^i and driver set D. Else, proceed to Step 6.

Step 6: For all the remaining nodes in C^n, add them one by one to the driven set D and, at each new addition to D, perform the check from Step 3.2.1, i.e., pruning the existing controlling path for each new addition in D.

Step 7: Output D as the set of driven nodes.

4.3 Heuristically Optimized Target Control Algorithms (HeTarCo1-3)

In Step 3 (resp. 3.1) of the previous two algorithms, at each iteration of the search process we find a maximum matching in between the nodes of G and the current target C^i. However, such maximum matchings might not be unique, in which case some of these maximum matchings might be more suitable to be chosen. Let us assume the algorithm is at some iteration i in its search procedure. Let $C^1, ..., C^i, D^1, ..D^{i-1}$ and D be the already computed sets of targets and driven nodes. Let G_{bi} be the bipartite graph constructed in iteration i, with nodes $L \cup R$, where $L = V_A$, $R = C^i$, and any node appearing both in V_A and in C^i is treated distinctly in L and R. When searching for a maximum matching (M_L, M_R) in G_{bi}, $M_L \subseteq L$ and $M_R \subseteq R$, we are setting the following heuristic criteria for guiding the process towards a minimum number of driven nodes. Note, not all criteria below can be followed in the same time.

– Criteria 1: When computing the maximum matching (M_L, M_R), maximize the use of already driven nodes in M_L.
– Criteria 2: When computing the maximum matching (M_L, M_R) try to avoid the creation of cyclic controlling path. That is, avoid selecting nodes $x \in M_L$ such that there exists $j \leq i$ and a sequence $u_{i+1}, ..., u_j$ such that $u_k \in C^k$ for all $j \leq k \leq i$, $u_{i+1} = u_j = x$, and for all $j \leq k \leq i$, u_j is matched to u_{j+1} in the corresponding bipartite graph.
– Criteria 3: When computing the maximum matching (M_L, M_R), maximize the use of nodes in M_L which have appeared in some previous $C^j, j < i$, on a path that is already controlled (ends with a driven node).
– Criteria 4: When computing the maximum matching (M_L, M_R), maximize the use of nodes in M_L which have appeared in some previous $C^j, j < i$, on a path that is not controlled yet.
– Criteria 5: When computing the maximum matching (M_L, M_R), maximize the use of edges (u, v) (with $u \in M_L$ and $v \in M_R$) which have been used in some previous matching and are part of at least one path that is already controlled.

– Criteria 6: When computing the maximum matching (M_L, M_R), maximize the use of edges (u, v) (with $u \in M_L$ and $v \in M_R$) which have been used in some previous matching, but are not part of any path that is already controlled.

Following subsets of the above selection criteria, as well as the previously introduced optimized control algorithm $(OpTarCo)$ as a base algorithm, we define in the following a series of three heuristically optimized target control algorithms, as follows.

Algorithm $HeTarCo1$: Within Step 3.1 of the $OpTarCo$ algorithm select a maximum matching (M_L, M_R) following Criteria 2.

Algorithm $HeTarCo2$: Within Step 3.1 of the $OpTarCo$ algorithm select a maximum matching (M_L, M_R) following Criteria 1, 2, 3, and 4, in this exact order of importance.

Algorithm $HeTarCo3$: Within Step 3.1 of the $OpTarCo$ algorithm select a maximum matching (M_L, M_R) following Criteria 2, 5, 6, 1, 3, and 4, in this exact order of importance.

5   Results: A Comparative Analysis of the Four Algorithms

We analyzed the performance of the four approximation algorithms against both randomly generated networks and targets, as well as against a human protein-protein interaction network, using as target a set of Breast Cancer specific essential genes.[2]

We predict the performance of all four algorithms to be highly dependent on the size of the network, i.e., the number of nodes and edges, the average degree, the size and the choice of the target set, as well as the overall control-affinity of the network, i.e., the size of its minimal driven set controlling the entire set of nodes. Thus, in order to perform a fair analysis of the algorithms against one-another we impose several conditions for our test cases.

We generated randomly a set of 100 networks, each over 1000 nodes and having exactly 4000 directed edges, i.e., all networks have equal average node (in/out) degree 4.[3] For each network, we randomly selected 10 target sets of size 100, 200, ..., up to 1000 (i.e., all the network's nodes), respectively. We performed 10 (independent) runs for each of the target sets (and networks) with each of the four algorithms; all runs were performed on the same Xeon 6/3 GHz core computer.

---

[2] Note that there is a one-to-one correspondence between genes and proteins; thus, having as target a set of essential genes means the equivalent set of essential proteins.

[3] Similar analyses were performed for networks of average degree from 2 to 6, but due to space limitations we concentrate here over average degree 4 networks; similar results were obtained in all cases, with more pronounced differences (for the normalized values) in the case of higher degree networks.

In order to compare the overall performance of all algorithms, as well as the performance of each individual algorithm applied over different test-cases, we normalize the performance of each run of an algorithm against the minimal total driven control set of the network, i.e., against the size of the minimum set of driven nodes controlling the entire network, as obtained over all runs with all algorithm (40 runs for each of the complete 1000 target node set).[4] Thus, a reported value of 1 for an algorithm's run (for a target and a network) signifies that the size of that solution is equal to the minimal total driven control set of that respective network.

In Fig. 2 we report the comparative analysis of the four algorithms with regards to the average (normalized) size of solutions after each run of the algorithm, Fig. 2(a), by taking to minimum of 10 runs for the same algorithm over the same target, Fig. 2(b), as well as the average time complexity of each individual run, Fig. 2(c). Our analysis shows that in terms of minimality of the driven set solution, all three heuristic algorithms, $HeTarCo1 - 3$ perform slightly better than $OpTarCo$, when the target set is proportionally small compared to the total number of nodes, i.e., less than 15 %. When the target size increases on the other hand, two out of three heuristically improved algorithms perform considerably worse, whereas the third one has a very similar performance as $OpTarCo$. However, in terms of average time taken by each of the algorithms to perform a run, there are up to 10-fold decreases for all heuristically improved algorithms.

In order to better analyze the importance of multiple runs over the solution size decrease we considered testing the four algorithms by doing multiple runs over a fixed time period, namely 12 h. For that we have selected a human protein-protein interaction network consisting of approx. 3000 nodes (i.e., proteins) and 1000 directed edges (i.e., protein interactions); the network was obtained from the SIGNOR (SIGnaling Network Open Resource) database [12]. For targets, we have intersected the set of nodes from the previous network with the list of Breast Cancer essential genes taken from the COLT-Cancer database [7]. In particular, we considered the MDA-MBD-231 cell line and followed the GARP (Gene Activity Rank Profile) and GARP-P values of corresponding proteins mentioned in the database, selecting those entries with negative GARP score and GARP-P value less than 0.05. The rationale behind this particular test-case can be found in network pharmacology: Identifying a relative small set of proteins which could control larger proportions of target essential genes would be advantageous for the development of efficient drugs. By cascading effects, these drugs could target several Breast Cancer essential genes in the same time, with minimal effect over the healthy cells (by definition, disease-specific essential genes are not taken among those genes which are essential for normal, i.e. healthy, cell survival).

---

[4] Note that computing the driven target control is different than computing the driver control, as for the latter one there is a known polynomial time algorithm computing the size of the minimal (total) driver control set, see [10]. In practice however, we observed that the driver and driven control values are very close to one-another in the case of randomly generated networks and real-life bio-medical networks.

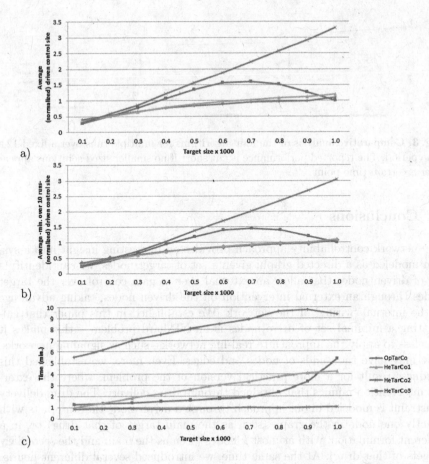

Fig. 2. Comparative analysis of the four algorithms over randomly generated networks. (a) The average (normalized) size of the driven set per algorithm and per target size; (b) The average (normalized) size of the minimal size (over 10 runs) of the driven set per algorithm and per target size; (c) the average time required for a single run, per algorithm and per target size.

The above procedure provided us with a set of 145 essential genes which could be used as target pool. We selected 3 target sets containing 30, 72, and 145 nodes, respectively; the choice of nodes for the smaller/incomplete targets was done according to an increasing ordering of their corresponding GARP values. The results of the 12 h runs of the algorithms are presented in Fig. 3. As it can be seen from this analysis, the heuristically improved algorithms, especially *HeTarCo2* and *HeTarCo3*, performed much better and faster.

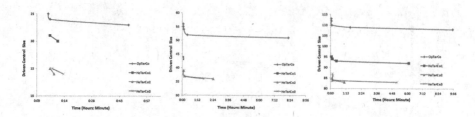

Fig. 3. Comparative analysis of the four algorithms for multiple runs over a fixed 12 h time period. The reported performance is cut short if no smaller sized solutions appear after a certain time point.

6   Conclusions

The network controllability approach provides an interesting insight into a system modeled as a directed graph: given a set of target nodes, we can identify a set of driven nodes that allow an external user to gain control over the target nodes through an external intervention on the driven nodes, taking advantage of the internal 'wiring' of the network. We established in this paper that calculating a minimal set of drive nodes is an NP-hard problem – this makes it hopeless to apply the approach to real-life networks, such as signaling networks, that may have thousands of nodes and edges. Even more, we established this hardness result for a more practical version of the problem, where the external intervention mimics that obtained through drug delivery. The drug delivery constraint is modeled in our approach through a driver node that interacts with exactly one node in the graph (seen as the main target of that drug) or, in a different formulation, with at most k nodes (seen as the main and the secondary targets of that drug). At the same time, we introduced several different heuristics for approximated target control; these algorithms find a set of driven nodes (perhaps not the smallest one) that control a given set of target nodes. Our algorithms improve significantly the currently known algorithm for the problem and we demonstrated in this paper that they are efficiently applicable even to real-life-size networks.

Judging according to our computer simulations, the suitability of each of our heuristically optimized algorithms depends greatly on the size of the target, proportional to the size of the network, and the type of the network, i.e., sparse or dense, homogeneous or non-homogeneous, etc. Determining which particular heuristics works best for some type of networks, and thus designing problem-specific heuristics, remains a topic of investigation. Meanwhile, given also the time efficiency of these heuristic algorithms, we recommend the parallel use of all three of them in determining the minimum target control of any particular network and target.

There are several other highly interesting research avenues that may be explored in this area. On the theoretical side, an open problem is to establish the approximation threshold of our heuristics. Another one, on which almost nothing is known, is on the general, rather than on the structural controllability of networks; in other words, this is the problem in which we also ask about the timing and the level of the external intervention, in addition to identifying the driver nodes where it should be applied. On the applied side, an interesting problem is to connect the network controllability approach with data on FDA-approved drug targets, and with data on gene-essentiality for different types of diseases; this has the potential of helping in the design of more diverse therapeutic strategies using currently known drugs.

References

1. Ashworth, A., Lord, C.J., Reis-Filho, J.S.: Genetic interactions in cancer progression and treatment. Cell **145**(1), 30–38 (2011)
2. Blomen, V.A., et al.: Gene essentiality and synthetic lethality in haploid human cells. Science **350**(6264), 1092–1096 (2015)
3. Brough, R., et al.: Searching for synthetic lethality in cancer. Curr. Opin. Genet. Dev. **21**, 34–41 (2011)
4. Gao, J., Liu, Y., D'Souza, M.R., Barabsi, L.A.: Target control of complex networks. Nat Comm., Article no. 5415 (2014)
5. Hopkins, A.L.: Network pharmacology: the next paradigm in drug discovery. Nat. Chem. Biol. **4**, 682–690 (2008)
6. Kalman, R.E.: Mathematical description of linear dynamical systems. J. Soc. Indus. Appl. Math. Ser. A **1**, 152–192 (1963)
7. Koh, Y.L.J., Brown, R.K., Sayad, A., Kasimer, D., Ketela, T., Moffat, J.: COLT-Cancer: functional genetic screening resource for essential genes in human cancer cell lines. Nucl. Acids Res. **40**(Database issue), D957–D963 (2012). doi:10.1093/nar/gkr959
8. Kolch, W., et al.: The dynamic control of signal transduction networks in cancer cells. Nat. Rev. **15**, 515–525 (2015)
9. Lin, C.-T.: Structural controllability. IEEE Trans. Autom. Control **19**, 201–208 (1974)
10. Liu, Y., Slotine, J.J., Barabsi, L.A.: Controllability of complex networks. Nature **473**, 167–173 (2011)
11. Marcotte, R., Brown, R.K., Suarez, F., et al.: Essential gene profiles in breast, pancreatic, and ovarian cancer cells. Cancer Discov. **2**(2), 172–189 (2012). doi:10.1158/2159-8290
12. Perfetto, L., Briganti, L., Calderone, A.: SIGNOR: a database of causal relationships between biological entities. Nucl. Acids Res. **44**(D1), D548–554 (2016). doi:10.1093/nar/gkv1048
13. Poljak, S.: On the generic dimension of controllable subspaces. IEEE Trans. Autom. Control **35**(3), 367–369 (1990). doi:10.1109/9.50361
14. Poljak, S., Murota, K.: Note on a graph-theoretic criterion for structural output controllability. IEEE Trans. Autom. Control **35**, 939–942 (1990)
15. Shields, R.W., Pearson, J.B.: Structural controllability of multi-input linear systems. IEEE Trans. Autom. Control **21**, 203–212 (1976)
16. Wang, T., et al.: Identification and characterization of essential genes in the human genome. Science **350**(6264), 1096–1101 (2015)
17. Zañudo, J.G.T., Albert, R.: Cell fate reprogramming by control of intracellular network dynamics. PLoS Comput. Biol. **11**(4), e1004193 (2015)
18. Zhan, T., Boutros, M.: Towards a compendium of essential genes - from model organisms to synthetic lethality in cancer cells. Crit. Rev. Biochem. Mol. Biol. **51**(2), 74–85 (2016)

High-Performance Symbolic Parameter Synthesis of Biological Models: A Case Study

Martin Demko, Nikola Beneš, Luboš Brim, Samuel Pastva,
and David Šafránek[✉]

Systems Biology Laboratory, Faculty of Informatics, Masaryk University,
Botanická 68a, 602 00 Brno, Czech Republic
{xdemko,xbenes3,brim,xpastva,xsafran1}@fi.muni.cz

Abstract. Complex behaviour arising in biological systems is described by highly parameterised dynamical models. Most of the parameters are mutually dependent and therefore it is hard and computationally demanding to find admissible parameter values with respect to hypothesised constraints and wet-lab measurements. Recently, we have developed several high-performance techniques for parameter synthesis that are based on parallel coloured model checking. These methods allow to obtain parameter values that guarantee satisfaction of a given set of dynamical properties and parameter constraints. In this paper, we review the applicability of our techniques in the context of biological systems. In particular, we provide an extended analysis of a genetic switch controlling the regulation in mammalian cell cycle phase transition and a synthetic pathway for biodegradation of a toxic pollutant in *E. coli*.

1 Introduction

Dynamical models in systems and synthetic biology are typically represented in terms of systems of ordinary differential equations (ODE). These models are quantitative and their functionality relies on adequate selection of parameter values. Practical parametric identification [19] is in many cases the only solution to obtain parameter values that fit with experimental observations. Due to the limited resolution of wet-lab experimentation, the task is hard. On the other hand, formal methods provide techniques that allow global analysis of model dynamics under uncertainty of parameter values. *Parameter synthesis* is a technique that allows identification of parameter values with respect to a given set of *a priori* known hypotheses (or requirements) on systems dynamics and parameter constraints (e.g., correlations of parameter values, constraints on production/degradation ratio, etc.).

There are several levels of complexity that significantly affect the tractability of parameter synthesis for biological models. First, the procedure requires consideration of all possible settings of parameters – points in the parameter space. The size of the parameter space grows exponentially with the number of

---

This work has been supported by the Czech Science Foundation grant GA15-11089S.

E. Bartocci et al. (Eds.): CMSB 2016, LNBI 9859, pp. 82–97, 2016.
DOI: 10.1007/978-3-319-45177-0_6

unknown parameters. Second, the complexity of the procedure increases with the number of dependencies among individual parameters. Parameter synthesis methods based on model checking or monitoring systems dynamics are capable to deal with parameter dependencies to some extent. On the contrary, traditional parameter estimation methods that search for an optimal fit of the parameters wrt the given data do not directly deal with dependent parameters [19] (manual work is required – model reduction or reformulation).

Several methods and tools based in computer science have been developed to tackle advanced analysis of biological models [4]. Given the complexity of the problem and the need for comprehensive large-scale models, there is a natural call for development of techniques prepared to perform efficiently on high-performance computing platforms [1]. The complexity caused by the state space and parameter space size can be reduced by parallel techniques. The achieved efficiency is again highly dependent on the modelling approach, the nature of models, and the properties considered.

In this paper, we formulate a general framework for parameter synthesis using several recently developed techniques based on model checking [2,7,9]. In particular, these techniques work with a given Computational Tree Logic (CTL) specification of behaviour constraints and a given initial parameter space (potentially refined with *a priori* known constraints on parameter values). In this context, parameter synthesis is solved by the coloured model checking technique extended with symbolic encoding of parameter valuations. The paper focuses primarily on case studies and practical evaluation of the workflow. As a part of the workflow, we apply a recently developed parallel algorithm published in [7].

The contribution of this paper is a general workflow for parameter synthesis based on model checking and application of this workflow to two different biological case studies that are of high importance in systems and synthetic biology. In particular, we provide an extended analysis of a genetic switch controlling the regulation in mammalian cell cycle phase transition and a synthetic pathway for the biodegradation of a toxic pollutant in *E. coli*.

Related Work. Existing techniques for parameter synthesis from temporal specification are either based on model checking performed directly over a qualitative finite quotient of systems dynamics [2,6,9,22], on techniques from hybrid systems [8], and on polytopic set representations employed with on-the-fly refinement of the parameter space [11]. Our technique can be treated as an extension to the first group of techniques. By employing SMT, we obtain a general framework that supports all parameterisations and constraints that can be encoded in a first-order logic over reals. Techniques in [6,15] also fit our workflow well, they encode parameter sets symbolically in terms of polytopes allowing to capture linear dependencies. To the best of our knowledge, the mentioned related techniques have not been parallelised yet.

Monitoring-based techniques [12,20] have an advantage that the function defining the systems dynamics is considered as a black box provided that there is basically no limitation on the form of parameterisation. A significant advantage of these techniques is the ability of quantitative analysis giving detailed insights on the sampled parameter space. In our workflow, we employ these methods in the post-processing phase. Additionally, a promising fact is that numerical solvers can be replaced with SMT solvers for non-linear functions and real domains [14,18]. However, these techniques currently focus on reachability analysis only and their extension to general temporal specifications is a non-trivial task yet to be explored as well as their inclusion into the overall model-checking-based parameter synthesis framework.

2   Methods

The overall procedure of parameter synthesis based on model checking consists of several tasks. The general overview of the workflow is depicted in Fig. 1 (left). In particular, the input of the method is a set of *behaviour constraints* specifying requirements on the systems dynamics, a set of *parameter constraints* collecting all *a priori* known restrictions, dependencies and correlations of individual parameter values, and finally, a dynamical system represented as a potentially non-linear system of ordinary differential equations (*ODE model*).

The workflow can be realised with any parameter synthesis method based on model checking and discretisation of the input ODE model by means of qualitative finite quotient rigorously abstracting its dynamics. In our case, we employ the technique of *coloured model checking* that provides a single-run heuristics

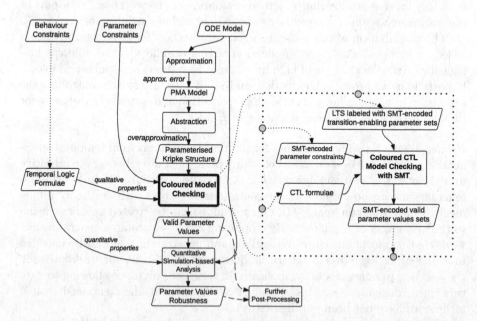

Fig. 1. Overall workflow of parameter synthesis for ODE models based on model checking (left). Refined description of the coloured model checking procedure for the case of SMT-based symbolic representation of parameter sets (right).

to exploration of the parameter space wrt the given behaviour constraints. The output of the method is a collection of sets of parameter values for which the given behaviour and parameter constraints are guaranteed to be satisfied. This result can be further post-processed. First, the SMT-based optimisation tools such as Symba [17] can be used to find approximate parameter values optimal with respect to a given objective function. Second, sampling or statistical model checking can provide detailed exploration of the valid parameter space including quantitative measures such as satisfaction degree [20]. All the steps starting from approximation, abstraction and finally the parameter synthesis are automatised by the respective algorithms. The only input required from the user (in addition to the models and constraints) is a parameter setting the precision of the approximation step. Finally, to automatise processing and visualisation of the resulting data, we have implemented a set of scripts written in the R language.

An important fact is that simulation results performed on the ODE models can be aligned with model checking results only approximately due to the different nature of approximation errors occurring in both methods. To explore the resulting space by monitoring while correctly aligning the results with model checking, the simulation has to be performed on the approximated (PMA) model.

Model. Let $\mathbb{P} \subseteq \mathbb{R}_{\geq 0}^m$ denote the *continuous parameter space* of dimension m. A *biological model* has to be given as a system of ODEs of the form $\dot{x} = f(x, \mu)$ where $x = (x_1, \ldots, x_n) \in \mathbb{R}_{\geq 0}^n$ is a vector of variables, $\mu = (\mu_1, \ldots, \mu_m) \in \mathbb{P}$ is a vector of parameters, and $f = (f_1, \ldots, f_n)$ is a vector where each component is a function constructed as a sum of reaction rates where every sum member represents an affine or bi-linear function of x, or a sigmoidal function of x. In particular, this restriction covers mass action kinetics with stoichiometric coefficients not greater than one and any sigmoidal kinetics such as all significant variants of enzyme or Hill kinetics. An important additional requirement is that each f_i must be affine in μ.

Constraints. To express *behaviour constraints* of interest, we employ the standard branching time logic CTL. The formulae of CTL are defined by the following abstract syntax:

$$\varphi ::= q \mid \neg\varphi \mid \varphi_1 \wedge \varphi_2 \mid \mathbf{AX}\,\varphi \mid \mathbf{EX}\,\varphi \mid \mathbf{A}(\varphi_1\,\mathbf{U}\,\varphi_2) \mid \mathbf{E}(\varphi_1\,\mathbf{U}\,\varphi_2)$$

where q ranges over the atomic propositions from the set AP. We use the standard abbreviations such as $\mathbf{EF}\,\varphi \equiv \mathbf{E}(\mathtt{tt}\,\mathbf{U}\,\varphi)$ and $\mathbf{AG}\,\varphi \equiv \neg\mathbf{EF}\,\neg\varphi$.

To express initial *parameter constraints* on the admissible parameter space, we employ a quantifier-free formula with parameters at the place of variables and inequalities at the place of predicates, denoted as Φ_I.

Approximation. We consider dynamical systems $\dot{x} = f(x, \mu)$ satisfying the criterion that every f_i is piecewise multi-affine (PMA) in x. To achieve that we employ the approach defined in [15]. In particular, each sigmoidal function member in f_i is approximated with an optimal sequence of piecewise affine ramp functions.

In this procedure, a finite number of thresholds is introduced for every component of x. The crucial factor of the approximation error is the number of piecewise affine segments.

Abstraction. We employ the rectangular abstraction [6,15]. We assume that we are given a set of thresholds $\{\theta_1^i, \ldots, \theta_{n_i}^i\}$ for each variable x_i satisfying $\theta_1^i < \theta_2^i < \cdots < \theta_{n_i}^i$. Each f_i is assumed to be multi-affine on each n-dimensional interval $[\theta_{j_1}^1, \theta_{j_1+1}^1] \times \cdots \times [\theta_{j_n}^n, \theta_{j_n+1}^n]$. We call these intervals rectangles. Each rectangle is uniquely identified via an n-tuple of numbers: $R(j_1, \ldots, j_n) = [\theta_{j_1}^1, \theta_{j_1+1}^1] \times \cdots \times [\theta_{j_n}^n, \theta_{j_n+1}^n]$, where the range of each j_i is $\{1, \ldots, n_i - 1\}$. We also define $VR(j_1, \ldots, j_n)$ to be the set of all vertices of $R(j_1, \ldots, j_n)$.

It has been shown that rectangular abstraction is conservative with respect to almost all trajectories of the original (continuous) PMA model [5]. The rectangular abstraction results in a state-transition system, a so-called *parameterised Kripke structure*, defined as $\mathcal{K} = (\mathbb{P}, S, \rightarrow, L)$ with $S = \{(j_1, \ldots, j_n) \mid \forall i : 1 \leq j_i \leq n_i\}$ where each $\alpha \in S$ represents the rectangle $R(\alpha)$. The atomic propositions representing concentration inequalities are assigned to adequate states by means of the labelling L. In particular, $L : S \rightarrow 2^{AP}$ where $AP = \{x_i \odot \theta_j^i \mid 1 \leq i \leq n, 1 \leq j \leq n_i\}, \odot \in \{\leq, \geq\}\}$. The transition relation $\rightarrow$ is defined between any two states representing adjacent rectangles. Each transition is associated with a subset of parameter values under which it is allowed.

Symbolic Encoding of Parameters. Let now $\alpha = (j_1, \ldots, j_n) \in S$, $1 \leq i \leq n$ and $d \in \{-1, +1\}$. We define $\alpha^{i,d} = (j_1, \ldots, j_i + d, \ldots, j_n)$ (if $j_i + d$ is in the valid range). Thus $\alpha^{i,d}$ describe all the neighbouring rectangles of α. We further define $v^{i,+1}(\alpha) = VR(\alpha) \cap \{(\ldots, j_i + 1, \ldots)\}$ and $v^{i,-1}(\alpha) = VR(\alpha) \cap \{(\ldots, j_i, \ldots)\}$. To define the transition relation $\rightarrow$ of the parameterised Kripke structure $\mathcal{K}$, every pair of states $\alpha, \alpha^{i,d} \in S$, $1 \leq i \leq n$, $d \in \{-1, 1\}$, is associated with a formula $\Phi_{\alpha, \alpha^{i,d}}$ symbolically encoding the set of parameter values $\mu \in \mathbb{P}$ for which the transition $\alpha \rightarrow \alpha^{i,d}$ is valid:

$$\Phi_{\alpha, \alpha^{i,d}} := \bigvee_{v \in v^{i,d}(\alpha)} d \cdot f_i(v, \mu) > 0$$

Additionally, the rectangular abstraction approximates the potential existence of a fixed point in any rectangle $\alpha \in S$. This is achieved conservatively by introducing a self-transition for every state provided that every self-transition $\alpha \rightarrow \alpha$ is labelled with the following formula:

$$\Phi_{\alpha, \alpha} := \neg \bigvee_{1 \leq i \leq n} ((\Phi_{\alpha^{i,-1}, \alpha} \wedge \Phi_{\alpha, \alpha^{i,+1}} \wedge \neg \Phi_{\alpha, \alpha^{i,-1}} \wedge \neg \Phi_{\alpha^{i,+1}, \alpha}) \\ \vee (\neg \Phi_{\alpha^{i,-1}, \alpha} \wedge \neg \Phi_{\alpha, \alpha^{i,+1}} \wedge \Phi_{\alpha, \alpha^{i,-1}} \wedge \Phi_{\alpha^{i,+1}, \alpha}))$$

The formula is true just if there is either a pair of transitions $\alpha^{i,-1} \rightarrow \alpha \rightarrow \alpha^{i,+1}$ or a pair of transitions $\alpha^{i,+1} \rightarrow \alpha \rightarrow \alpha^{i,-1}$ provided that the respective two transitions are the only transitions allowed in ith dimension through the rectangle α. According to [6], this situation implies that the zero vector is not included in the convex hull of the vectors in rectangle vertices. That makes a necessary condition for non-existence of a fixed point inside the rectangle. We have found this representation of fixed points as satisfactory for our biological case studies [2, 9].

Remark 1 (Interval-Based Encoding). In case every $f_i(x, \mu)$ depends on at most a single component in μ (there is at most one unknown parameter per equation), encoding of parameter values can be significantly simplified. In particular, in such case all parameters in μ are *mutually independent* and therefore any set of parameter values can be represented as a Cartesian product of closed intervals describing ranges of individual parameters values. The reason for that comes from inherent properties of rectangular abstraction. In consequence, the representation of parameter sets as well as the overall parameter synthesis procedure can be significantly simplified. In [9] we have presented an efficient solution for the respective class of parameterised models that avoids SMT encoding.

Coloured Model Checking Integrated with SMT Solver. The procedure of coloured model checking with SMT-encoded parameter values is briefly illustrated in Fig. 1 (right). Formally, let $\mathcal{K} = (\mathcal{P}, S, I, \rightarrow, L)$ be the parameterised Kripke structure with symbolic description as explained above. Let further φ be a CTL formula over AP. The goal of *coloured CTL model checking* is, given $\mathcal{K}$, Φ_I, and φ, to find the function $\mathcal{F}$ that assigns to every state of the Kripke structure the set of parameters that ensure the satisfaction of the CTL formula. Formally, the function is described as $\mathcal{F}(s) = \{p \in \mathcal{P} \mid p \models \Phi_I, s \models_{\mathcal{K}_p} \varphi\}$.

We have adapted the CTL model checking algorithm to perform on the parameterised Kripke structure with symbolic encoding of parameter values. Briefly, the states are iteratively labelled with all subformulae of φ and the respective SMT formulae representing the parameter values for which the particular subformula is valid. This is realised starting from atomic propositions and back-propagating the information while unfolding the formula structure until the fixed point is reached. In every phase the SMT solver is executed to decide on which transitions are valid for the propagation. The distributed algorithm is presented in [7].

Remark 2 (Consequences of Overapproximation). It is important to note how the overapproximative abstraction affects model checking results. For a formula in the universal fragment of CTL (ACTL), the abstraction causes parameter values synthesised by model checking to be under-approximated wrt the entire set of parameter values for which the formula is exactly valid in the PMA model [5]. For the existential fragment (ECTL), we obtain over-approximation of the exact set of parameter values.

3   Case Study

3.1   Biodegradation of 1,2,3-Trichloropropane in *E. Coli*

Anthropogenic halogenated compounds were unknown to nature until the industrial revolution, and microorganisms have not had sufficient time to evolve

$$\text{TCP} \xrightarrow{\text{DhaA}} \text{DCP} \xrightarrow{\text{HheC}} \text{ECH} \xrightarrow{\text{EchA}} \text{CPD} \xrightarrow{\text{HheC}} \text{GDL} \xrightarrow{\text{EchA}} \text{GLY}$$

$$\frac{d[TCP]}{dt} = -\frac{k_1 \cdot DhaA \cdot [TCP]}{K_{m,1}+[TCP]} \qquad\qquad \frac{d[CPD]}{dt} = \frac{k_3 \cdot EchA \cdot [ECH]}{K_{m,3}+[ECH]} - \frac{k_4 \cdot HheC \cdot [CPD]}{K_{m,4}+[CPD]}$$

$$\frac{d[DCP]}{dt} = \frac{k_1 \cdot DhaA \cdot \cdot [TCP]}{K_{m,1}+[TCP]} - \frac{k_2 \cdot HheC \cdot [DCP]}{K_{m,2}+[DCP]} \qquad \frac{d[GDL]}{dt} = \frac{k_4 \cdot HheC \cdot [CPD]}{K_{m,4}+[CPD]} - \frac{k_5 \cdot HheC \cdot [GDL]}{K_{m,5}+[GDL]}$$

$$\frac{d[ECH]}{dt} = \frac{k_2 \cdot HheC \cdot [DCP]}{K_{m,2}+[DCP]} - \frac{k_3 \cdot EchA \cdot [ECH]}{K_{m,3}+[ECH]} \qquad \frac{d[GLY]}{dt} = \frac{k_5 \cdot HheC \cdot [GDL]}{K_{m,5}+[GDL]}$$

$$k_1 = 1.05, \; k_2 = 0.751, \; k_3 = 14.37, \; k_4 = 2.38, \; k_5 = 3.96,$$
$$K_{m,1} = 1.79, \; K_{m,2} = 1.00, \; K_{m,3} = 0.09, \; K_{m,4} = 0.86, \; K_{m,5} = 3.54$$

Fig. 2. (Up) An abstract scheme of the original system. Note that enzymes HheC and EchA are employed twice on the pathway. The reverse mass flow is considered negligible and abstracted away. (Down) The mathematical model. Enzyme concentrations are considered as constant (and unknown) parameters. Units: $k_x(s^{-1})$, $K_{m,x}(mM)$.

enzymes for their degradation. A synthetic route for conversion of the highly toxic 1,2,3-trichloropropane (TCP) to glycerol (GLY) in Escherichia coli was assembled and the research was published in [16].

TCP is an emerging toxic groundwater pollutant and suspected carcinogen which spreads to the environment mainly due to improper waste management. According to [16] no naturally occurring bacterial pathway capable of degradation of TCP has been found so far. However, a synthetic pathway compound of five intermediates with harmless glycerol as a final product and utilising enzymes from other bacterial species was assembled (Fig. 2).

These enzymes are haloalkane dehalogenase (DhaA) from *Rhodococcus rhodochrous* and haloalcohol dehalogenase (HheC), epoxide hydrolase (EchA) from *Agrobacterium radiobacter*. They have the major role in this pathway. In order to achieve an efficient implementation of the pathway it is important to quantitatively characterise mutual interplay and optimal concentration levels of these enzymes. In general, the higher is the enzyme concentration the higher is the flux rate. Especially, if a substrate and its intermediates are more or less toxic to a host cell such a requirement becomes critical.

Unfortunately, the solution is not straightforward because each of the enzymes has a distinct rate and some of the intermediate products are much more toxic than the others. Additionally, since these enzymes are not natural proteins in *E. coli*, they have to be produced at the expense of other substances. This is called *metabolic burden*. In other words, there must be a balance in concentrations of these enzymes in order to degrade TCP as fast as possible while not killing the host by enervation. Therefore, we employ our workflow to answer the question on optimal enzyme concentration levels.

The original model taken from [13] was reduced in order to minimise the dimensionality of the system. Redundant reactions are eliminated based on their rates and catalytic efficiency (defined as $\frac{k_x}{K_{m,x}}$, see Table 1). In general, greater catalytic efficiency means a faster reaction flux towards generation of the product. Since we need to preserve the number of unknown parameters, the very first

Table 1. Model reactions including enzymes, reaction constants and additional information about catalytic efficiency.

Reaction	Enzyme	Rate (k)	Michaelis const. (K_m)	Cat. efficiency $(\frac{k}{K_m})$
TCP→DCP	DhaA	1.050	1.79	0.587
DCP→ECH	HheC	0.751	1.00	0.751
ECH→CPD	EchA	14.370	0.09	159.670
CPD→GDL	HheC	2.380	0.86	2.767
GDL→GLY	EchA	3.960	3.54	1.119

reaction cannot be omitted. Reaction towards CPD is undeniably the fastest reaction of the model not just due to the best catalytic efficiency but also because of the highest affinity which is an alternative interpretation of the reciprocal Michaelis constant. Therefore this reaction can be omitted in our model. The reaction towards GDL has the second fastest flux and since it is much faster than the last reaction it can be omitted as well. Finally, we have reduced the model to only three reactions which significantly helps to reduce the model state space while making the investigation of the three uncertain parameters tractable.

The desired property is defined verbally as *"complete degradation of TCP as fast as possible with the least accumulated toxicity"*. The notion of toxicity is based on inhibitory concentration of particular molecules. Our framework is designed for manipulation with differential expressions rather than with numerical assignments. Hence we are not able to directly observe actual amount of toxicity. But the toxicity has a direct connection to the concentrations of intermediates. To this end, we translate the desired property as *"TCP completely degrades and the concentration of intermediates does not exceed given bounds"*. The bounds are based on experimental data of the original model (Fig. 3) with the default setting of parameters (DhaA = 0.003, HheC = 0.0036, EchA = 0.0029 (mM)) and initial concentrations ($[TCP] = 2\ mM$, $[other\ species] = 0\ mM$).

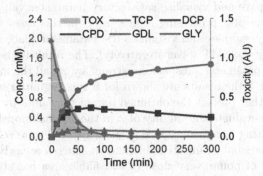

Fig. 3. Experimental data from the original model [16]. We are interested just in the progress of TCP, DCP, GDL and GLY taken as variables of our reduced model. Note the time scale and the maximal reached concentration of DCP and GDL.

Constants are shown in Fig. 2. The presented data reveal that the considered boundary is reasonable for the concentration 0.5 mM or less. In consequence, we proceed by testing various combinations of bounds for GDL and DCP in the interval $[0, 0.5]$ of mM.

It has been mentioned that concentration of enzymes cannot be unlimited due to the metabolic burden (which is not the object of investigation in this paper). According to the default values the initial constraints for these parameters are therefore set to the interval $[0.0, 0.02]$ of mM. The parameter synthesis workflow is employed in order to find parameter values satisfying the desired property. The template of CTL formula expressing the property (denoted as φ) is a combination of several smaller subformulae:

$$\varphi_1 = (\mathbf{AG}\ [TCP] < y), \quad \varphi_2 = (\mathbf{A}([TCP] > x)\mathbf{U}(\mathbf{AF}\ \varphi_1)),$$

$$\varphi_3 = (\mathbf{AG}\ [GLY] > x), \quad \varphi_4 = (\mathbf{A}([GLY] < y)\mathbf{U}(\mathbf{AF}\ \varphi_3)),$$

$$\varphi_6 = (\mathbf{AG}\ [DCP] < v), \quad \varphi_7 = (\mathbf{AG}\ [GDL] < w),$$

$$\varphi_5 = (\varphi_2 \wedge \varphi_4), \quad \varphi_8 = (\varphi_5 \wedge \varphi_6), \quad \varphi = (\varphi_8 \wedge \varphi_7),$$

where x, y, v and w are estimated values making a particular instance of this property. Here $x = 1.9$ (according to [16] where authors use the value 2 mM), $y = 0.01$ (cannot be zero), $v \in \{0.5, 0.3, 0.1\}$ and $w \in \{0.5, 0.25, 0.1\}$ (variations based on an observation of the experimental data in Fig. 3).

The resulting formula is quite large. However, due to the global nature of enumerative CTL model checking algorithm all the subformulae are investigated during the process. This feature can be very convenient in many cases. The computation took more than one day on one computing node while less than 2 h on twenty nodes (each node equipped with common HW – Intel Xeon quad-core 2 GHz and 16 GB of RAM).

The result of parameter synthesis is the set of initial states (satisfying φ) each accompanied with a set of respective values of the parameters (DhaA, HheC, EchA). Results are encoded as a formula in the SMT-LIB format 2.5 [3]. Consequently, to compare and visualise satisfactory parameter values in a human-readable form some post-processing is necessary. In this case, we run a script that systematically samples and visualises the parameter space encoded by the formula (by calling the SMT solver iteratively). The result is the graphical representation of the parameter subspace constraint by φ and the initial parameter constraints. In Fig. 4 the results are shown for a specific initial state.

Finally, we further process the obtained parameter values by numerical simulation in order to evaluate the validity of φ in the original model (Fig. 5). Some points of the resulting parameter space (Fig. 4 (up left)) were selected as representatives of the satisfiable, unsatisfiable and *in-between* area. By *in-between* it is meant the layer of points very close to satisfiable area but still being unsatisfiable. Since we operate on an overapproximated system, the result represents an underapproximation of the exact solution. Hence the points in the *in-between* area might satisfy a given property in the original ODE model even though our framework has refused them.

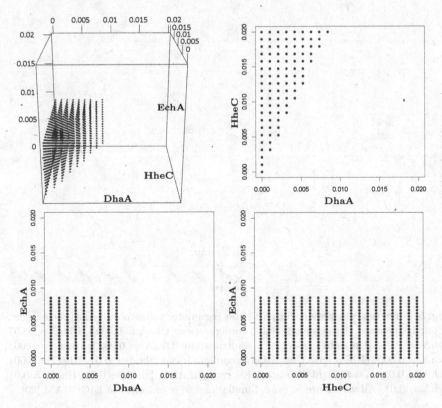

Fig. 4. A sample of the resulting parameter space for a particular initial state: TCP $\in$ [1.9, 1.9586], DCP $\in$ [0.448898, 0.5], GDL $\in$ [0.0, 0.0669138], GLY $\in$ [0.0, 0.01]. Dotted area corresponds to φ ($v = 0.5$, $w = 0.25$). (Up left figure) The 3D space sampled with 400 points per a layer. (Other figures) Projection of the 3D plot for every combination of unknown parameters. All values are shown in mM.

Note that due to the global nature of our algorithm all states satisfying the property have been found. Concentration of all variables in this case study has been restricted to the interval [0.0, 5.0] mM. Our framework reveals parameter values satisfying φ also for initial states beyond the singular initial concentration of particular species considered in [16]. The most interesting are the initial states that increase the upper limit of TCP concentration wrt φ (Fig. 6).

3.2 Regulation of G_1/S Cell Cycle Transition

This case study is focused on extension of previous analysis from [9] of the model describing regulatory interactions controlling a transition between two phases of a mammalian cell cycle [21]. In particular, the model explains the core mechanism behind the irreversible decision for cell division described by a two-gene regulatory network of interactions between the tumour suppressor protein pRB

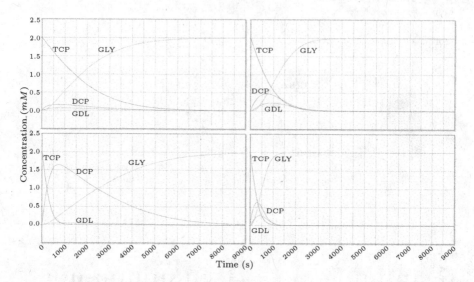

Fig. 5. Numerical simulations for particular parameter values obtained as the outcome of our framework. (Up left) Satisfiable configuration: DhaA = 0.0015, HheC = 0.007, EchA = 0.01. (Up right) *In-between* configuration: DhaA = 0.0035, HheC = 0.005, EchA = 0.005. (Down left) Unsatisfiable configuration: DhaA = 0.01, HheC = 0.001, EchA = 0.01. (Down right) Unsatisfiable configuration: DhaA = 0.01, HheC = 0.01, EchA = 0.01. All values are in mM. Simulations were obtained in BIOCHAM [20].

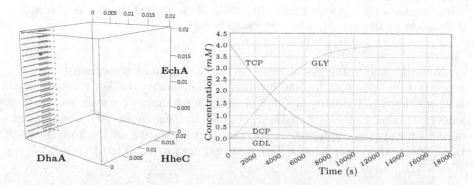

Fig. 6. (Left) Resulting parameter space for a specific initial state: TCP $\in$ [3.84186, 5.0], DCP $\in$ [0.0, 0.448898], GDL $\in$ [0.0, 0.0669138], GLY $\in$ [0.0, 0.01]. The red dot shows the selected point for parameters values: DhaA = 0.001, HheC = 0.005, EchA = 0.015. (Right) Numerical simulation for the selected point. All values are in mM. Simulation was obtained in BIOCHAM [20]. (Color figure online)

and the central transcription factor $E2F1$ (Fig. 7 (left)). For suitable parameter values, two distinct stable attractors may exist (the so-called *bistability*). In [21] a numerical bifurcation analysis of $E2F1$ stable concentration depending on the degradation parameter of pRB (ϕ_{pRB}) has been provided. Note that traditional methods for bifurcation analysis hardly scale to more than a single model parameter.

In this paper we demonstrate that by employing our algorithm we can provide bifurcation analysis for more than one parameter. In particular, we focus on the synthesis of values of two interdependent parameters. We show how the new results complement the results obtained with the algorithm employing the interval-based representation of mutually independent parameters [9]. Additionally, we compare the results achieved within our workflow with the numerical analysis provided in [21].

The property of *bistability* expresses that the system is able to settle in two distinct stable states (i.e., levels of concentration) for specific initial conditions and particular parameter values. It implies existence of a decision-making point (or area) in the system.

The main outcome of the original analysis is shown in Fig. 8 (left) (produced by numerical analysis) displaying the dependency of stable concentration of $E2F1$ on value of ϕ_{pRB} (degradation rate). The most interesting area called *unstable* (for $\phi_{pRB} \in [0.007, 0.027]$) determines feasible values of ϕ_{pRB} wrt the above property. For $\phi_{pRB} < 0.007$ the system converges to a lower-concentration stable equilibrium whereas for $\phi_{pRB} > 0.027$ it converges to a higher-concentration stable equilibrium.

The CTL representation of the property in consideration is $\varphi_1 = (\mathbf{EF\,AG}\,low \land \mathbf{EF\,AG}\,high)$ where $low = (0.5 < E2F1 < 2.5)$ (representing safe cell behaviour) and $high = (4 < E2F1 < 7.5)$ (representing excessive cell division). During the single run of our algorithm all subformulae of φ_1 have been analysed. Let $\varphi_2 = (\mathbf{AG}\,low)$ and $\varphi_3 = (\mathbf{AG}\,high)$ as the most interesting.

In [9] we have investigated perturbations of a single parameter ϕ_{pRB} with the initial constraint $\phi_{pRB} \in [0.001, 0.025]$. According to the Sect. 2 we have first created the PMA approximation of the original ODE model (Fig. 7 (right)) by approximating each non-linear function in the right-hand side of ODEs with a sum of optimal sequence of piecewise affine ramp functions (the precision has been set to 70 automatically generated segments per each non-linear function). For such a setting the verification process took less than 10 seconds on twenty nodes. The results were processed by a Python script (Fig. 8 (right)). The plot

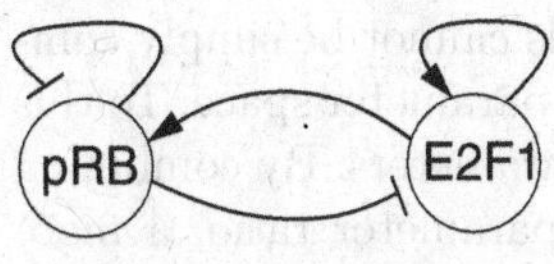

$$\frac{d[pRB]}{dt} = k_1 \frac{[E2F1]}{K_{m1}+[E2F1]} \frac{J_{11}}{J_{11}+[pRB]} - \phi_{pRB}[pRB]$$

$$\frac{d[E2F1]}{dt} = k_p + k_2 \frac{a^2+[E2F1]^2}{K_{m2}^2+[E2F1]^2} \frac{J_{12}}{J_{12}+[pRB]} - \phi_{E2F1}[E2F1]$$

$a = 0.04$, $k_1 = 1$, $k_2 = 1.6$, $k_p = 0.05$, $\phi_{pRB} = 0.005$
$\phi_{E2F1} = 0.1$, $J_{11} = 0.5$, $J_{12} = 5$, $K_{m1} = 0.5$, $K_{m2} = 4$

Fig. 7. G_1/S transition regulatory network (left) and its ODE model (right).

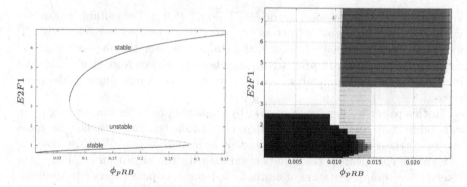

Fig. 8. (Left) Equilibrium curve for $E2F1$ in proportion to ϕ_{pRB} as the result of bifurcation analysis [21] (the authors confirmed the scale of ϕ_{pRB} in the figure should be 0.005-0.035 according to the text). (Right) Model checking results. Red and blue are the *high* and *low* stable regions, respectively. Yellow are the states where φ_1 holds. (Color figure online)

intentionally depicts the same space as the Fig. 8 (left) to show obvious similarities of these results. The blue area stands for stable concentration of $E2F1$ (y-axis) with particular value of ϕ_{pRB} (x-axis) satisfying the property φ_2, whereas the red area satisfies the property φ_3. The yellow area (in the middle) stands for possibility of reaching both stable concentrations. Due to mixing of existential and universal quantifiers (see Sect. 2), the results achieved for φ_1 cannot be exactly interpreted. On the contrary, the results for φ_2 and φ_3 are guaranteed due to the conservativeness of the abstraction.

Although the algorithm based on interval-based encoding performs fast, it is limited to independent parameters only. To overcome this limitation, we have employed the SMT-based algorithm to explore two uncertain mutually dependent parameters. The method is computationally more demanding (about one order of magnitude for each pair of dependent parameters). The goal of our extended analysis is to explore the mutual effect of the degradation parameter of pRB (ϕ_{pRB}) and the production parameter of pRB (k_1) on the bistability. Additionally, we perform post-processing of achieved results by employing additional constraints on the parameter space (e.g., imposing a lower and upper bound on the production/degradation parameter ratio) and show an alternative way of presenting the results.

In particular, we involve the SMT-based tool Symba [17] to obtain an approximated interval of the bounds on valid parameter values. Since the considered parameters are linearly dependent, the resulting intervals cannot be simply combined to display the two-dimensional validity area in the parameter space. To this end, we employ Symba to explore the ratio of the two parameters. By combining initial parameter constraints with the bounds on the parameter ratio, a more accurate parameter subspace is acquired. Such an outcome has been used with the initial constraint $\phi_{pRB} \in [0.001, 0.1]$ and $k_1 \in [0.001, 10]$ (Fig. 9 (up left)).

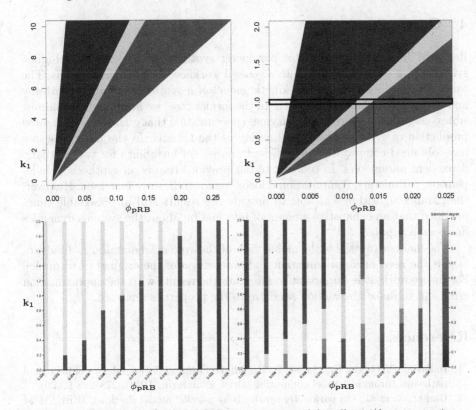

Fig. 9. (Up left) The resulting parameter space merged for all initial concentrations. Each area corresponds to a different property: φ_1 (yellow), φ_2 (blue) and φ_3 (red). (Up right) The same parameter space magnified and projected to ϕ_{pRB}-axis. The framed region agrees with the original numerical bifurcation analysis performed in [21] for ϕ_{pRB}. (Down) Landscapes of the parameter space according to the quantitative satisfaction degree computed by BIOCHAM for φ_2 (left) and φ_3 (right), respectively. (Color figure online)

Additionally, we have explored a refined parameter space ($\phi_{pRB} \in [0.001, 0.025]$ and $k_1 \in [0.001, 2]$) where a one-dimensional projection on the ϕ_{pRB}-axis is highlighted for $k_1 \approx 1$, the default value of k_1 (Fig. 9 (up right)).

The analysis took 8 min on twenty nodes (excluding post-processing). The obtained results can be used as a base for further analysis. We employ the feature of BIOCHAM [10] to compute the *landscape* function that allows investigation of quantitative satisfaction degree of the properties explored (Fig. 9 (down)). LTL reformulation of φ_2 and φ_3 has been used (φ_1 cannot be expressed in LTL). The lighter is the colour the higher the satisfaction degree.

4   Conclusions

Recently developed methods for parameter synthesis of piecewise multi-affine systems have been embedded into a general workflow for biological models. The workflow has been applied to a kinetic model of a synthetic metabolic pathway and to a model of biological switch. In the former case, we have predicted admissible configurations of required enzymes concentration that guarantee the desired production of glycerol under elimination of the toxicity. In the latter case, we have obtained computationally efficient analysis of bistability for two mutually dependent parameters. In contrast to our previous results on synthesis of independent parameters, computational loads were significantly increased. However, the parallel algorithm was able to provide the results still in reasonable times provided that an exhaustive amount of information about the systems dynamics has been computed.

The main advantage is the global view of the systems dynamics. A disadvantage is the need for approximation and abstraction of the original ODE model. For future work, it is important to integrate the results with the approximation error and to make abstraction sensitive to the properties analysed.

References

1. Ballarini, P., Guido, R., Mazza, T., Prandi, D.: Taming the complexity of biological pathways through parallel computing. Brief. Bioinform. **10**(3), 278–288 (2009)
2. Barnat, J., et al.: On parameter synthesis by parallel model checking. IEEE/ACM Trans. Comput. Biol. Bioinform. **9**(3), 693–705 (2012)
3. Barrett, C., Fontaine, P., Tinelli, C.: The SMT-LIB Standard: Version 2.5. Technical report, Department of Computer Science, The University of Iowa (2015)
4. Bartocci, E., Lió, P.: Computational modeling, formal analysis, and tools for systems biology. PLoS Comput. Biol. **12**(1), 1–22 (2016)
5. Batt, G., Belta, C., Weiss, R.: Model checking genetic regulatory networks with parameter uncertainty. In: Bemporad, A., Bicchi, A., Buttazzo, G. (eds.) HSCC 2007. LNCS, vol. 4416, pp. 61–75. Springer, Heidelberg (2007)
6. Batt, G., Yordanov, B., Weiss, R., Belta, C.: Robustness analysis and tuning of synthetic gene networks. Bioinformatics **23**(18), 2415–2422 (2007)
7. Beneš, N., Brim, L., Demko, M., Pastva, S., Šafránek, D.: Parallel SMT-based parameter synthesis with application to piecewise multi-affine systems. In: ATVA 2016. LNCS. Springer (2016) (to appear)
8. Bogomolov, S., Schilling, C., Bartocci, E., Batt, G., Kong, H., Grosu, R.: Abstraction-based parameter synthesis for multiaffine systems. In: Piterman, N., et al. (eds.) HVC 2015. LNCS, vol. 9434, pp. 19–35. Springer, Heidelberg (2015)
9. Brim, L., Češka, M., Demko, M., Pastva, S., Šafránek, D.: Parameter synthesis by parallel coloured CTL model checking. In: Roux, O., Bourdon, J. (eds.) CMSB 2015. LNCS, vol. 9308, pp. 251–263. Springer, Heidelberg (2015)
10. Calzone, L., Fages, F., Soliman, S.: BIOCHAM: an environment for modeling biological systems and formalizing experimental knowledge. Bioinformatics **22**(14), 1805–1807 (2006)

11. Dang, T., Dreossi, T., Piazza, C.: Parameter synthesis through temporal logic specifications. In: Bjørner, N., de Boer, F. (eds.) FM 2015. LNCS, vol. 9109, pp. 213–230. Springer, Heidelberg (2015)

12. Donzé, A., Fanchon, E., Gattepaille, L.M., Maler, O., Tracqui, P.: Robustness analysis and behavior discrimination in enzymatic reaction networks. PLoS ONE **6**(9), e24246 (2011)

13. Dvořák, P.: Engineering of the synthetic metabolic pathway for biodegradation of environmental pollutant. Ph.D. thesis, Masaryk University (2014)

14. Gao, S., Kong, S., Clarke, E.M.: dReal: an SMT solver for nonlinear theories over the reals. In: Bonacina, M.P. (ed.) CADE 2013. LNCS, vol. 7898, pp. 208–214. Springer, Heidelberg (2013)

15. Grosu, R., Batt, G., Fenton, F.H., Glimm, J., Le Guernic, C., Smolka, S.A., Bartocci, E.: From cardiac cells to genetic regulatory networks. In: Gopalakrishnan, G., Qadeer, S. (eds.) CAV 2011. LNCS, vol. 6806, pp. 396–411. Springer, Heidelberg (2011)

16. Kurumbang, N.P., et al.: Computer-assisted engineering of the synthetic pathway for biodegradation of a toxic persistent pollutant. ACS Synth. Biol. **3**(3), 172–181 (2013)

17. Li, Y., Albarghouthi, A., Kincaid, Z., Gurfinkel, A., Chechik, M.: Symbolic optimization with SMT solvers. In: POPL 2014, pp. 607–618. ACM (2014)

18. Madsen, C., Shmarov, F., Zuliani, P.: BioPSy: an SMT-based tool for guaranteed parameter set synthesis of biological models. In: Roux, O., Bourdon, J. (eds.) CMSB 2015. LNCS, vol. 9308, pp. 182–194. Springer, Heidelberg (2015)

19. Raue, A., et al.: Comparison of approaches for parameter identifiability analysis of biological systems. Bioinformatics **30**, 1440–1448 (2014)

20. Rizk, A., Batt, G., Fages, F., Soliman, S.: A general computational method for robustness analysis with applications to synthetic gene networks. Bioinformatics **25**(12), i169–i178 (2009)

21. Swat, M., Kel, A., Herzel, H.: Bifurcation analysis of the regulatory modules of the mammalian G1/S transition. Bioinformatics **20**(10), 1506–1511 (2004)

22. Yordanov, B., Belta, C.: Parameter synthesis for piecewise affine systems from temporal logic specifications. In: Egerstedt, M., Mishra, B. (eds.) HSCC 2008. LNCS, vol. 4981, pp. 542–555. Springer, Heidelberg (2008)

Influence Systems vs Reaction Systems

François Fages[1]([⊠]), Thierry Martinez[2], David A. Rosenblueth[1,3], and Sylvain Soliman[1]

[1] Inria Saclay-Île-de-France, Team Lifeware, Palaiseau, France
{Francois.Fages,Sylvain.Soliman}@inria.fr
[2] Inria Paris, SED, Paris, France
Thierry.Martinez@inria.fr
[3] Instituto de Investigaciones en Matemáticas Aplicadas y en Sistemas (IIMAS), Universidad Nacional Autónoma de México (UNAM), Mexico, D.F., Mexico
drosenbl@unam.mx

Abstract. In Systems Biology, modelers develop more and more reaction-based models to describe the mechanistic biochemical reactions underlying cell processes. They may also work, however, with a simpler formalism of influence graphs, to merely describe the positive and negative influences between molecular species. The first approach is promoted by reaction model exchange formats such as SBML, and tools like CellDesigner, while the second is supported by other tools that have been historically developed to reason about boolean gene regulatory networks. In practice, modelers often reason with both kinds of formalisms, and may find an influence model useful in the process of building a reaction model. In this paper, we introduce a formalism of influence systems with forces, and put it in parallel with reaction systems with kinetics, in order to develop a similar hierarchy of boolean, discrete, stochastic and differential semantics. We show that the expressive power of influence systems is the same as that of reaction systems under the differential semantics, but weaker under the other interpretations, in the sense that some discrete behaviours of reaction systems cannot be expressed by influence systems. This approach leads us to consider a positive boolean semantics which we compare to the asynchronous semantics of gene regulatory networks *à la* Thomas. We study the monotonicity properties of the positive boolean semantics and derive from them an efficient algorithm to compute attractors.

1 Introduction

In Systems Biology, modelers develop more and more reaction models to describe the biochemical reactions underlying cell processes. This approach is promoted by reaction-model exchange formats such as SBML [18] and by the subsequent creation of large reaction-based model repositories such as BioModels [25], without prejudging of their interpretation by differential equations, Markov chains, Petri nets, or boolean transition systems [12].

Modelers can also work, however, with a simpler formalism of *influence systems* to merely describe the positive and negative influences between molecular

© Springer International Publishing AG 2016
E. Bartocci et al. (Eds.): CMSB 2016, LNBI 9859, pp. 98–115, 2016.
DOI: 10.1007/978-3-319-45177-0_7

species, without fixing their implementation with biochemical reactions. In particular, boolean influence systems have been popularized in the 70's by Glass, Kauffman [15] and Thomas [30,31] to reason about gene regulatory networks, represented by ordinary graphs between genes given with a boolean transition table which defines their synchronous or asynchronous boolean transition semantics. Necessary conditions for multi-stability (cell differentiation) and oscillations (homeostasis) have been given in terms of positive or negative circuits in the influence graph [27,29]. Several tools such as GINsim [22], GNA [4] or Griffin [28], use these properties and powerful graph-theoretic and model-checking techniques to automate reasoning about the boolean state transition graph, compute attractors and verify various reachability and path properties. The representation of boolean influence systems by Petri nets was described in [6] but leads to complicated encodings. It is also worth mentioning that influence systems with spatial information have been nicely developed in [7] as a formalism particularly suitable for describing natural algorithms in life sciences and social dynamics.

In Systems Biology, modelers often reason with both kinds of formalisms, and may find it useful to use and maintain an influence model in the process of building a reaction model, for instance in order to reduce it while preserving the essential influence circuits [23]. One reason is that it is easier to visualize influence systems, rather than reaction systems for which complicated graphical conventions such as SBGN [26] have been developed. While it is clear that the influence graph is an abstraction of the reaction hypergraph [12], and perhaps more surprisingly that the Jacobian influence system derived from the differential semantics of a reaction system is largely independent of the kinetics [13], influence models are mostly used for their graphical representation and their boolean semantics, but more rarely as a modeling paradigm for systems biology with quantitative semantics using differential equations, or stochastic semantics.

In this paper, we introduce a formalism of influence systems with forces, which we put in parallel with reaction systems with kinetics, in order to develop a similar hierarchy of boolean, discrete, stochastic and differential semantics for influence systems, similarly to what is done for reasoning about programs in the framework of abstract interpretation [10,12]. We show that the expressive power of influence systems is the same as that of reaction systems under the differential semantics, but is weaker under the other interpretations, in the sense that some formal discrete behaviours of reaction systems cannot be expressed by influence systems. This approach provides an influence model with a hierarchy of possible interpretations related by precise abstraction relationships, so that, for instance, if a behavior is not possible in the boolean semantics, it is surely not possible in the stochastic semantics whatever the influence forces are.

This leads us to consider a positive boolean semantics which we compare to the asynchronous semantics of gene regulatory networks à la Thomas. We study the monotonicity properties of the positive boolean semantics and derive from them an efficient algorithm to compute attractors. These concepts are illustrated with models from the literature.

2 Preliminaries on Reaction Systems with Kinetics

In this article, unless explicitly noted, we will denote by capital letters (e.g. S) sets or multisets, by bold letters (e.g., x) vectors and by small roman or Greek letters elements of those sets or vectors (e.g. real numbers, functions). For a multiset M, let $\mathrm{Set}(M)$ denote the set obtained from the support of M, and brackets like $M(i)$ denote the multiplicity in the multiset (usually the stoichiometry). $\geq$ will denote the pointwise order for vectors, multisets and sets (i.e. inclusion).

2.1 Syntax

We recall here definitions from [11,13] for directed reactions with inhibitors:

Definition 1. *A reaction* over molecular species $S = \{x_1, \ldots, x_s\}$ *is a quadruple* (R, M, P, f), *also noted* f *for* $R/M \Rightarrow P$, *where* R *is a multiset of reactants, M a set of inhibitors, P a multiset of products, all composed of elements of S, and $f : \mathbb{R}^s \to \mathbb{R}$, called* kinetic expression, *is a mathematical function over molecular species concentrations. A reaction system is a finite set of reactions.*

It is worth noting that a molecular species in a reaction can be both a reactant and a product (i.e. a catalyst), or both a reactant and an inhibitor (e.g. Botts–Morales enzymes [19]). Such molecular species are not distinguished in SBML and both are called reaction *modifiers*. Unlike SBML, we find it useful to consider only directed reactions (reversible reactions being represented here by two reactions) and to enforce the following compatibility conditions between the kinetic expression and the structure of a reaction.

Definition 2 ([11,13]). *A reaction (R, M, P, f) over molecular species $\{x_1, \ldots, x_s\}$ is* well formed *if the following conditions hold:*

1. $f(x_1, \ldots, x_s)$ *is a partially differentiable function, non-negative on $\mathbb{R}_+^s$;*
2. $x_i \in R$ *if and only if $\partial f / \partial x_i(x) > 0$ for some value $x \in \mathbb{R}_+^s$;*
3. $x_i \in M$ *if and only if $\partial f / \partial x_i(x) < 0$ for some value $x \in \mathbb{R}_+^s$.*

A reaction system is well formed *if all its reactions are well formed.*

Example 1. The classical prey-predator model of Lotka–Volterra can be represented by the following well-formed reaction system (without reaction inhibitors) between a proliferating prey A and a predator B:

```
k1*A*B for A+B=>2*B.
k2*A for A=>2*A.
k3*B for B=>_.
```

2.2 Hierarchy of Semantics

As detailed in [12], a reaction system can be interpreted with different formalisms that are formally related by abstraction relationships in the framework of abstract interpretation [10] and form a hierarchy of semantics. We simply recall here the definitions of the different semantics of a reaction system.

The *differential* semantics corresponds to the association of an Ordinary Differential Equation (ODE) system with the reactions in the usual way:

$$\frac{dx_j}{dt} = \sum_{(R_i, M_i, P_i, f_i)} (P_i(j) - R_i(j)) \times f_i$$

It is worth noting that in this interpretation, the inhibitors are supposed to decrease the reaction rate but do not prevent the reaction from proceeding with effects on the products and reactants. For instance, in Example 2, we get the classical Lotka–Volterra equations $dB/dt = k1 * A * B - k3 * B$, $dA/dt = k2 * A - k1 * A * B$, and the well-known oscillations between the concentrations of the prey and the predator.

The *stochastic semantics* for reaction systems defines transitions between discrete states describing numbers of each molecule, i.e. vectors x of $\mathbb{N}^s$. A transition is enabled if there are enough reactants, and the reaction propensity is defined by the kinetics:

$$\forall(R_i, M_i, P_i, f_i), x \xrightarrow{f_i}_S x' \text{ with propensity } f_i \text{ if } x \geq R_i, x' = x - R_i + P_i$$

Transition probabilities between discrete states are obtained through normalization of the propensities of all enabled reactions, and the time of next reaction can be computed from the rates *à la* Gillespie [14]. In this interpretation, the inhibitors are supposed to decrease the reaction propensity but do not prevent the reaction from occurring. They are thus ignored here by the stochastic transition conditions as in the differential semantics. In Example 1, the stochastic interpretation can exhibit some oscillations similar to the differential interpretation, and (almost surely) the extinction of the predator.

The *discrete*, or *Petri Net*, semantics is similar but ignores the kinetics and is thus a trivial abstraction of the stochastic semantics by a forgetful functor:

$$\forall(R_i, M_i, P_i, f_i), x \longrightarrow_D x' \text{ if } x \geq R_i, x' = x - R_i + P_i$$

The *boolean semantics* is similar to the *discrete* one but on boolean vectors x of $\mathbb{B}^s$, obtained by the "zero, non-zero" abstraction of integers. With this abstraction, when the number of a molecule is decremented, it can still remain present, or become absent. It is thus necessary to take into account all the possible complete consumption or not of the reactants in order to obtain a correct boolean abstraction of the discrete and stochastic semantics [12]. The *boolean transition system* $\longrightarrow_B$ is thus defined by:

$$\forall(R_i, M_i, P_i, f_i), \forall C \in \mathcal{P}(\text{Set}(R_i)), x \longrightarrow_B x' \text{ if } x \supseteq \text{Set}(R_i), x' = x \setminus C \cup \text{Set}(P_i)$$

It is worth remarking that in Example 2 under this boolean interpretation, one can observe either the stable coexistence of the prey and the predator, or the extinction of the predator with or without the preceding extinction of the prey.

As proven in [12], the last three of these semantics are related by successive Galois connections, which means that *if a behaviour is not possible in the boolean semantics, it is not possible in the stochastic semantics* whatever the reaction kinetics are. On the other hand, the first differential semantics is not an abstraction but rather a limit of the first one for high number of molecules, as shown for instance in [14].

It is worth noticing that the set of inhibitors of a reaction is just a syntactical annotation which has not been used to define the different semantics of the hierarchy. One can also consider a *boolean semantics with negation* where the set of inhibitors of a reaction is seen as a conjunction of negative conditions for the transition (disjunctions can be represented with several reactions). The *boolean with negation transition system* $\longrightarrow_{BN}$ is then defined by:

$$\forall(R_i, M_i, P_i, f_i)\forall C \in \mathcal{P}(\mathrm{Set}(R_i))\boldsymbol{x} \longrightarrow_{BN} \boldsymbol{x}'$$
$$\text{if } \boldsymbol{x} \supseteq \mathrm{Set}(R_i), \boldsymbol{x} \cap M_i = \emptyset, \boldsymbol{x}' = \boldsymbol{x} \setminus C \cup \mathrm{Set}(P_i)$$

However, this strict interpretation of inhibitors by negations restricts the set of possible boolean transitions and is not compatible with the differential semantics, since in that interpretation an inhibitor may just slightly decrease the rate of a reaction without preventing it from proceeding.

2.3   Influence Graph of a Reaction System

Here we recall two definitions of the influence graph associated with a reaction system, and their equivalence under general assumptions [11,13]. The first definition is based on the Jacobian matrix J formed of the partial derivatives $J_{ij} = \partial\dot{x}_i/\partial x_j$, where $\dot{x}_i$ is defined by the differential semantics.

Definition 3. *The* differential influence graph *associated with a reaction system is the graph having for vertices the molecular species, and for edge-set the following two kinds of edges:*

$$\{A \rightarrow^+ B \mid \partial\dot{x}_B/\partial x_A > 0 \text{ for some value } \boldsymbol{x} \in \mathbb{R}^s_+\}$$
$$\cup\{A \rightarrow^- B \mid \partial\dot{x}_B/\partial x_A < 0 \text{ for some value } \boldsymbol{x} \in \mathbb{R}^s_+\}$$

Definition 4. *The* syntactical influence graph *associated with a reaction system M is the graph having for vertices the molecular species, and for edges the following set of positive and negative influences:*

$$\{A \rightarrow^+ B \mid \exists(R_i, M_i, P_i, f_i) \in M, (R_i(A) > 0 \text{ and } P_i(B) - R_i(B) > 0)$$
$$\text{or } (A \in M_i \text{ and } P_i(B) - R_i(B) < 0)\}$$
$$\cup\{A \rightarrow^- B \mid \exists(R_i, M_i, P_i, f_i) \in M, (R_i(A) > 0 \text{ and } P_i(B) - R_i(B) < 0)$$
$$\text{or } (A \in M_i \text{ and } P_i(B) - R_i(B) > 0)\}$$

The syntactical graph is trivial to compute, in linear time, by browsing the syntax of the rules. Both definitions are equivalent under general assumptions:

Theorem 1 ([11, 13]). *For any well-formed reaction system such that the syntactical influence graph contains no conflict (i.e. no pair of the form $A \rightarrow^+ B$ and $A \rightarrow^- B$), the syntactical and differential influence graphs are identical.*

3 Influence Systems with Forces

Reaction systems allow the description of mechanistic models of cell processes, but modelers can also work with a simpler formalism of *influence systems* to merely describe the positive and negative influences between molecular species, without fixing their implementation with biochemical reactions. In this section, we propose a syntax for influence systems with forces which allows us to define a hierarchy of differential, stochastic, discrete and positive boolean semantics, similarly to reaction systems. We then focus on different boolean semantics, and compare them to Thomas's setting for gene regulatory networks.

3.1 Syntax

The idea is to syntactically distinguish conjunctive from disjunctive conditions by writing influences with several sources for representing a conjunction of conditions, while the different influences on a same target express a disjunction of conditions. These syntactical conventions are a particular case of the concept of multiplexes introduced in [5] restricted here to disjunctive normal forms.

Definition 5. *Given $S = \{x_1, \ldots, x_s\}$ a set of species, an influence system I is a set of quintuples (P, N, t, σ, f) called* influences, *where $P \subset S$ is called the positive sources of the influence, $N \subset S$ the negative sources, $t \in S$ is the target, sign $\sigma \in \{+, -\}$ is the sign of the influence, and f is a real-valued mathematical function of $\mathbb{R}^s$, called the* force *of the influence.*

Influences of sign $+$ will be called *positive influences* and those of sign $-$, *negative influences*. In addition, we distinguish the positive sources from the negative sources in an influence (positive or negative), in order to annotate the fact that in the differential semantics, the source increases or decreases the force of the influence, and in the boolean semantics with negation whether the source or the negation of the source is a condition for a change in the target.

For practical reasons we provide an ASCII syntax for influence systems which is used in Biocham[1] v4: they will be written as sequences of lines, where each set is written as a comma-separated sequence of the corresponding species, where the sign is represented as an arrow separating sources from the target, -> for positive influences, and -< for negative influences, and where the positive and negative sources are separated by a / that can be omitted if there is no negative source.

Let us now define the concept of *well-formed* influence systems similarly to the above Definition 2 for reaction systems, with a particular condition for the target of a negative influence, as follows:

---

[1] http://lifeware.inria.fr/biocham.

Definition 6. *An influence (P, N, t, σ, f) over molecular species $\{x_1, \ldots, x_s\}$ is well formed if the following conditions hold:*

1. *$f(x_1, \ldots, x_s)$ is a partially differentiable function, non-negative on $\mathbb{R}_+^s$;*
2. *$x_i \in P$ if and only if $\sigma = +$ (resp. $-$) and $\partial f / \partial x_i(\boldsymbol{x}) > 0$ (resp. < 0) for some value $\boldsymbol{x} \in \mathbb{R}_+^s$;*
3. *$x_i \in N$ if and only if $\sigma = +$ (resp. $-$) and $\partial f / \partial x_i(\boldsymbol{x}) < 0$ (resp. > 0) for some value $\boldsymbol{x} \in \mathbb{R}_+^s$;*
4. *$t \in P$ if $\sigma = -$.*

An influence system is well formed if all its influences are well formed.

Example 2. The prey-predator model of Lotka–Volterra of Example 1 can also be represented by the following well-formed influence system

```
k1*A*B for A,B -< A.
k1*A*B for A,B -> B.
k2*A for A -> A.
k3*B for B -< B.
```

composed of four influences with positive sources only, $(\{A, B\}, \emptyset, A, -, k1 * A * B)$, $(\{A, B\}, \emptyset, B, +, k1 * A * B)$, $(\{A\}, \emptyset, A, +, k2 * A)$ and $(\{B\}, \emptyset, B, -, k3 * B)$.

3.2 Hierarchy of Semantics

Given a set of species $S = \{x_1, \ldots, x_s\}$ and an influence system I over S, the *differential* semantics associates the following ODE system with I:

$$\frac{dx_k}{dt} = \sum_{(P_i, N_i, x_k, +, f_i) \in I} f_i - \sum_{(P_j, N_j, x_k, -, f_j) \in I} f_j$$

Intuitively, it adds up all the forces of the positive influences on x_k and subtracts all forces of negative influences on x_k in the derivative of x_k over time. For instance, in Example 2, we get the same equations as for Example 1.

It is worth noticing that the negative sources in a well-formed influence decrease the force of the influence but do not disable it. Consequently, we define similarly the *stochastic* semantics of an influence system with forces, by a transition system, noted $\longrightarrow_S$, between discrete states, i.e. vectors $\boldsymbol{x}$ of $\mathbb{N}^s$, with the condition that the positive sources are present in sufficient number, with a transition propensity defined by the force, and the target updated as follows:

$$\forall (P_i, N_i, A_i, \sigma_i, f_i), \boldsymbol{x} \xrightarrow{f_i}_S \boldsymbol{x}' \text{ with propensity } f_i \text{ if } \boldsymbol{x} \geq P_i, \boldsymbol{x}' = \boldsymbol{x} \, \sigma_i \, A_i$$

Transition probabilities between discrete states are obtained through normalization of the propensities of all enabled transitions, and the time of next reaction can also be given *à la* Gillespie [14]. In this interpretation, the negative sources are supposed to decrease the influence propensity but do not prevent the influence from proceeding. They are thus ignored here by the stochastic transition conditions.

The *discrete* (or *Petri Net*) semantics simply ignores the forces:

$$\forall (P_i, N_i, A_i, \sigma_i, f_i), x \longrightarrow_D x' \text{ if } x \geq P_i, x' = x \ \sigma_i \ A_i$$

The *positive boolean semantics* is defined on boolean vectors x of $\mathbb{B}^s$, by the "zero, non-zero" abstraction of integers of the discrete semantics. Unlike reaction systems, this boolean semantics associates one transition with one influence:

$$\forall (P_i, N_i, A_i, \sigma_i, f_i), x \longrightarrow_B x' \text{ if } x \geq P_i, x' = x \ \sigma_i \ A_i$$

This boolean semantics is positive in the sense that it ignores the negative sources of an influence and contains no negation in the influence enabling condition. In Example 2 we get the same boolean transitions as in Example 1, although in general one can expect to get more transitions (as shown below in Sect. 3.4).

With these definitions, one can obtain as in [12], a hierarchy of semantics related by simple abstraction relationships (Galois connections in the framework of abstract interpretation [10]) which allows us to state, for instance, that if a behaviour is not possible in the positive boolean semantics, it is also not possible in the discrete or stochastic semantics for any forces.

3.3   Influence Graph of an Influence System

One can define the differential influence graph of an influence system as in Definition 3 for reaction systems, and get a similar equivalence result with the following

Definition 7. *The* syntactical influence graph *associated with an influence system M is the graph having for vertices the molecular species, and for edges the following set of positive and negative influences:*

$$\{A \rightarrow^+ B \mid \ \exists (P_i, N_i, B, \sigma_i, f_i) \in M, \ (A \in P_i \text{ and } \sigma_i = +)$$
$$\text{or } (A \in N_i \text{ and } \sigma_i = -)\}$$
$$\cup \{A \rightarrow^- B \mid \exists (P_i, N_i, B, \sigma_i, f_i) \in M, \ (A \in P_i \text{ and } \sigma_i = -)$$
$$\text{or } (A \in N_i \text{ and } \sigma_i = +)\}$$

Proposition 1. *For a well-formed influence system such that the syntactical influence graph contains no conflict, the syntactical and differential influence graphs are identical.*

Proof. Recall that $\dot{x}_B = \sum_{(P_i, N_i, x_B, +, f_i) \in I} f_i - \sum_{(P_j, N_j, x_B, -, f_j) \in I} f_j$

Hence $\frac{\partial \dot{x}_B}{\partial x_A} = \sum_{(P_i, N_i, x_B, +, f_i) \in I} \frac{\partial f_i}{\partial x_A} - \sum_{(P_j, N_j, x_B, -, f_j) \in I} \frac{\partial f_j}{\partial x_A}$

Since the SIG does not have any conflict, $A \rightarrow^+ B$ is in the SIG (a similar reasoning can be made for $A \rightarrow^- B$) iff

$$\frac{\partial \dot{x}_B}{\partial x_A} = \sum_{(P_i, N_i, x_B, +, f_i) \in I, A \in P_i, A \notin N_i} \frac{\partial f_i}{\partial x_A} - \sum_{(P_j, N_j, x_B, -, f_j) \in I, A \notin P_j, A \in N_j} \frac{\partial f_j}{\partial x_A}$$

Now, since the influence system is well formed, all terms of the left-hand sum are non-negative ($A \notin N_i$) and strictly positive for some points x_i and all terms of the right-hand sum are non-positive ($A \notin P_j$) and strictly negative for some x_j.

We have that $A \rightarrow^+ B$ in the SIG *iff* the above sum has at least one term, which is equivalent to the existence of some x in the state space where one of the terms above is non-null, and therefore $\frac{\partial x_B}{\partial x_A} > 0$, i.e., $A \rightarrow^+ B$ is in the DIG.   □

3.4   Expressive Power Compared to Reaction Systems

Theorem 2. *Any (well-formed) influence system with forces can be represented by a (well-formed) reaction system with kinetics, with the same boolean, discrete, stochastic and differential semantics.*

Proof. Let us represent a positive influence $f\,forP/N \rightarrow t$ by a catalytic synthesis reaction $f \text{ for } P/N \Rightarrow P + t$.

Similarly, let us represent a negative influence $f \text{ for } P/N \prec t$, by an active degradation reaction $f \text{ for } P + t/N \Rightarrow P$.

It is straightforward to verify that the boolean, discrete, stochastic as well as differential semantics recalled and defined above are the same.

Furthermore, the well-formedness condition is preserved. Indeed, this property only depends on the forces/kinetic expressions and on the reactants/inhibitors, which do not change through that transformation thanks to the condition that in *well-formed* influence systems. In addition, $t \in P$ in the case of the negative influence.   □

This theorem shows that an influence system can be simulated by a reaction system for the different semantics. The converse does not hold for the boolean semantics, for instance. Indeed, let us consider boolean semantics of the reaction $C \Rightarrow A + B$ (the kinetics is omitted). We have a transition from the state $(A, B, C) = (0, 0, 1)$ to $(1, 1, 0)$ which is obviously not possible in an influence system since only one variable can change in one transition. However, the converse holds for the differential semantics:

Theorem 3. *Under the differential semantics, (well-formed) influence and reaction systems have the same expressive power.*

Proof. For each reaction (R, M, P, f) of a given reaction system, let us add the following influences:

$$(\text{Set}(R), \text{Set}(M), x_i, +, (P(i) - R(i)) \times f) \text{ when } P(i) - R(i) > 0$$
$$(\text{Set}(R), \text{Set}(M), x_i, -, (R(i) - P(i)) \times f) \text{ when } P(i) - R(i) < 0$$

The associated ODE system collects all $(P_i - R_i) \times f$ exactly as in the differential semantics of the original reaction system. Furthermore, it is easy to check that these influences are well formed since the original reaction is well formed and the force is only a positive integer multiplied by the original kinetic expression.   □

This theorem shows that as far as the differential semantics is concerned, the influence systems have the same expressive power as reaction systems and there is no theoretical reason to develop a reaction model. This does not mean that there is a canonical reaction system associated with an influence system. Generally, different implementations with reactions are possible without changing the differential semantics. They represent extra information that is irrelevant to the analysis or simulation of the differential equations, but could lead to different stochastic simulations, for instance.

3.5   Boolean Semantics with Negation

One can also consider a boolean semantics with negation for influence systems, where the negative sources are interpreted as negations in the enabling condition. Formally, the *boolean with negation semantics* of an influence system is then defined by the following transition system:

$$\forall (P_i, N_i, A_i, \sigma_i, f_i), x \longrightarrow_B x' \text{ if } x \geq P_i, \ x \cap N_i = \emptyset, \ x' = x \ \sigma_i \ A_i$$

This interpretation allows us to represent more boolean transition semantics. Let us call a *unitary transition system*, a transition system that updates *at most one* variable of x in each transition. It is worth remarking that in this case, the state transition graph lives on a hypercube (e.g. Fig. 2 of Sect. 5).

Proposition 2. *Any unitary boolean transition system can be represented by an influence system under the boolean semantics with negation.*

Proof. It is sufficient to notice that since a unitary boolean transition $s \longrightarrow_{BN} s'$ changes at most one species, say x_i, from s to s', it can be represented by either a positive influence, $(P, N, x_i, +)$, if $s'(x_i) = 1$, or a negative influence, $(P, N, x_i, -)$, if $s'(x_i) = 0$, with $P = \{x \mid s(x) = 1\}$ and $N = \{x \mid s(x) = 0\}$.   □

Let us call a *positive boolean transition system* one that contains no negation in the conditions for enabling a transition, i.e. if a transition is enabled when $s(x_i) = 0$ then it is also enabled when $s(x_i') = 1$.

Corollary 1. *Any unitary positive boolean transition system can be represented by an influence system under the positive boolean semantics.*

Proof. In the influence system associated by Proposition 2 with a positive unitary transition system, any influence that has negative sources is doubled by a counterpart influence where such sources are positive (by definition of positive boolean transition system). Therefore, in the associated influence system, the negative sources can be simply ignored, as done by the positive boolean semantics.   □

3.6   Functional Boolean Semantics with Negation *à la* Thomas

The boolean semantics defined by René Thomas originally for gene networks [31], is *functional*, in the sense that the next boolean state x' is defined by a boolean function $\phi(x)$, not a relation. In this setting, the synchronous semantics is deterministic, and the non-deterministic asynchronous semantics is obtained by interleaving, by considering all the possible transitions that change the boolean value of *exactly one* of the genes at a time. A truly non-deterministic influence system such as $\{(A, \emptyset, B, +, f), (A, \emptyset, B, -, g)\}$, for which the transition relation is not a function, cannot be represented. Thomas's setting excludes self-loops in the state transition graph and all steady states are stable (i.e. terminal states):

Proposition 3. *The boolean transition systems definable by Thomas's regulatory networks are the unitary boolean transition systems without self-loops.*

Proof. A Thomas's transition graph is necessarily unitary and without self-loops since each transition changes the boolean value of exactly (not at most) one variable at a time. The converse follows from Proposition 2 by excluding the possibility of having self-loop transitions which change no variable. □

This restriction to transition functions is even more striking in Thomas's multilevel setting, where the above system can (in the discrete semantics) have transitions from $(1, 1)$ both to $(1, 0)$ and to $(1, 2)$. That would necessitate the corresponding logical parameter for B to be at the same time < 1 and > 1. It is worth noting that despite this restriction, the logical formalism of Thomas is successfully used in a wide variety of models [16,17,24,32] in systems biology.

4   Properties of the Positive Boolean Semantics

In this section, we focus on the positive boolean semantics of influence systems and study its properties. Recall that $\leq$ denotes the pointwise order on $\{0, 1\}$ coordinates of vectors representing states.

Proposition 4 (Monotonicity). *The positive boolean semantics of influence systems is monotonic: let I be an influence system over $S = \{x_1, \ldots, x_s\}$ and v_1, v_2 be two boolean states, i.e., vectors of $\mathbb{B}^s$*

$$
\begin{array}{ccc}
 & v_1 \longrightarrow v_1' & \\
 & \leq \Big\downarrow \qquad \vdots \leq & \\
 & v_2 \dashrightarrow v_2' &
\end{array}
$$

if $v_1 \leq v_2$ then $\forall v_1', v_1 \longrightarrow v_1', \exists v_2', v_1' \leq v_2'$ and $v_2 \longrightarrow v_2'$

Proof. One can simply notice that since there are no negations in the enabling conditions, any influence that is enabled in v_1 is also enabled in v_2. □

It is worth noticing that this monotonicity property for transitions is fundamentally different from that of monotone dynamical systems [3] which are deterministic, and therefore impose the monotonicity property on the *unique* image of v_1 and v_2. In our setting, Proposition 4 states that there exists some $v_2' \geq v_1'$, but the existence of negative influences in the system permits that some other images of v_2 might not be greater than v_1'. Nevertheless, we have

Proposition 5 (Greatest element). *Let C be a Terminal Strongly Connected Component (TSCC) of the state transition graph of a positive influence system, then C has a greatest element: $\exists v_0 \in C, \forall v \in C, v \leq v_0$.*

Proof. Let us prove this proposition by contradiction: assume that there are two incomparable maximal elements v_1 and v_2 in C. Since C is strongly connected there is a path from v_1 to v_2 and along that path a state v_3 and its successor in the path v_4 such that $v_3 \leq v_1$ and $v_4 \not\leq v_1$, as $v_2 \not\leq v_1$. Now, using Proposition 4 we get that $v_1 \longrightarrow v_1'$ with $v_4 \leq v_1'$ and $v_1' \in C$ since C is terminal. However v_1' is either greater or less than v_1 since it is the result of applying a single influence. If $v_1 < v_1'$ we have a contradiction as we supposed v_1 maximal. If $v_1' \leq v_1$ we get $v_4 \leq v_1$ by transitivity and that is also contradictory. $\square$

Corollary 2. *To enumerate the attractors, i.e., TSCCs, of a positive influence system, it is enough to check the strongly connected components (SCCs) of states that have no strictly increasing transition.*

Proof. This is an immediate consequence of Proposition 5 as each TSCC can be represented by its greatest element which has no strictly increasing transition. $\square$

Notice that stable states are a particular case with no strictly decreasing transition either. Moreover, any strictly decreasing transition should be "reversible" for the SCC to be a TSCC. This allows us to rule out potential TSCC candidates without exploring their whole SCC in Algorithm 1 (implementation available in Biocham v4).

Algorithm 1. TSCC maximal elements candidates enumeration algorithm

---

procedure LIST_TSCC_CANDIDATES
    $Constraints \leftarrow \{P \wedge \neg N \implies t \mid (P, N, +, t, f) \in I\}$
                                ▷ Enabled positive influences must not change the state
    $Candidates \leftarrow$ ENUMERATESOLUTIONS($Constraints$)
    for $C \in Candidates$ **do**
        if C has no strictly decreasing transition **then**
            C is a *stable* steady state
        else if C has a non-reversible strictly decreasing transition **then**
            C is *not* in a TSCC
        else
            C's SCC must be explored to check if it is a TSCC
        end if
    end for
end procedure
function ENUMERATESOLUTIONS($Constraints$)
    Iteratively solve by SAT/CP the CSP defined by $Constraints$
    return The set of solutions
end function

Proposition 6. *Given an influence system, there is at least one TSCC of the original influence system in each TSCC of its positive semantics.*

Proof. The positive semantics only adds transitions by enabling more influences, it can therefore only merge TSCCs. $\square$

This result suggests finding complex attractors of non-positive systems, such as logical models à la Thomas [27,29], by enumerating the greatest elements of the TSCCs of their positive boolean semantics, and then looking for attractors of the original system. This approach provides an over-approximation of the attractors and is complementary to recent works which provide lower-bounds on their number [20].

5   Examples

5.1   Influence Model of p53/Mdm2 DNA Damage Repair System [1]

The p53/Mdm2 DNA damage repair system is an interesting oscillatory system which has been first modeled in [8] by a reaction system with differential equations, and then in [1,2] by simplified influence systems with discrete and differential semantics respectively.

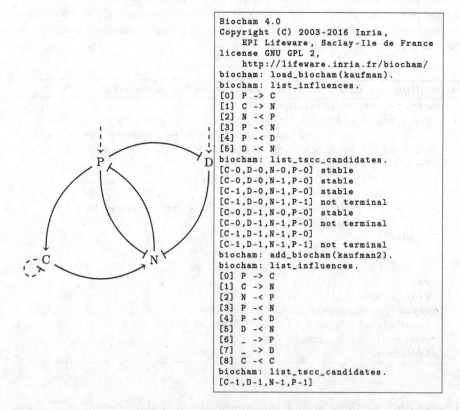

```
Biocham 4.0
Copyright (C) 2003-2016 Inria,
    EPI Lifeware, Saclay-Ile de France
license GNU GPL 2,
    http://lifeware.inria.fr/biocham/
biocham: load_biocham(kaufman).
biocham: list_influences.
[0] P -> C
[1] C -> N
[2] N -< P
[3] P -< N
[4] P -< D
[5] D -< N
biocham: list_tscc_candidates.
[C-0,D-0,N-0,P-0]  stable
[C-0,D-0,N-1,P-0]  stable
[C-1,D-0,N-1,P-0]  stable
[C-1,D-0,N-1,P-1]  not terminal
[C-0,D-1,N-0,P-0]  stable
[C-0,D-1,N-1,P-0]  not terminal
[C-1,D-1,N-1,P-0]
[C-1,D-1,N-1,P-1]  not terminal
biocham: add_biocham(kaufman2).
biocham: list_influences.
[0] P -> C
[1] C -> N
[2] N -< P
[3] P -< N
[4] P -< D
[5] D -< N
[6] _ -> P
[7] _ -> D
[8] C -< C
biocham: list_tscc_candidates.
[C-1,D-1,N-1,P-1]
```

Fig. 1. Left: Influence graph displayed in Fig. 4 of [1], without the activation multi-levels. The dashed influences are those added in our second version of the model. Right: Biocham v4 session for computing the TSCC in both influence systems.

We illustrate here the search for TSCCs with two versions of the influence model of [1]. In the first model, we simply transcribe the graph of Fig. 4 of the authors as a boolean influence system. We therefore ignore the multi-level aspect they developed. In the second model, we add some activations on p53 and DNA-damage, and an inhibition on cytoplasmic Mdm2, in order to take into account some basal state of the model. The influence systems and the computed TSCCs are listed in the Biocham session depicted in Fig. 1.

Our algorithm shows that there is in each case a single complex attractor (i.e. not marked as stable or not terminal), accordingly to [1], and four stable steady states in the first case. Note that in [2], this influence model was further extended with differential and stochastic dynamics which could be represented in our setting by influence forces.

5.2 Influence Model of the Mammalian Circadian Clock [9]

A good example of the use of logical models *à la* Thomas is the recent paper by Comet et al. [9] studying different variants of small models of the circadian rhythms in mammals. A direct import in Biocham v4 of the logical model of Sect. 5 of [9] gives the following influence system with negative sources:

```
_ / L -> L.
L -< L.
_ / G, PC -> G.
G, PC -< G.
G / PC, L -> PC.
PC / G -< PC.
PC, L -< PC.
```

The positive semantics of this system is close to the original boolean semantics with negation *à la Thomas* of the model. They both have a single TSCC: the vector $(1, 1, 1)$ that is found by the command list_tscc_candidates as sole candidate. Furthermore, only a few state transitions become reversible in the positive boolean semantics, while they are irreversible in the original boolean semantics with negation *à la Thomas* of the model, as depicted in Fig. 2.

The approximation introduced by the positive boolean semantics can be explained by quantitative dynamics considerations. For instance, when G is on, the transcription leading to the PER-CRY complexes is stimulated, however [9] explains that these complexes can only migrate to the nucleus in absence of light. This *absence* cannot be checked in a positive semantics model, however the consensus mechanistic process is rather thought to be a modulation of PER transcription by light (see for instance [21] for the mammalian case). Being purely quantitative, it is not easy to take into account such a regulation in a boolean model except with the reversible activation of PC when G is on, whether L is on or not. This is what happens in our positive model as can be seen in the right panel of Fig. 2, and it is similar to what happens for the light in the original model.

The same reasoning explains the reversible inactivation of G when PC is active. Indeed there is a basal synthesis of G that cannot check, in a positive

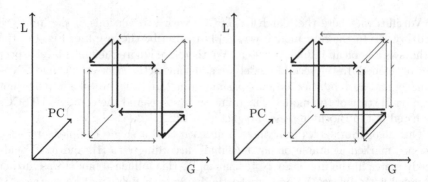

Fig. 2. State transition graphs of the model under, **Left:** the boolean semantics with negation *à la* Thomas, similar to Fig. 7 of [9], **Right:** the positive boolean semantics, where some state transitions have become reversible.

setting, that PC is inactive in order to activate the genes. Once again, the mechanistic process is a quantitative inhibition of the CLOCK-BMAL1 complexes by PER-CRY and a conservative boolean approximation of that process is reflected by the reversible activation of G in presence of PC.

In [9], the authors also restrict the possible behaviours by introducing delays for the boolean transitions which could be considered as a further expansion of the formalism.

6   Discussion

In this paper, we hope to have clarified some differences between influence systems and reaction systems, and especially some subtle discrepancies between the precise boolean semantics that have been considered in the literature. As far as the modeling of one biological system is concerned, the modeler can work with one formalism and one tool to answer the questions about their model. Nevertheless, as soon as different modeling tools are to be used, or the model has to be communicated and reused for another purpose, understanding and mastering these discrepancies in the semantics of the interactions become crucial.

We have shown that, for influence systems and reaction systems with inhibitors, one can obtain a hierarchy of semantics which goes from the concrete stochastic semantics to a discrete Petri net, and then a positive boolean semantics in which the inhibitors of the reactions or influences are just ignored. This is consistent with the fact that the inhibitors decrease the rate or force in the quantitative semantics, but do not really prevent the reaction or influence from proceeding. This convention thus ensures that all discrete behaviours are approximated when we go up in the abstractions of the hierarchy of semantics, and that if a behaviour is not possible in the positive boolean semantics (which can be checked by model-checking methods for instance) it is not possible in the stochastic semantics for any forces. Furthermore, we have shown that in the positive boolean semantics, the monotonicity of the transition relation allows us to enumerate the complex attractors more efficiently by restricting the search to the greatest elements candidates.

On the other hand, the boolean semantics à la Thomas of influence systems, interprets inhibitors as negations, and contains a restriction on the definition of the transition relation by a function, not a relation, which limits the sources of non-determinism. We have shown that the boolean semantics with negation leads to a more expressive formalism in which any unitary boolean transition system can be encoded, but does not correspond to an abstraction of the stochastic semantics, unless the stochastic transitions interprets inhibitors as negative conditions which does not correspond to the differential semantics. With the functional restriction, we have proven that each TSCC in the positive semantics contains at least one TSCC of the semantics à la Thomas, and thus that our algorithm can be used to prune the search space in this setting also.

We have also shown that reaction systems and influence systems have the same expressive power under the differential semantics. This means that, as far as the differential equations are concerned, the details given in the reactant-product structure of a reaction system are not necessary, and that the same differential equations can be derived from an influence system with forces. Several reaction systems can be associated with an influence system with the same differential semantics. This leaves open the design of canonical forms for reaction systems, and computer tools for automatically maintaining the implementation of an influence system by a reaction system.

Acknowledgements. We are grateful to Paul Ruet for interesting discussions on Thomas's framework, and to the reviewers for their comments. This work was partially supported by ANR project Hyclock under contract ANR-14-CE09-0011, and PASPA-DGAPA-UNAM, Conacyt grants 221341 and 261225.

References

1. Abou-Jaoudé, W., Ouattara, D.A., Kaufman, M.: From structure to dynamics: frequency tuning in the p53-Mdm2 network: I. logical approach. J. Theor. Biol. **258**(4), 561–577 (2009)
2. Abou-Jaoudé, W., Ouattara, D.A., Kaufman, M.: From structure to dynamics: frequency tuning in the p53-Mdm2 network: II differential and stochastic approaches. J. Theor. Biol. **264**, 1177–1189 (2010)
3. Angeli, D., Sontag, E.D.: Monotone control systems. IEEE Trans. Autom. Control **48**(10), 1684–1698 (2003)
4. Batt, G., et al.: Genetic Network Analyzer: a tool for the qualitative modeling and simulation of bacterial regulatory networks. In: van Helden, J., Toussaint, A., Thieffry, D. (eds.) Bacterial Molecular Networks. Methods in Molecular Biology, vol. 804, pp. 439–462. Springer, New York (2012)
5. Bernot, G., Comet, J.P., Khalis, Z.: Gene regulatory networks wih multiplexes. In: Proceedings of European Simulation and Modelling Conference, ESM 2008. pp. 423–432 (2008)
6. Chaouiya, C.: Petri net modelling of biological networks. Brief. Bioinform. **8**, 210 (2007)

7. Chazelle, B.: Natural algorithms and influence systems. Commun. ACM **55**(12), 101–110 (2012)
8. Ciliberto, A., Capuani, F., Tyson, J.J.: Modeling networks of coupled enzymatic reactions using the total quasi-steady state approximation. PLOS Comput. Biol. **3**(3), e45 (2007)
9. Comet, J.P., Bernot, G., Das, A., Diener, F., Massot, C., Cessieux, A.: Simplified models for the mammalian circadian clock. Procedia Comput. Sci. **11**, 127–138 (2012)
10. Cousot, P., Cousot, R.: Abstract interpretation: a unified lattice model for static analysis of programs by construction or approximation of fixpoints. In: Proceedings of the 6th ACM Symposium on Principles of Programming Languages, POPL1977, pp. 238–252, Los Angeles. ACM, New York (1977)
11. Fages, F., Gay, S., Soliman, S.: Inferring reaction systems from ordinary differential equations. Theor. Comput. Sci. **599**, 64–78 (2015)
12. Fages, F., Soliman, S.: Abstract interpretation and types for systems biology. Theor. Comput. Sci. **403**(1), 52–70 (2008)
13. Fages, F., Soliman, S.: From reaction models to influence graphs and back: a theorem. In: Fisher, J. (ed.) FMSB 2008. LNCS (LNBI), vol. 5054, pp. 90–102. Springer, Heidelberg (2008)
14. Gillespie, D.T.: Exact stochastic simulation of coupled chemical reactions. J. Phys. Chem. **81**(25), 2340–2361 (1977)
15. Glass, L., Kauffman, S.A.: The logical analysis of continuous, non-linear biochemical control networks. J. Theor. Biol. **39**(1), 103–129 (1973)
16. González, A.G., Chaouiya, C., Thieffry, D.: Qualitative dynamical modelling of the formation of the anterior-posterior compartment boundary in the drosophila wing imaginal disc. Bioinformatics **24**, 234–240 (2008)
17. Grieco, L., Calzone, L., Bernard-Pierrot, I., Radvanyi, F., Kahn-Perlès, B., Thieffry, D.: Integrative modelling of the influence of mapk network on cancer cell fate decision. PLOS Comput. Biol. **9**(10), e1003286 (2013)
18. Hucka, M., et al.: The systems biology markup language (SBML): a medium for representation and exchange of biochemical network models. Bioinformatics **19**(4), 524–531 (2003). http://sbml.org/
19. Katsumata, M.: Graphic representation of Botts-Morales equation for enzyme-substrate-modifier system. J. Theor. Biol. **36**(2), 327–338 (1972)
20. Klarner, H., Bockmayr, A., Siebert, H.: Computing symbolic steady states of Boolean networks. In: Wąs, J., Sirakoulis, G.C., Bandini, S. (eds.) ACRI 2014. LNCS, vol. 8751, pp. 561–570. Springer, Heidelberg (2014)
21. Leloup, J.C., Goldbeter, A.: Toward a detailed computational model for the mammalian circadian clock. Proc. Nat. Acad. Sci. **100**, 7051–7056 (2003)
22. Naldi, A., Berenguier, D., Fauré, A., Lopez, F., Thieffry, D., Chaouiya, C.: Logical modelling of regulatory networks with GINsim 2.3. Biosystems **97**(2), 134–139 (2009)
23. Naldi, A., Remy, E., Thieffry, D., Chaouiya, C.: A reduction of logical regulatory graphs preserving essential dynamical properties. In: Degano, P., Gorrieri, R. (eds.) CMSB 2009. LNCS, vol. 5688, pp. 266–280. Springer, Heidelberg (2009)
24. Naldi, A., Carneiro, J., Chaouiya, C., Thieffry, D.: Diversity and plasticity of th cell types predicted from regulatory network modelling. PLoS Comput. Biol. **6**(9), e1000912 (2010)

25. le Novère, N., Bornstein, B., Broicher, A., Courtot, M., Donizelli, M., Dharuri, H., Li, L., Sauro, H., Schilstra, M., Shapiro, B., Snoep, J.L., Hucka, M.: BioModels database: a free, centralized database of curated, published, quantitative kinetic models of biochemical and cellular systems. Nucleic Acid Res. 1(34), D689–D691 (2006)

26. le Novere, N., Hucka, M., Mi, H., Moodie, S., Schreiber, F., Sorokin, A., Demir, E., Wegner, K., Aladjem, M.I., Wimalaratne, S.M., Bergman, F.T., Gauges, R., Ghazal, P., Kawaji, H., Li, L., Matsuoka, Y., Villeger, A., Boyd, S.E., Calzone, L., Courtot, M., Dogrusoz, U., Freeman, T.C., Funahashi, A., Ghosh, S., Jouraku, A., Kim, S., Kolpakov, F., Luna, A., Sahle, S., Schmidt, E., Watterson, S., Wu, G., Goryanin, I., Kell, D.B., Sander, C., Sauro, H., Snoep, J.L., Kohn, K., Kitano, H.: The systems biology graphical notation. Nat. Biotechnol. 27(8), 735–741 (2009)

27. Remy, E., Ruet, P., Thieffry, D.: Graphic requirements for multistability and attractive cycles in a Boolean dynamical framework. Adv. Appl. Math. 41(3), 335–350 (2008)

28. Rosenblueth, D.A., Muñoz, S., Carrillo, M., Azpeitia, E.: Inference of Boolean networks from gene interaction graphs using a SAT solver. In: Dediu, A.-H., Martín-Vide, C., Truthe, B. (eds.) AlCoB 2014. LNCS, vol. 8542, pp. 235–246. Springer, Heidelberg (2014)

29. Ruet, P.: Local cycles and dynamical properties of Boolean networks. Math. Found. Comput. Sci. 26(4), 702–718 (2016)

30. Thomas, R.: Boolean formalisation of genetic control circuits. J. Theor. Biol. 42, 565–583 (1973)

31. Thomas, R., D'Ari, R.: Biological Feedback. CRC Press, Boca Raton (1990)

32. Traynard, P., Fauré, A., Fages, F., Thieffry, D.: Logical model specification aided by model- checking techniques: application to the mammalian cell cycle regulation. Bioinformatics, Special issue of ECCB (2016)

Local Traces: An Over-Approximation of the Behaviour of the Proteins in Rule-Based Models

Jérôme Feret[(✉)] and Kim Quyên Lý[(✉)]

DI-ENS (INRIA/ÉNS/CNRS/PSL), Paris, France
feret@ens.fr, quyen@di.ens.fr

Abstract. Thanks to rule-based modelling languages, we can assemble large sets of mechanistic protein-protein interactions within integrated models. Our goal would be to understand how the behaviour of these systems emerges from these low-level interactions. Yet this is a quite long term challenge and it is desirable to offer intermediary levels of abstraction, so as to get a better understanding of the models and to increase our confidence within our mechanistic assumptions.

In this paper, we propose an abstract interpretation of the behaviour of each protein, in isolation. Given a model written in Kappa, this abstraction computes for each kind of protein a transition system that describes which conformations this protein can take and how a protein can pass from one conformation to another one. Then, we use simplicial complexes to abstract away the interleaving order of the transformations between conformations that commute. As a result, we get a compact summary of the potential behaviour of each protein of the model.

1 Introduction

Thanks to rule-based modelling languages, as Kappa, one can model accurately the biochemical interactions between proteins involved for instance in signalling pathways, without abstracting away *a priori*, when they are available, the mechanistic details about these interactions. For example, one can describe faithfully the formation of dimmers, scaffold proteins, and the phosphorylation of proteins on multiple sites, in a very compact way. Yet, understanding how the behaviour of the systems may emerge from these interactions remains a challenge. Moreover, when models become large, no matter they have been humanly written, or automatically assembled from the literature, as suggested in [14], it becomes crucial to get some automatic tools to understand the content of the models and to check that what is modelled matches with what the modeller has in mind.

This material is based upon works partially sponsored by the Defense Advanced Research Projects Agency (DARPA) and the U. S. Army Research Office under grant number W911NF-14-1-0367, and by the ITMO Plan Cancer 2014. The views, opinions, and/or findings contained in this article are those of the authors and should not be interpreted as representing the official views or policies, either expressed or implied, of DARPA, the U. S. Department of Defense, or ITMO.

© Springer International Publishing AG 2016
E. Bartocci et al. (Eds.): CMSB 2016, LNBI 9859, pp. 116–131, 2016.
DOI: 10.1007/978-3-319-45177-0_8

We use the abstract interpretation framework [3,4] to systematically derive automatic static analyses for Kappa models. Applications range from model debugging, to the abstraction of complex properties offering new insights to investigate the system overall behaviour. In this paper, we propose to study the behaviour of each protein in isolation. Starting from a formal definition of the trace semantics, we collect the behaviour of each kind of protein independently, and summarise the potential steps to reach these conformations within a transition system. When proteins have too many interaction sites, it is crucial to take benefit of the potential independence between some conformation changes in some protein states. Taking inspiration from simplicial complexes [8], we introduce the notion of macrotransition systems, in which the behaviour of different subsets of sites can be described independently, abstracting away the potential interleaving between their behaviour. The result is a scalable and convenient way to visualise both the different conformations that each protein may take and the causal relations among the different conformation changes.

Related Works. A qualitative analysis is proposed in [6,9]. This abstraction captures all the conformations an agent may take in a Kappa model. In the present paper, we go further and compute, for each agent, a transition system that describes the causal relationships among its potential conformational changes.

Causality plays an important role in the understanding and the verification of concurrent systems, as found in Systems Biology. Several frameworks are available to study and understand causality, and to reduce the combinatorial complexity of the models, by exploiting pair of commutative transitions. Partial order reduction is broadly used in model checking [10]. It consists in restricting the transitions of a concurrent system so as to force its computation to follow a canonical order for the interleaving of commutative transitions. Event structures [13] focus on the causal relations between events in a concurrent system. In [5], they provide a compact description of trace samples, in which the events which are not necessary, are discarded. Yet, it is worth noting that these discarded events may have a kinetics impact. An application of event structures in static analysis can be found in [2]. Since they focus on accumulating the effect of causally related transformations, event structures somehow obfuscate the notion of states. Our notion of macrotransition systems is inspired from simplicial complexes. Simplicial complexes can be used for describing concurrent systems up to the interleaving order of commutative transitions [8]. They describe the state of the system as a point moving along a geometrical object, in which commutative transitions are denoted by higher dimension faces. Our formalism offers a convenient compact abstraction of all the potential conformation changes of a protein, without discarding any transition.

Outline. In Sect. 2, we introduce two case studies to motivate our framework. In Sect. 3, we describe Kappa. In Sect. 4, we define its finite trace semantics, that we abstract in Sect. 5, by over-approximating the behaviour of each kind of agent thanks to *local* transition systems. Lastly in Sect. 6, we explain how to abstract away the interleaving order of the transitions that commute in these local transition systems.

2   Case Studies

So as to motivate our goal, we introduce two models as case studies.

The first model describes the formation of some dimmers. Two kinds of proteins are involved: ligands and membrane receptors. When activated by ligands, receptors can form stable dimmers, as described by the means of the interaction rules in Fig. 1. We are interested in one particular binding site in ligand proteins, and in four sites in receptor proteins. Ligand proteins are depicted as circles, whereas receptor proteins are depicted as rectangles. Their binding sites are drawn as smaller circles. Some sites are connected pair-wisely. For the others, we use the symbol '⊣' to specify a free site and the symbol '−' to specify a site that is bound to an unspecified site. By convention, the site alone on its side in a receptor protein is the one that can bind to a ligand protein; the three sites on the other side can form bonds with other receptors (their order matters).

Let us now give more details about the interactions between these proteins. A ligand protein and a receptor protein may bind to each other provided that the sites that are dedicated to this binding are both free (e.g. see Fig. 1(a)), or detach from each other, provided that the receptor protein is not yet involved in a dimmer (e.g. see Fig. 1(b)). Two activated receptor proteins can form a symmetric bond

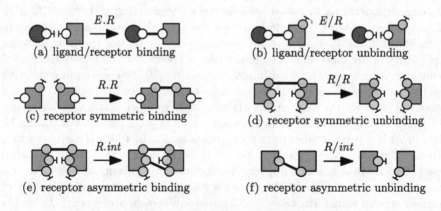

(a) ligand/receptor binding

(b) ligand/receptor unbinding

(c) receptor symmetric binding

(d) receptor symmetric unbinding

(e) receptor asymmetric binding

(f) receptor asymmetric unbinding

Fig. 1. Rules for dimmer formation.

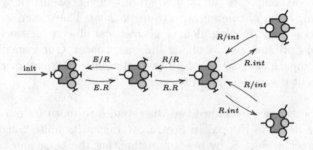

Fig. 2. The local transition system of membrane receptors.

by connecting their respective top-most site (e.g. see Fig. 1(c)), or break this bond unless an asymmetric bond has been formed already (e.g. see Fig. 1(d)). To gain stability, a dimmer with a symmetric link can form an asymmetric one by connecting one of its free site in the first receptor protein to the free site of the other kind in the second receptor protein (e.g. see Fig. 1(e)), or break this connection (e.g. see Fig. 1(f)).

Writing interaction rules can be error prone. Especially, which amount of information should be put in rules, is often not so clear. So as to gain confidence in our modelling process, we propose to compute, for each kind of protein, a *local* transition system. The goal is to abstract the different conformations that each protein may take, and how a given protein may pass from one conformation to another one. As an example, the local transition system for receptor proteins is given in Fig. 2 (there are two transitions for the rule R/Int, since it operates differently on the first and on the second receptort of its left hand side; the same remark holds for the rule $R.Int$). We claim that it provides a helpful summary of the effect of the rules on the behaviour of each protein instance.

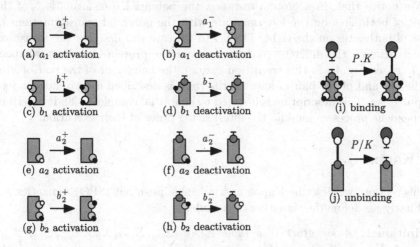

(a) a_1 activation   (b) a_1 deactivation

(c) b_1 activation   (d) b_1 deactivation   (i) binding

(e) a_2 activation   (f) a_2 deactivation

(g) b_2 activation   (h) b_2 deactivation   (j) unbinding

Fig. 3. Rules for the protein with four phosphorylation sites.

When proteins have too many interaction sites, we can no longer describe extensively their sets of potential conformations. Our second model deals with a protein with four phosphorylation sites and a single binding site. The lower left (resp. lower right) site can be phosphorylated without any condition (e.g. see Figs. 3(a) and (e)). The upper left (resp. upper right) site can get phosphorylated, if the lower left (resp. lower right) site is still phosphorylated (e.g. see Fig. 3(c) and (g)). When the four sites are all phosphorylated, the conformation of the protein changes which reveals the binding site. Then the protein can bind to another kind of protein (e.g. see Fig. 3(i)). This bond can be released with no condition (e.g. see Fig. 3(j)). Phosphorylated sites can be dephosphorylated

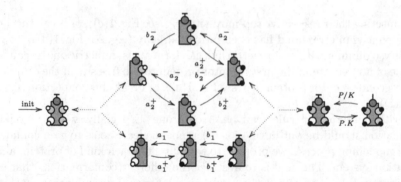

Fig. 4. Local transition system for the protein with four phosphorylation sites.

under the following conditions: as long as a protein is bound, none of its site can be dephosphorylated; as long as the upper left site is phosphorylated, the lower left site cannot be dephosphorylated (e.g. see Figs. 3(b), (d), (f), and (h)).

We notice that, in a protein instance, the potential transformations of the states of both sites on the left *commute* with the potential transformations of those of both sites on the right. Thanks to this, we can describe the transition system between the different conformations of the protein in a more compact way (e.g. see Fig. 4). In this transition system, the behaviour of the pair of sites on the left and of the pair of sites on the right is described as two independent subprocesses. This description is inspired by simplicial complexes [8]. It describes independent processes modulo the interleaving order of their execution.

3   Kappa

In this section, we describe Kappa and its single push-out (SPO) semantics.

Firstly we define the signature of a model.

Definition 1. *A signature is a tuple* $\Sigma = (\Sigma_{ag}, \Sigma_{site}, \Sigma_{int}, \Sigma_{ag-st}^{int}, \Sigma_{ag-st}^{lnk})$ *where: 1. Σ_{ag} is a finite set of agent types, 2. Σ_{site} is a finite set of site identifiers, 3. Σ_{int} is a finite set of internal state identifiers, 4. and $\Sigma_{ag-st}^{lnk} : \Sigma_{ag} \to \wp(\Sigma_{site})$ and $\Sigma_{ag-st}^{int} : \Sigma_{ag} \to \wp(\Sigma_{site})$ are site maps.*

Agent types in Σ_{ag} denote agents of interest, as kinds of proteins for instance. A site identifier in Σ_{site} represents an identified locus for capability of interactions. Each agent type $A \in \Sigma_{ag}$ is associated with a set of sites which can bear an internal state $\Sigma_{ag-st}^{int}(A)$ and a set of sites which can be linked $\Sigma_{ag-st}^{lnk}(A)$. We assume without any loss of generality that $\Sigma_{ag-st}^{lnk}(A) \cap \Sigma_{ag-st}^{int}(A) = \emptyset$, for any $A \in \Sigma_{ag}$ and we write $\Sigma_{ag-st}(A)$ for the set of sites $\Sigma_{ag-st}^{lnk}(A) \uplus \Sigma_{ag-st}^{int}(A)$.

Example 1. We define the signature for the model in the second case study as $\Sigma := (\Sigma_{ag}, \Sigma_{site}, \Sigma_{int}, \Sigma_{ag-st}^{int}, \Sigma_{ag-st}^{lnk})$ where: $\Sigma_{ag} := \{P, K\}$; $\Sigma_{site} := \{a_1, a_2, b_1, b_2, x\}$; $\Sigma_{int} := \{\circ, \bullet\}$; $\Sigma_{ag-st}^{int} := [P \mapsto \{a_1, a_2, b_1, b_2\}, K \mapsto \emptyset]$;

$\Sigma_{ag-st}^{lnk} := [P \mapsto \{x\}, K \mapsto \{x\}]$. The agent type P denotes the first kind of proteins and K the second one; the site identifier x denotes the binding site (both in P and K), and the site identifiers a_1, a_2, b_1, b_2 denote respectively the lower left, upper left, lower right, and upper right sites in the protein P.

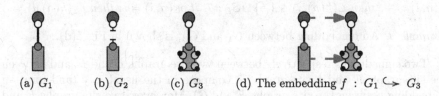

(a) G_1 (b) G_2 (c) G_3 (d) The embedding $f : G_1 \hookrightarrow G_3$

Fig. 5. Three site-graphs G_1, G_2, and G_3, and an embedding f.

Site-graphs describe both patterns and chemical mixtures. Their nodes are typed agents with some sites which can bear internal states and binding states.

Definition 2. *A site-graph is a tuple $G = (\mathcal{A}, type, \mathcal{S}, \mathcal{L}, p\kappa)$ where: 1. $\mathcal{A} \subseteq \mathbb{N}$ is a finite set of agents, 2. type : $\mathcal{A} \to \Sigma_{ag}$ is a function mapping each agent to its type, 3. $\mathcal{S}$ is a set of sites such that $\mathcal{S} \subseteq \{(n, i) \mid n \in \mathcal{A}, i \in \Sigma_{ag-st}(type(n))\}$, 4. $\mathcal{L}$ is a function between the sets $\{(n, i) \in \mathcal{S} \mid i \in \Sigma_{ag-st}^{lnk}(type(n))\}$ and $\{(n, i) \in \mathcal{S} \mid i \in \Sigma_{ag-st}^{lnk}(type(n))\} \cup \{\dashv, -\}$, such that for any two sites $(n, i), (n', i') \in \mathcal{S}$, we have $(n', i') = \mathcal{L}(n, i)$ if and only if $(n, i) = \mathcal{L}(n', i')$; 5. and $p\kappa$ is a function between the sets $\{(n, i) \in \mathcal{S} \mid i \in \Sigma_{ag-st}^{int}(type(n))\}$ and Σ_{int}.*

A site $(n, i) \in \mathcal{S}$ such that $i \in \Sigma_{ag-st}^{int}(type(n))$ is called a property site, whereas a site $(n, i) \in \mathcal{S}$ such that $i \in \Sigma_{ag-st}^{lnk}(type(n))$ is called a binding site. Whenever $\mathcal{L}(n, i) = \dashv$, the binding site (n, i) is free. Various levels of information can be given about the sites that are bound. Whenever $\mathcal{L}(n, i) = -$, the binding site (n, i) is bound to an unspecified site. Whenever $\mathcal{L}(n, i) = (n', i')$ (and hence $\mathcal{L}(n', i') = (n, i)$), the sites (n, i) and (n', i') are bound together.

For a site-graph G, we write as $\mathcal{A}_G$ its set of agents, $type_G$ its typing function, $\mathcal{S}_G$ its set of sites, $\mathcal{L}_G$ its set of links, and $p\kappa_G$ its set of the internal states.

A mixture is a site-graph in which the state of each site in each agent is documented. Formally, a site-graph G is a chemical mixture, if and only if, $\mathcal{S}_G = \{(n, i) \mid n \in \mathcal{A}_G, i \in \Sigma_{ag-st}(type_G(n))\}$.

Example 2. Three site-graphs G_1, G_2, and G_3 are drawn in Figs. 5(a), (b), and (c). For the sake of brevity, we only give the explicit definition of the first one: 1. $\mathcal{A}_{G_1} = \{1, 2\}$, 2. $type_{G_1} = [1 \mapsto P, 2 \mapsto K]$, 3. $\mathcal{S}_{G_1} = \{(1, x), (2, x)\}$, 4. $\mathcal{L}_{G_1} = [(1, x) \mapsto (2, x), (2, x) \mapsto (1, x)]$, 5. $p\kappa_{G_1} = []$; Among these three site-graphs, we notice that only G_3 is a chemical mixture.

Two site-graphs can be related by structure-preserving injective functions, which are called embeddings. the notion of embedding is defined as follows:

Definition 3. *An embedding* $h : G \hookrightarrow H$ *between two site-graphs* G *and* H *is a function of agents* $h : \mathcal{A}_G \to \mathcal{A}_H$ *satisfying, for all agent identifiers* m, $n \in \mathcal{A}_G$, *for all site identifiers* $i \in \Sigma_{ag-st}(type_G(n))$, $i' \in \Sigma_{ag-st}(type_G(n'))$, *and for all internal state identifier* $\iota \in \Sigma_{int}$: 1. *if* $m \neq n$, *then* $h(m) \neq h(n)$; 2. $type_G(n) = type_H(h(n))$; 3. *if* $(n, i) \in \mathcal{S}_G$, *then* $(h(n), i) \in \mathcal{S}_H$; 4. *if* $\mathcal{L}(n, i) = (n', i')$, *then* $\mathcal{L}(h(n), i) = (h(n'), i')$; 5. *if* $\mathcal{L}(n, i) = \dashv$, *then* $\mathcal{L}(h(n), i) = \dashv$; 6. *if* $\mathcal{L}(n, i) = -$, *then* $\mathcal{L}(h(n), i) \in \{-\} \cup \mathcal{S}_H$; 7. *if* $p\kappa(n, i) = \iota$, *then* $p\kappa(h(n), i) = \iota$.

Example 3. An embedding between G_1 and G_3, is shown in Fig. 5(d).

Two embeddings respectively between two site-graphs E and F, and between the site-graph F and a site-graph G, compose in the usual way (and form an embedding between the site-graphs E and G). Moreover, two site-graphs E and F, such that there exists an embedding between E and F, and an embedding between F and E, are said isomorphic. An embedding between two isomorphic site-graphs, is called an isomorphism. Given three site-graphs L, R, and D, a couple of embeddings respectively between the site-graphs D and L, and between the site-graphs D and R, is called a span between L and R. Besides, a couple of embeddings respectively between the site-graphs L and D, and between the site-graphs R and D is called a cospan between L and R.

Transformations between site-graphs are described by rules (e.g. see Figs. 1 and 3). For the sake of simplicity, we assume that rules can break and create bonds between sites, and can change the internal states of sites, but we consider neither agent degradation, nor agent creation.

These requirements are formalised in the following definition:

Definition 4. *A rule is a span of embeddings* $L \xleftarrow{h_L} D \xrightarrow{h_R} R$ *such that: 1.* $\mathcal{A}_D = \mathcal{A}_L$ *and* $\mathcal{A}_D = \mathcal{A}_R$; 2. *for all agents* $n \in \mathcal{A}_D$, $h_L(n) = n$ *and* $h_R(n) = n$; 3. $\mathcal{S}_D = \mathcal{S}_L'$ *and* $\mathcal{S}_D = \mathcal{S}_R$; 4. *for all sites* $(n, i) \in \mathcal{S}_D$, *if* $\mathcal{L}_R(n, i) = -$, *then* $\mathcal{L}_L(n, i) = -$.

Since we do consider neither agent creation, nor agent degradation, we can assume that the agents in the left hand side and in the right hand side of a rule are the same (constraint 1) and that both embeddings preserve agent identifiers (constraint 2). The constraint 3. ensures that, in a rule, sites cannot be removed or added. Lastly, the constraint 4. ensures that, when the binding state of a site is modified, it is not replaced with the state $-$.

A rule $L \hookleftarrow D \hookrightarrow R$ is usually denoted as $L \to R$.

Rules can be applied to site-graphs via an embedding, by the means of a push-out construction.

Definition 5 ([5]). *Let* r *be a rule* $L \to R$, L' *be a site-graph, and* h_L *be an embedding between the site-graphs* L *and* L'. *Then, there exists a rule* r' *between the site-graph* L' *and a site-graph* R' *and an embedding* h_R *between the site-graphs* R *and* R' *such that both following properties are satisfied: 1.* $h_R r = r' h_L$; 2. *for all rule* r'' *between the site-graph* L' *and a site-graph* R'' *and all embedding* h_R' *between* R *and* R'' *such that:* $h_R' r = r'' h_L$, *there exists a unique embedding* h

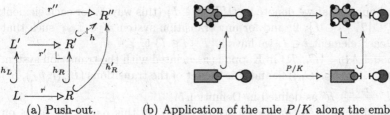

(a) Push-out.   (b) Application of the rule P/K along the embedding f.

Fig. 6. Rule application.

between R' and R'' such that $r'' = hr'$ and $h_{R''} = hh_{R'}$. With these notations, we say that there is a transition from the state L' into the state R' via a computation step with the label (r, h_L), and we write $L' \xrightarrow{(r, h_L)} R'$. Moreover, with the same notations, whenever the site-graph L' is a chemical mixture, the site-graph R' is a chemical mixture as well.

In Definition 5, the compositions between rules and embeddings are defined by the means of a pull-back construction.

Example 4. The embedding f between the site-graph G_1 and the left hand side of the rule P/K induces a computation step as described in Fig. 6(b).

In Kappa, a model $\mathcal{M}$ (over a given signature Σ) is defined as a pair $(G_0, \mathcal{R})$ where: 1. G_0 is a chemical mixture; 2. and $\mathcal{R}$ is a set of rules. The chemical mixture G_0 denotes the initial state. Since we focus only on qualitative properties, we do not associate rules with kinetic rates.

4   Trace Semantics

In this section, we define the semantics of a model (written in Kappa) as the set of the traces that is induced by the underlying transition system.

We assume that we are given $\mathbb{Q}$ a set of *states* and $\mathbb{L}$ a set of *labels*. We call a *transition* any triple (q, λ, q') in the set $\mathbb{Q} \times \mathbb{L} \times \mathbb{Q}$. In Kappa, states are chemical mixtures whereas transition labels are pairs composed of a rule and an embedding between the left hand side of this rule and a chemical mixture.

A transition system is given by a set of initial states and a set of transitions, as formalised in the following definition:

Definition 6 (Transition system). *A transition system is a pair* $(\mathcal{Q}_0, T)$ *where: 1.* $\mathcal{Q}_0 \subseteq \mathbb{Q}$*; 2.* $T \subseteq \mathbb{Q} \times \mathbb{L} \times \mathbb{Q}$.

We denote as $\mathbb{T}_{\mathbb{Q}, \mathbb{L}}$ the set of all the transition systems over $\mathbb{Q}$ and $\mathbb{L}$. Transition systems can be ordered by the relation $\sqsubseteq$ that is defined as $(\mathcal{Q}_0, T) \sqsubseteq (\mathcal{Q}_0', T')$ if and only if 1. $\mathcal{Q}_0 \subseteq \mathcal{Q}_0'$, 2. and $T \subseteq T'$. The pair $(\mathbb{T}_{\mathbb{Q}, \mathbb{L}}, \sqsubseteq)$ is indeed a complete lattice. This means that any family $(T_i)_{i \in I}$ of transition systems has a least upper bound, that we denote by $\sqcup\{T_i \mid i \in I\}$ (this way, 1. for each element $i \in I$, $T_i \sqsubseteq \sqcup\{T_i \mid i \in I\}$, 2. and for any transition system $Y \in \mathbb{T}_{\mathbb{Q},\mathbb{L}}$ such that $T_i \sqsubseteq Y$ for each element $i \in I$, we have $\sqcup\{T_i \mid i \in I\} \sqsubseteq Y$).

Each model $\mathcal{M} := (G_0, \mathcal{R})$ in Kappa is associated with the transition system $(\mathcal{Q}_0, T)$ where 1. $\mathcal{Q}_0 = \{G_0\}$; 2. and T is the set of the transitions $(L', (r, h_L), R')$ such that $L' \xrightarrow{(r,h_L)} R'$ as defined in Definition 5.

Each transition system induces a set of traces. In this paper, we focus on finite traces, that are made of an initial state followed by a (potentially empty) finite sequence of transitions, each of them starting from the state the previous transition had ended in. This is formalised in the following definition:

Definition 7 (Finite traces). *A* finite trace *is a pair* $\tau = (q_0', (q_i, \lambda_i, q_i')_{1 \le i \le p})$, *where* q_0' *is a state (in* $\mathbb{Q}$*), and* $(q_i, \lambda_i, q_i')_{1 \le i \le p}$ *is a family of transitions (in* $\mathbb{Q} \times \mathbb{L} \times \mathbb{Q}$*), such that* $q_{i-1}' = q_i$ *for each integer* i *between 1 and* p.

We denote as $\mathcal{T}_{\mathbb{Q},\mathbb{L}}^\star$ the set of all the traces over the sets $\mathbb{Q}$ and $\mathbb{L}$.

With the notations of Definition 7, we call the state q_0' (resp. q_p') the initial (resp. the final) state of the trace τ and we denote it as $\mathrm{fst}(\tau)$, (resp. as $\mathrm{last}(\tau)$). When a trace is made of a single state, we write it as q instead of $(q, ())$. Besides, any transition $t := (q_{p+1}, \lambda_{p+1}, q_{p+1}')$ such that the state q_{p+1} is equal to the final state q_p' of the trace τ, can be concatenated to the trace τ. In such a case, we write $\tau \frown (q_{p+1}, \lambda_{p+1}, q_{p+1}')$ for the finite trace $(q_0', (q_i, \lambda_i, q_i')_{1 \le i \le p+1})$.

A transition system $(\mathcal{Q}_0, T)$ induces a set of traces $\gamma_{\mathbb{Q},\mathbb{L}}(\mathcal{Q}_0, T)$, that is defined as the set of the traces $(q_0', (q_i, \lambda_i, q_i')_{1 \le i \le p})$ such that: 1. the state q_0' belongs to the set $\mathcal{Q}_0$; 2. and for each integer i between 1 and p, the transition (q_i, λ_i, q_i') belongs to the set T. The set of traces $\gamma_{\mathbb{Q},\mathbb{L}}(\mathcal{Q}_0, T)$ can also be defined as the least fix-point of the operator $\mathbb{F}_{\mathcal{Q}_0, T}$ that maps any set of traces $X \in \wp(\mathcal{T}_{\mathbb{Q},\mathbb{L}}^\star)$ into the set of traces $\{q \mid q \in \mathcal{Q}_0\} \cup \{\tau \frown (q, \lambda, q') \mid \tau \in X \wedge \mathrm{last}(\tau) = q \wedge (q, \lambda, q') \in T\}$. The operator $\mathbb{F}_{\mathcal{Q}_0, T}$ is monotonic (that is to say that, for any two sets of traces $X, Y \in \wp(\mathcal{T}_{\mathbb{Q},\mathbb{L}}^\star)$, if $X \subseteq Y$, then $\mathbb{F}_{\mathcal{Q}_0, T}(X) \subseteq \mathbb{F}_{\mathcal{Q}_0, T}(Y)$). By [12], it follows that $\mathbb{F}_{\mathcal{Q}_0, T}$ has a fix-point that is included in any other fix-point. We denote this fix-point $lfp\ \mathbb{F}_{\mathcal{Q}_0, T}$ (thus, we have 1. $lfp\ \mathbb{F}_{\mathcal{Q}_0, T} = \mathbb{F}_{\mathcal{Q}_0, T}(lfp\ \mathbb{F}_{\mathcal{Q}_0, T})$, 2. and for each set $X' \subseteq \wp(\mathcal{T}_{\mathbb{Q},\mathbb{L}}^\star)$ such that $X' = \mathbb{F}_{\mathcal{Q}_0, T}(X')$, we have: $lfp\ \mathbb{F}_{\mathcal{Q}_0, T} \subseteq X'$).

Conversely, a set $X \subseteq \mathcal{T}_{\mathbb{Q},\mathbb{L}}^\star$ of finite traces can be abstracted by the transition system $\alpha_{\mathbb{Q},\mathbb{L}}(X)$ that is defined as the pair $(\mathcal{Q}_0, T)$ with: 1. $\mathcal{Q}_0 = \{\mathrm{fst}(\tau) \mid \tau \in X\}$; 2. $T = \{t_i \mid \exists (q_0, (t_i)_{1 \le i \le p}) \in X, i \in [\![1, p]\!]\}$. The pair $(\alpha_{\mathbb{Q},\mathbb{L}}, \gamma_{\mathbb{Q},\mathbb{L}})$ forms a Galois connection between the complete lattices $(\wp(\mathcal{T}_{\mathbb{Q},\mathbb{L}}^\star), \subseteq)$ and $(\mathbb{T}_{\mathbb{Q},\mathbb{L}}, \sqsubseteq)$. This means that for any transition system $(\mathcal{Q}_0', T')$, we have $\alpha_{\mathbb{Q},\mathbb{L}}(X) \sqsubseteq (\mathcal{Q}_0', T')$, if and only if, $X \subseteq \gamma_{\mathbb{Q},\mathbb{L}}(\mathcal{Q}_0', T')$. It follows (e.g. see [4]), that: 1. functions $\alpha_{\mathbb{Q},\mathbb{L}}$ and $\gamma_{\mathbb{Q},\mathbb{L}}$ are both monotonic; 2. the function $\alpha_{\mathbb{Q},\mathbb{L}}$ maps each set of finite traces to the smallest (for $\sqsubseteq$) transition system which induces this set of traces.

5   Local Transition Systems

In Sect. 4, we have associated each model with a transition system, that describes the set of the finite traces of this model. However, such transition systems are usually too large to be computed, or even if they could, they are too complex to help understanding the behaviour of the models. We propose to simplify these transition systems by focusing on the behaviour of each protein independently, abstracting away which proteins are bound together. This abstraction had already been applied in [6,9], to infer the relationships among the state of sites in protein instances. Here we extend this static analysis to traces.

Firstly we explain how to track the behaviour of each protein independently while forgetting about the bonds between pairs of binding sites. For any agent identifier $n \in \mathbb{N}$, we denote by β_n the function that, when applied to a site-graph G containing an agent with identifier n: 1. replaces any bond between two sites by two occurrences of the symbol '$-$'; 2. restricts the site-graph G to the agent with identifier n; 3. renames the identifier n with 1. More formally, the site-graph $\beta_n(G)$ is defined by: 1. $\mathcal{A}_{\beta_n(G)} := \{1\}$; 2. $type_{\beta_n(G)} := [1 \mapsto type_G(n)]$; 3. $\mathcal{S}_{\beta_n(G)} := \{(1, i) \mid (n, i) \in \mathcal{S}_G\}$; 4. the function $\mathcal{L}_{\beta_n(G)}$ maps each site $(1, i)$ such that $i \in \Sigma^{lnk}_{ag-st}(type_G(n))$ and $(n, i) \in \mathcal{S}_G$, to the symbol '$\dashv$' whenever $\mathcal{L}_G(n, i) = \dashv$, and to the symbol '$-$' otherwise; 5. the function $p\kappa_{\beta_n(G)}$ maps each site $(1, i)$ such that $i \in \Sigma^{int}_{ag-st}(type_G(n))$ and $(n, i) \in \mathcal{S}_G$, to $p\kappa_G(n, i)$.

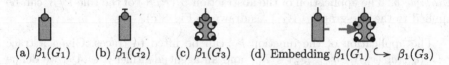

(a) $\beta_1(G_1)$ (b) $\beta_1(G_2)$ (c) $\beta_1(G_3)$ (d) Embedding $\beta_1(G_1) \hookrightarrow \beta_1(G_3)$

Fig. 7. Abstraction of the site-graphs G_1, G_2, G_3, and of the embedding f.

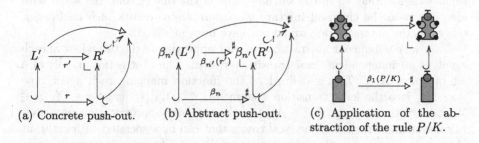

(a) Concrete push-out.   (b) Abstract push-out.   (c) Application of the abstraction of the rule P/K.

Fig. 8. The abstraction of a push-out (in the concrete) is a push-out (in the abstract), and application of rule (in the abstract).

Example 5. The restriction of the site-graphs G_1, G_2, and G_3 to their agent with identifier 1, are depicted in Figs. 7(a), (b), and (c). For instance, the site-graph $\beta_1(G_1)$ is equal to the tuple $(\{1\}, [1 \mapsto P], \{(1, x)\}, [(1, x) \mapsto -], [])$. Moreover, the function $[1 \mapsto 1]$ induces an embedding between the site-graph $\beta_1(G_1)$ and the site-graph $\beta_1(G_3)$, as depicted in Fig. 7(d).

For each agent type A, we define the set of the local views $\mathbb{Q}_A$, for the agents of type A, as the set of the site-graphs G with only one agent of type A and identifier 1, that documents the state of all the sites in $\Sigma_{ag-st}(A)$ (that is to say such that $\mathcal{S}_G = \{(1,i) \mid i \in \Sigma_{ag-st}(A)\}$). Besides, we define the set of the local transition labels $\mathbb{L}_A$ as the set of pairs (r,n) where r is a rule, and n is an agent identifier. Intuitively, the local transition label (r,n) denotes the fact that a rule r is applied along an embedding that matches the agent with identifier n in the left hand side of the rule r, to the local view to which we want to apply the rule. Given a rule $r : L \hookleftarrow D \hookrightarrow R$ and an identifier $n \in \mathcal{A}_L$ of an agent in the left hand side L of the rule r, we define the abstraction $\beta_n(r)$ of the rule r as the span $\beta_n(L) \hookleftarrow \beta_n(D) \hookrightarrow \beta_n(R)$. In this span, both embeddings are induced by the function $[1 \mapsto 1]$. The copsan $\beta_n(r)$ is not a rule in general, yet it can be applied along any embedding between its left hand side $\beta_n(L)$ and a site-graph L' thanks to a push-out construction (e.g. see Fig. 8). In such a case, we write $L' \xrightarrow{(r,n)}_{\sharp} R'$ for the corresponding abstract computation step. Moreover, we notice that in an abstract computation step $q^{\sharp} \xrightarrow{(r,n)}_{\sharp} q^{\sharp\prime}$ if $q^{\sharp}$ is a local view in $\mathbb{Q}_A$, then $q^{\sharp\prime}$ is a local view in $\mathbb{Q}_A$ as well.

Example 6. The application of the abstraction $\beta_1(P/K)$ of the rule P/K can be applied to the site-graph $\beta_1(G_1)$, as drawn in Fig. 8(c).

The application of the function β_n can be lifted to traces. Given a trace $\tau := (q_0', (q_i, (r_i, f_i), q_i')_{1 \leq i \leq p}) \in \mathcal{T}_{\mathbb{Q},\mathbb{L}}^{\star}$ and an agent identifier $n \in \mathcal{A}_{q_0'}$, we define the local trace $\beta_n(\tau)$ as the trace: $(\beta_n(q_0'), (\beta_n(q_{\sigma(i)}), (r_{\sigma(i)}, n_i), \beta_n(q_{\sigma(i)}'))_{1 \leq i \leq p'})$ where $1. \sigma_1, \ldots, \sigma_{p'}$ is the sequence (in increasing order) of the integers i between 1 and p such that $\beta_n(q_i) \neq \beta_n(q_{i+1})$; 2. and for each integer i between 1 and p', the integer n_i is the unique identifier such that the embedding f_{σ_i} maps the agent with identifier n_i in the left hand side of the rule r_{σ_i} into the agent with identifier n in the chemical mixture q_{σ_i} (there always exists such an integer, otherwise the abstract state would not have been modified).

Now we combine our abstractions to over-approximate the behaviour of each agent as an independent *local* transition system. The abstraction $\alpha_\pi(X)$ of a set of traces $X \subseteq \mathcal{T}_{\mathbb{Q},\mathbb{L}}^{\star}$ is defined as the function mapping each agent type $A \in \Sigma_{ag}$ into the local transition system $\alpha_{\mathbb{Q},\mathbb{L}}[\mathbb{Q}_A, \mathbb{L}_A](\{\beta_n(\tau) \mid \tau \in X, n \in \mathcal{A}_{\mathrm{fst}(\tau)}$ such that $type_{\mathrm{fst}(\tau)}(n) = A\}$) (i.e. the best over-approximation, as a transition system, of the set of the local traces that can be associated with an agent of type A). Conversely, the concretization $\gamma_\pi(Y)$ of a function Y between agent types and local transition systems, is defined as the set of the traces $\tau \in \mathcal{T}_{\mathbb{Q},\mathbb{L}}^{\star}$ such that for each identifier n of an agent in the initial state $\mathrm{fst}(\tau)$ of the trace τ, the local trace $\beta_n(\tau)$ belongs to the set $\gamma_{\mathbb{Q},\mathbb{L}}[\mathbb{Q}_A, \mathbb{L}_A](Y(type_{\mathrm{fst}(\tau)}(n)))$.

By [4], the function $\alpha_\pi \circ \mathbb{F}_{\mathcal{Q}_0,T} \circ \gamma_\pi$ is the best abstract counterpart to the function $\mathbb{F}_{\mathcal{Q}_0,T}$. The function $\alpha_\pi \circ \mathbb{F}_{\mathcal{Q}_0,T} \circ \gamma_\pi$ is monotonic and its least fix-point satisfies the inclusion: $lfp\, \mathbb{F}_{\mathcal{Q}_0,T} \subseteq \gamma_\pi(lfp\, \alpha_\pi \circ \mathbb{F}_{\mathcal{Q}_0,T} \circ \gamma_\pi)$. The least fix-point in the right hand side of the inclusion can be computed in a finite number of iterations, since the domain of $\alpha_\pi \circ \mathbb{F}_{\mathcal{Q}_0,T} \circ \gamma_\pi$ is finite. However computing these iterations is quite cumbersome, because it intertwines the computation of the local transition systems that are associated to each agent type. We propose to desynchronise these computations. To do this, we introduce, for each agent type $A \in \Sigma_{ag}$, the function $\mathbb{F}^{\sharp}_{\mathcal{Q}_0,T,A}$ that maps any local transition system $(\mathcal{Q}^{\sharp}_0, T^{\sharp}) \in \mathbb{T}_{\mathcal{Q}_A, \mathbb{L}_A}$ to the local transition system $(\mathcal{Q}^{\sharp}_0{}', T^{\sharp}{}') \in \mathbb{T}_{\mathcal{Q}_A, \mathbb{L}_A}$ where: 1. $\mathcal{Q}^{\sharp}_0{}' = \mathcal{Q}^{\sharp}_0 \cup \{\beta_n(q_0) \mid q_0 \in \mathcal{Q}_0,\ n \in \mathcal{A}_{q_0},\ type_{q_0}(n) = A\}$; 2. and $T^{\sharp}{}'$ is the union between the set $T^{\sharp}$ and the set of the transitions $(q^{\sharp}, (r, n), q^{\sharp}{}')$ such that the local view $q^{\sharp}$ is reachable in the transition system $(\mathcal{Q}^{\sharp}_0, T^{\sharp})$ and such that $q^{\sharp} \xrightarrow{(r,n)}{}^{\sharp} q^{\sharp}{}'$.

For any function Y mapping each agent type $A \in \Sigma_{ag}$ to a local transition system over the states $\mathbb{Q}_A$ and the transition labels $\mathbb{L}_A$, we have $[\alpha_{\pi} \circ \mathbb{F}_{\mathcal{Q}_0,T} \circ \gamma_{\pi}](Y) \sqsubseteq \mathbb{F}^{\sharp}_{\mathcal{Q}_0,T,A}(Y)$. Moreover, for each agent type $A \in \Sigma_{ag}$, the function $\mathbb{F}^{\sharp}_{\mathcal{Q}_0,T,A}$ is monotonic. By [12], for each agent type $A \in \Sigma_{ag}$, the function $\mathbb{F}^{\sharp}_{\mathcal{Q}_0,T,A}$ has a least fix-point. By [4], the inclusion $lfp\, \mathbb{F}_{\mathcal{Q}_0,T} \subseteq \gamma_{\pi}([A \in \Sigma_{ag} \mapsto lfp\, \mathbb{F}^{\sharp}_{\mathcal{Q}_0,T,A}(A)])$ is satisfied, which ensures the soundness of our approach.

Thus, we have derived an abstraction of the finite trace semantics, as a family of local transition systems, that describes the behaviour of each particular type of agent, in isolation. Each such local transition system can be computed independently iteratively. We can apply our framework to abstract the local transition system associated to the membrane receptors in the model of dimmer formation (e.g. see Fig. 1). As expected, the result is the abstract transition system that is given in Fig. 2.

6   Macrotransition Systems

The analysis described in Sect. 5 can also be applied to our second case study (e.g. see Fig. 3). But there are too many conformations for the result to be visualisable in practice. In this section, we propose to identify which transitions commute and provide a more compact symbolic representation of local transition systems, which abstracts the interleaving order of commutative transitions.

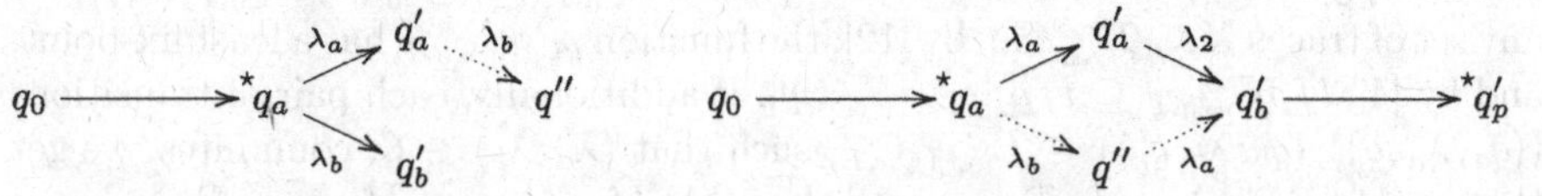

(a) Concurrent transitions.   (b) Consecutive transitions.

Fig. 9. Pairs of commutative transitions.

We firstly define the notion of pairs of commutative transitions in a set of traces. Given a set $X \subseteq T^{*}_{\mathbb{Q}, \mathbb{L}}$ of traces, We denote by T_X the set of the transitions that occur in X, i.e. $T_X := \{(q_i, \lambda_i, q'_i) \mid (q_0, (q_i, \lambda_i, q'_i))_{1 \le i \le p}, 1 \le i \le p\}$. For the sake of simplicity, and since this is the case in Kappa, we assume that any transition is fully defined by the state it is starting from and the label of the transition. Formally, we assume that for any two transitions $(q_1, \lambda_1, q'_1), (q_2, \lambda_2, q'_2) \in T_X$, the states q'_1 and q'_2 are equal whenever the pairs (q_1, λ_1) and (q_2, λ_2) are equal.

Definition 8. *We say that transitions with labels $\lambda_a \in \mathbb{L}$ and $\lambda_b \in \mathbb{L}$ commute in the set of traces $X \subseteq T^\star_{\mathbb{Q},\mathbb{L}}$, if and only if both following properties are satisfied:
1. for any trace $\tau \in T^\star_{\mathbb{Q},\mathbb{L}}$, and any states $q, q'_a, q'_b \in \mathbb{Q}$ such that $\tau \frown (q, \lambda_a, q'_a)$ and $\tau \frown (q, \lambda_b, q'_b)$ are well defined traces that both belong to the set X, there exists a state $q'' \in \mathbb{Q}$, such that the trace $\tau \frown (q, \lambda_a, q'_a) \frown (q'_a, \lambda_b, q'')$ belongs to the set X; 2. and for any trace $(q_0, (t_i)_{1 \le i \le p})$ in X, if there exists three states $q_a, q'_a, q'_b \in \mathbb{Q}$ and an integer j between 1 and $p-1$ satisfying both $t_j = (q_a, \lambda_a, q'_a)$ and $t_{j+1} = (q'_a, \lambda_b, q'_b)$, then there exists a state $q'' \in \mathbb{Q}$ such that the trace $(q_0, (t'_i)_{1 \le i \le p})$, where t'_i is defined as (q_a, λ_b, q'') whenever $i = j$, as (q'', λ_a, q'_b) whenever $i = j+1$, and as t_i otherwise, is well defined and belongs to the set X.*

In Definition 8, the first property entails that whenever after a given prefix of trace, two transitions that commute are enable (they necessarily starts from the same state), each transition can be followed by the other one modulo the fact that the latter transition should now start from the ending state of the former one (e.g. see Fig. 9(a)), whereas the second property entails that two consecutive transitions that commute can always be performed in the reverse order, modulo the fact that the intermediary state has to be modified (e.g. see Fig. 9(b)).

Given $\mathcal{C} \in \wp(\mathbb{L}^2)$ a set of pairs of transition labels, we define the operator $\rho_{\mathcal{C}}$ which maps each set of traces $X \subseteq T^\star_{\mathbb{Q},\mathbb{L}}$, into the smallest set of traces $\rho_{\mathcal{C}}(X) \subseteq T^\star_{\mathbb{Q},\mathbb{L}}$ that contains the set X and in which each pair of transitions $((q_a, \lambda_a, q'_a), (q_b, \lambda_b, q'_b)) \in T^2_X$ such that $(\lambda_a, \lambda_b) \in \mathcal{C}$, commutes. The function $\rho_{\mathcal{C}}$ is an upper closure operator, that is to say that: 1. $\rho_{\mathcal{C}}$ is monotonic; 2. $\rho_{\mathcal{C}}$ is idempotent (i.e. $\rho_{\mathcal{C}} \circ \rho_{\mathcal{C}} = \rho_{\mathcal{C}}$); 3. and $\rho_{\mathcal{C}}$ is extensive (i.e. $X \subseteq T^\star_{\mathbb{Q},\mathbb{L}}\rho_{\mathcal{C}}(X)$, $\forall X \subseteq T^\star_{\mathbb{Q},\mathbb{L}}$). We notice that the fix-points of the upper closure $\rho_{\mathcal{C}}$ are the set of traces $X \subseteq T^\star_{\mathbb{Q},\mathbb{L}}$ such that each pair of transitions $((q_a, \lambda_a, q'_a), (q_b, \lambda_b, q'_b)) \in T^2_X$ such that $(\lambda_a, \lambda_b) \in \mathcal{C}$, commutes. We can use the upper closure $\rho_{\mathcal{C}}$ to accelerate the computation of the set of traces $\gamma_{\mathbb{Q},\mathbb{L}}(\mathcal{Q}_0, T)$ that is induced by a given transition system $(\mathcal{Q}_0, T)$ provided that each pair of transitions $((q_a, \lambda_a, q'_a), (q_b, \lambda_b, q'_b)) \in T^2_{\gamma_{\mathbb{Q},\mathbb{L}}(\mathcal{Q}_0, T)}$ such that $(\lambda_a, \lambda_b) \in \mathcal{C}$, commutes in the set of traces $\gamma_{\mathbb{Q},\mathbb{L}}(\mathcal{Q}_0, T)$. The function $\rho_{\mathcal{C}} \circ \mathbb{F}_{\mathcal{Q}_0,T}$ is monotonic and satisfies $\mathbb{F}_{\mathcal{Q}_0,T}(X) \subseteq [\rho_{\mathcal{C}} \circ \mathbb{F}_{\mathcal{Q}_0,T}](X))$, for any set of traces $X \subseteq T^\star_{\mathbb{Q},\mathbb{L}}$. So, by [12], the function $\rho_{\mathcal{C}} \circ \mathbb{F}_{\mathcal{Q}_0,T}$ has a least fix-point, and by [4], $lfp\ \mathbb{F}_{\mathcal{Q}_0,T} \subseteq lfp\ [\rho_{\mathcal{C}} \circ \mathbb{F}_{\mathcal{Q}_0,T}]$. If additionally, each pair of transitions $((q_a, \lambda_a, q'_a), (q_b, \lambda_b, q'_b)) \in T_{\gamma_{\mathbb{Q},\mathbb{L}}(\mathcal{Q}_0, T)}$ such that $(\lambda_a, \lambda_b) \in \mathcal{C}$, commutes, we get that $\rho_{\mathcal{C}}(lfp\ \mathbb{F}_{\mathcal{Q}_0,T}) = lfp\ \mathbb{F}_{\mathcal{Q}_0,T}$, and thus that $lfp\ \mathbb{F}_{\mathcal{Q}_0,T} = lfp\ [\rho_{\mathcal{C}} \circ \mathbb{F}_{\mathcal{Q}_0,T}]$.

In Kappa, pairs of commutative local transitions can be identified syntactically. Let us consider an agent type $A \in \Sigma_{ag}$. Given the label $(r, n) \in \mathbb{L}_A$ of a local transition, we denote as $\text{TEST}(r, n)$ the set of the site identifiers i such that the site (n, i) occurs in the domain of the rule r, and as $\text{MOD}(r, n)$ the set of the site identifiers i such that the site (n, i) occurs in the domain of the rule r without occurring in its left hand side. Then, for any two labels $\lambda_1, \lambda_2 \in \mathbb{L}_A$ of local transitions such that both $\text{MOD}(\lambda_1) \cap \text{TEST}(\lambda_2) = \emptyset$ and $\text{MOD}(\lambda_2) \cap \text{TEST}(\lambda_1) = \emptyset$, any pair of local transitions with the labels λ_1 and λ_2 commutes [11].

Example 7. In the rules Fig. 3, we have: We have: $\text{TEST}(b_1^+, 1) = \{a_1, b_1\}$; $\text{MOD}(b_1^+, 1) = \{b_1\}$; $\text{TEST}(b_2^+, 1) = \{a_2, b_2\}$; $\text{MOD}(b_2^+, 1) = \{b_2\}$. As a consequence any two local transitions with respective labels $(b_1^+, 1)$ and $(b_2^+, 1)$ commute.

Given a local view $v \in \mathbb{Q}_A$, we consider the set $\Lambda(v)$ as the set of all the transitions labels λ such that there exists a local transition starting from the view v and with the label λ. We define the site-graph $\mathrm{FRAME}_\lambda(v)$ as the site-graph that is obtained by removing from the site-graph v any site that belongs to the set $\bigcup\{\mathrm{MOD}(\lambda') \mid \lambda' \in \Lambda(v) \setminus \{\lambda\}\}$. Intuitively, $\mathrm{FRAME}_\lambda(v)$ is the restriction of the local view v to the sites that cannot be modified by local transitions that commute with a local transition with the label λ.

We are left to provide a compact data-structure to represent transition systems with pairs of commutative transitions. We propose to use macrostates and macrotransitions. A *macrostate* is a symbolic representation of a set of (micro)states (that behave similarly), and *macrotransitions* are transitions between macrostates, that denote some transitions between the corresponding microstates. Let us assume that we are given a set of macrostates $\mathbb{Q}^\sharp$. Each macrostate $q^\sharp \in \mathbb{Q}^\sharp$ denotes a set of microstates $\Gamma_{\mathbb{Q},\mathbb{Q}^\sharp}(q^\sharp) \subseteq \mathbb{Q}$.

A macrotransition system is defined as follows:

Definition 9. *A macrotransition system is a pair $(\mathcal{Q}_0, T^\sharp)$ where: 1. $\mathcal{Q}_0 \subseteq \mathbb{Q}$; 2. $T^\sharp \subseteq \mathbb{Q}^\sharp \times \mathbb{L} \times \mathbb{Q}^\sharp$.*

A macrotransition system is made of a set of initial microstates, and a set of labelled transitions between macrostates. Each macrotransition is an implicit representation of a set of transitions between microstates. Formally, each macrotransition $(q^\sharp, \lambda, q^{\sharp\prime})$ denotes the set of the transitions $(q, \lambda, q') \in \mathbb{Q} \times \mathbb{L} \times \mathbb{Q}$ for which there exists a set of macrostates $X \subseteq \mathbb{Q}^\sharp$ satisfying: $q = \bigcap\{\Gamma_{\mathbb{Q},\mathbb{Q}^\sharp}(x) \mid x \in X \cup \{q^\sharp\}\}$ and $q' = \bigcap\{\Gamma_{\mathbb{Q},\mathbb{Q}^\sharp}(x) \mid x \in X \cup \{q^{\sharp\prime}\}\}$. We denote as $\Gamma'_{\mathbb{Q},\mathbb{Q}^\sharp}(q^\sharp, \lambda, q^{\sharp\prime})$ the set of the transitions denoted by the macrotransition $(q^\sharp, \lambda, q^{\sharp\prime})$. A macrotransition system $(\mathcal{Q}_0, T^\sharp)$ is an abstraction of the transition system $(\mathcal{Q}_0, \cup\{\Gamma'_{\mathbb{Q},\mathbb{Q}^\sharp}(t) \mid t \in T^\sharp\})$, that we denote as $\Gamma_{\mathcal{Q}_0,T,\mathbb{Q}^\sharp}(\mathcal{Q}_0, T^\sharp)$.

In Kappa, a macrostate is a site-graph that can be embedded in a local view. Each macrostate intentionally denotes the set of the local views it can be embedded in. Now we mimic the computation of $\mathbb{F}^\sharp_{\mathcal{Q}_0,T,A}$ in macrotransition systems. We define the function $\mathbb{G}^\sharp_{\mathcal{Q}_0,T^\sharp}$ mapping each macrotransition system $(\mathcal{Q}_0, T^\sharp)$ to the macrotransition system where the set of initial states is defined as the union between $\mathcal{Q}_0$ and the set $\{\beta_n(q_0) \mid n \in \mathcal{A}(q_0),\ type_{q_0}(n) = A\}$; and where the set of macrotransitions is obtained 1. by adding in the set $T^\sharp$ any transition of the form $(\mathrm{FRAME}_\lambda(v), \lambda, v')$ for any local view $v \in \mathbb{Q}_A$ such that: (a) the local view v is reachable in the transition system $\gamma_{\mathbb{Q}_A,\mathbb{L}_A}(\Gamma_{\mathcal{Q}_0,T,\mathbb{Q}^\sharp}(\mathcal{Q}_0, T^\sharp))$; (b) and $\mathrm{FRAME}_\lambda(v) \xrightarrow{\lambda}{}^\sharp v'$, 2. before removing any macrotransition t such that there exists a macrotransition t' satisfying $\Gamma'_{\mathbb{Q},\mathbb{Q}^\sharp}(t) \subset \Gamma'_{\mathbb{Q},\mathbb{Q}^\sharp}(t')$.

For any macrotransition system $X^\sharp$, both following inclusions are satisfied:
1. $\gamma_{\mathbb{Q}_A,\mathbb{L}_A}(\mathbb{F}^\sharp_{\mathcal{Q}_0,T,A}(\Gamma_{\mathcal{Q}_0,T,\mathbb{Q}^\sharp}(X^\sharp))) \subseteq \gamma_{\mathbb{Q}_A,\mathbb{L}_A}(\Gamma_{\mathcal{Q}_0,T,\mathbb{Q}^\sharp}(\mathbb{G}^\sharp_{\mathcal{Q}_0,T^\sharp}(X^\sharp)))$ (soundness);
2. $\gamma_{\mathbb{Q}_A,\mathbb{L}_A}(\Gamma_{\mathcal{Q}_0,T,\mathbb{Q}^\sharp}(\mathbb{G}^\sharp_{\mathcal{Q}_0,T^\sharp}(X^\sharp))) \subseteq \rho_C(\gamma_{\mathbb{Q}_A,\mathbb{L}_A}(\mathbb{F}^\sharp_{\mathcal{Q}_0,T,A}(\Gamma_{\mathcal{Q}_0,T,\mathbb{Q}^\sharp}(X^\sharp))))$ (relative completeness). It follows (since $\Gamma_{\mathcal{Q}_0,T,\mathbb{Q}^\sharp}(\emptyset, \emptyset) = (\emptyset, \emptyset)$) that, when $k \in \mathbb{N}$ increases, the sequence of the images by $[\gamma_{\mathbb{Q}_A,\mathbb{L}_A} \circ \Gamma_{\mathcal{Q}_0,T,\mathbb{Q}^\sharp}]$ of the iterates $\mathbb{G}^{\sharp(k)}_{\mathcal{Q}_0,T^\sharp}(\emptyset, \emptyset)$ of the function $\mathbb{G}^\sharp_{\mathcal{Q}_0,T^\sharp}$, starting from the macrotransition system

$(\emptyset, \emptyset)$, stations ultimately at the value $\gamma_{\mathbb{Q}_A, L_A}(lfp\ \mathbb{F}^{\sharp}_{\mathbb{Q}_0, T, A})$. The result of our analysis is defined as the macrotransition system $\mathbb{G}^{\sharp(l)}_{\mathbb{Q}_0, T^{\sharp}}(\emptyset, \emptyset)$, where $l \in \mathbb{N}$ is the rank at which this limit is reached.

Our analysis is integrated within an open-source static analyser [1]. It relies on binary decision diagrams, to describe implicitly which microtransitions are not yet covered by any macrotransition (as required both in the function $\mathbb{G}^{\sharp}_{\mathbb{Q}_0, T^{\sharp}}$ and to detect when to stop the iterations). We obtain, for the second case study in Fig. 3, the macrotransition system that is described in Fig. 4. We have also applied to a similar model with 5 pairs of phosphorylation sites and obtained the corresponding macrotransition system in around one second.

7   Conclusion

Kappa [7] allows for the description of highly combinatorial systems of interactions between proteins. But it is not always obvious to check the consistency of the models that are written in Kappa or to get an overview of how these models work. To cope with this issue, we have proposed an abstract interpretation framework to automatically over-approximate the behaviour of each agent of a model, independently, by the means of a *local* transition system.

Since these local transition systems may remain too combinatorial when proteins have many interaction sites, we have designed a coarser description of the local transition systems, inspired by simplicial systems. This latter description abstracts away the interleaving order between commutative transitions, by the means of transitions between macrostates, that denotes symbolically sets of transitions between microstates. Our tool computes these macrotransitions directly.

Acknowledgement. This work has been motivated by models written by Héctor Medina, and by Nathalie Théret and Jean Cocquet. We deeply thank them, as well as Pierre Boutillier, Ioana Cristescu, Vincent Danos, Walter Fontana, Russ Harmer, Jean Krivine, Jonathan Laurent, and Jean Yang, for fruitful discussions.

References

1. Boutillier, P., Feret, J., Krivine, J., Kim Lý, Q.: Kasim development homepage. http://kappalanguage.org
2. Chatain, T., Haar, S., Jezequel, L., Paulevé, L., Schwoon, S.: Characterization of reachable attractors using petri net unfoldings. In: Mendes, P., Dada, J.O., Smallbone, K. (eds.) CMSB 2014. LNCS, vol. 8859, pp. 129–142. Springer, Heidelberg (2014)
3. Cousot, P., Cousot, R.: Abstract interpretation: a unified lattice model for static analysis of programs by construction or approximation of fixpoints. In: Proceedings of POPL 1977 (1977)
4. Cousot, P., Cousot, R.: Systematic design of program analysis framework. In: Proceedings of POPL 1979 (1979)

5. Danos, V., Feret, J., Fontana, W., Harmer, R., Hayman, J., Krivine, J., Thompson-Walsh, C.D., Winskel, G.: Graphs, rewriting and pathway reconstruction for rule-based models. In: Proceedings of FSTTCS 2012. LIPIcs, vol. 18 (2012)
6. Danos, V., Feret, J., Fontana, W., Krivine, J.: Abstract interpretation of cellular signalling networks. In: Logozzo, F., Peled, D.A., Zuck, L.D. (eds.) VMCAI 2008. LNCS, vol. 4905, pp. 83–97. Springer, Heidelberg (2008)
7. Danos, V., Laneve, C.: Formal molecular biology. TCS **325**(1), 69–110 (2004)
8. Fajstrup, L., Goubault, É., Raußen, M.: Detecting deadlocks in concurrent systems. In: Sangiorgi, D., de Simone, R. (eds.) CONCUR 1998. LNCS, vol. 1466, pp. 332–347. Springer, Heidelberg (1998)
9. Feret, J.: Reachability analysis of biological signalling pathways by abstract interpretation. In: Proceedings of ICCMSE 2007, vol. 963(2). AIP (2007)
10. Godefroid, P. (ed.): Partial-Order Methods for the Verification of Concurrent Systems. LNCS, vol. 1032. Springer, Heidelberg (1996)
11. Laurent, J.: Causal analysis of rule-based models of signaling pathways (2015)
12. Tarski, A.: A lattice-theoretical fixpoint theorem and its applications. Pacific J. Math. **5**(2), 285–309 (1955)
13. Winskel, G.: Event structures. In: Brauer, W., Reisig, W., Rozenberg, G. (eds.) Advances in Petri Nets 1986. LNCS, vol. 255, pp. 325–392. Springer, Heidelberg (1987)
14. You, J.: Darpa sets out to automate research. Science **347**, 465 (2015)

Bifurcation Analysis of Cardiac Alternans Using δ-Decidability

Md. Ariful Islam[1]([✉]), Greg Byrne[2], Soonho Kong[1], Edmund M. Clarke[1],
Rance Cleaveland[3], Flavio H. Fenton[4], Radu Grosu[5,6], Paul L. Jones[2],
and Scott A. Smolka[5]

[1] Carnegie Mellon University, Pittsburgh, PA, USA
mdarifui@cs.cmu.edu
[2] US Food and Drug Administration, Silver Spring, MD, USA
[3] University of Maryland, College Park, MD, USA
[4] Georgia Institute of Technology, Atlanta, GA, USA
[5] Vienna University of Technology, Vienna, Austria
[6] Stony Brook University, Stony Brook, NY, USA

Abstract. We present a bifurcation analysis of electrical alternans in
the two-current Mitchell-Schaeffer (MS) cardiac-cell model using the theory of δ-decidability over the reals. Electrical alternans is a phenomenon
characterized by a variation in the successive *Action Potential Durations*
(APDs) generated by a single cardiac cell or tissue. Alternans are known
to initiate re-entrant waves and are an important physiological indicator
of an impending life-threatening arrhythmia such as ventricular fibrillation. The bifurcation analysis we perform determines, for each control
parameter τ of the MS model, the *bifurcation point* in the range of τ
such that a small perturbation to this value results in a transition from
alternans to non-alternans behavior. To the best of our knowledge, our
analysis represents the first formal verification of non-trivial dynamics
in a numerical cardiac-cell model.

Our approach to this problem rests on encoding alternans-like behavior in the MS model as a 11-mode, multinomial hybrid automaton (HA).
For each model parameter, we then apply a sophisticated, guided-search-
based reachability analysis to this HA to estimate parameter ranges
for both alternans and non-alternans behavior. The bifurcation point
separates these two ranges, but with an uncertainty region due to the
underlying δ-decision procedure. This uncertainty region, however, can
be reduced by decreasing δ at the expense of increasing the model exploration time. Experimental results are provided that highlight the effectiveness of this method.

1 Introduction

An important component of cardiac electrodynamic modeling is the ability to
understand and predict qualitative changes that take place in the dynamics as
model parameters are varied [1,9,29]. One well-known change involves a transition to *alternans*: a phenomenon characterized by a period-doubling bifurcation
where, while cells are paced at a constant period, their response has different

© Springer International Publishing AG 2016
E. Bartocci et al. (Eds.): CMSB 2016, LNBI 9859, pp. 132–146, 2016.
DOI: 10.1007/978-3-319-45177-0_9

dynamics between even and odd beats, with one long action potential following a short one [23]. Alternans are known to destabilize waves [15] and initiate re-entrant waves and represent an important physiological indicator of an impending life-threatening arrhythmia such as ventricular fibrillation [19,27].

About 100 mathematical models [10] have been developed to recreate and study, to varying degrees of complexity, the electrical dynamics of a cardiac cell (i.e., cardiomyocyte). A particularly appealing one in terms of its mathematical tractability is the model of Mitchell and Schaeffer [24], which represents the cellular electrodynamics using only two state variables: a voltage variable v that describes the trans-membrane potential, and a gating variable h that describes the internal ionic state of the cell.

In this paper, we present a bifurcation analysis of electrical alternans in the two-current Mitchell-Schaeffer (MS) cardiac-cell model[1] using the theory of δ-decidability over the reals [12]. The bifurcation analysis we perform determines, for each parameter τ of the MS model, the *bifurcation point* in the range of τ such that a small perturbation to this value results in a transition from alternans to non-alternans behavior; see Fig. 1. To the best of our knowledge, our analysis represents the first formal verification of non-trivial dynamics in a realistic cardiac-cell model.

Our approach to this problem rests on encoding alternans-like behavior in the MS model as an 10-mode, multinomial hybrid automaton (HA). For each MS model parameter, we then apply a sophisticated, guided-search-based reachability analysis to this HA to estimate ranges for both alternans and non-alternans behavior. The bifurcation point separates these two ranges, but with an uncertainty region due to the underlying δ-decision procedure. This uncertainty region, however, can be reduced by decreasing δ at the expense of increasing the model exploration time. Experimental results are provided that highlight the effectiveness of this method.

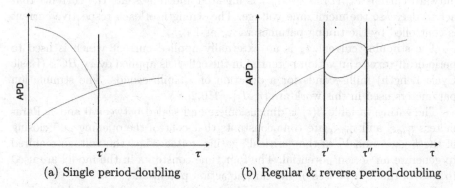

(a) Single period-doubling   (b) Regular & reverse period-doubling

Fig. 1. Bifurcation analysis of *alternans* for parameter τ. Parameter values τ' and τ'' are bifurcation points.

---

[1] A third current I_s, which is not intrinsic to the MS model, is used to stimulate the cell to produce an action potential.

This paper is organized as follows. Section 2 presents the MS model and gives a brief overview of the dReach tool that we use to perform reachability analysis. Section 3 represents the MS model as an HA and then extends the MS HA to encode alternans behavior. Section 4 formally defines *bifurcation analysis* of alternans, and outlines an approach to perform the analysis by reducing it to a parameter-synthesis problem for the HA that encodes alternans. Section 5 presents our results for the bifurcation analysis of all of the control parameters of the MS model. Section 6 considers related work. Section 7 offers our concluding remarks and directions for future work.

2   Background

2.1   Mitchell-Schaefer Model

The Mitchell-Schaefer model is an activator-inhibitor system that describes the electrical dynamics of a ventricular myocyte. The model involves two coupled, nonlinear ordinary differential equations of the form:

$$\dot{v} = I_{in}(v, h) + I_{out}(v) + I_s(t)$$
$$\dot{h} = \begin{cases} \frac{1-h}{\tau_{open}} & v < V_g \\ \frac{-h}{\tau_{close}} & v \geq V_g \end{cases} \qquad (1)$$

where $v(t)$ is the transmembrane voltage and $h(t)$ is a gating variable (as in a voltage-gated ion channel [10]). The voltage ranges from -85 to 20 mV in a real cardiac cell, but has been scaled to the range $[0, 1]$ in the MS model, and is expressed as the sum of three currents: an inward current, outward current and stimulation current. The inward current $I_{in}(v, h) = hv^2(1-v)/\tau_{in}$ is designed to replicate the behavior of fast-acting gates found in more complex models. The outward current $I_{out}(v) = -v/\tau_{out}$ is ungated and represents the currents that act to decrease the membrane voltage. The strength of each respective current is controlled by the timing parameters τ_{in} and τ_{out}.

The stimulus current I_s is an externally applied current which is used to periodically excite an action potential in the cell. It is applied every BCL (Basic Cycle Length) milliseconds for a duration of τ_s milliseconds. The stimulation parameters used in this work are $[\tau_s, I_s] = [1, 0.2]$.

The gating variable $h(t)$ is dimensionless and scaled between 0 and 1. Parameters τ_{close} and τ_{open} are time constants that control the opening and closing of the h-gate, and V_g is the "critical" gating voltage; i.e., the voltage required to generate an action potential. The four time constants in the model are used to control the four phases of the cardiac action potential.

For certain parameter values, the Mitchell-Schaeffer model can exhibit alternans, a state which successively exhibits alternating short-long values of the APD. An example of alternans and non-alternans behavior in the voltage time series is shown in Fig. 2.

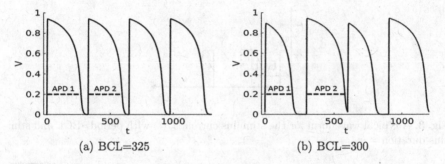

Fig. 2. The voltage time series for the Mitchell-Schaeffer model using parameter values $[\tau_{in}, \tau_{out}, \tau_{open}, \tau_{close}, V_g]=[0.3, 6, 20, 150, 0.1]$. The threshold value used to compute the APD is $V_T = 0.2$. (a) The time series does not meet the definition of alternans since APD1=APD2. (b) The time series meets the definition of alternans since the APDs alternate in length (short, long, short long).

2.2 The dReach Tool

dReach [21] is a bounded reachability analysis tool for nonlinear hybrid systems. It takes a hybrid automaton $\mathcal{H}$, reachability properties $\mathcal{P}$, a numerical error bound $\delta \in \mathbb{Q}^+$, and an unrolling depth $k \in \mathbb{N}$ as inputs. It then encodes a bounded-reachability problem for a hybrid automaton as a first-order formula over the reals and solves the formula using the delta-decision SMT solver dReal [14]. There are two possible outputs from the dReach tool:

- unreachable: dReach confirms that there is no trace satisfying the reachability properties up to k discrete jumps.
- δ-reachable: dReach shows that there exists a trace ξ satisfying the reachability properties if we consider a user-specified numerical perturbation $\delta \in \mathbb{Q}^+$ in $\mathcal{H}$. The tool also provides a feature to visualize this trace.

We note that the bounded-reachability problem for nonlinear hybrid automata is undecidable [3]. The tool is implemented in the framework of delta-complete analysis for bounded reachability of hybrid systems [13], which provides an algorithm for the originally undecidable problem by using approximation (the use of δ in the analysis).

3 Hybrid Automata for the MS Model and Alternans

In this section, we represent the MS model as a hybrid automaton and extend this automaton to encode alternans and non-alternans behavior.

3.1 Hybrid Automaton ($\mathcal{H}_M$) for the MS model

The stimulus current $I_s(t)$ in Eq. 1 is typically a periodic square-wave pulse of fixed duration (τ_s). An example of such a wave form is shown in Fig. 3.

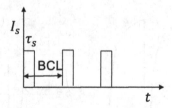

Fig. 3. A typical wave form for the stimulus current $I_s(t)$ with period=BCL and stimulus duration $= \tau_s$.

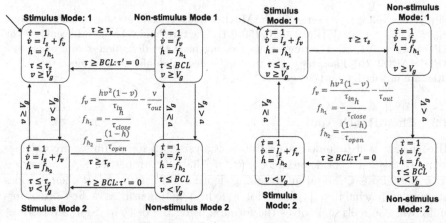

(a) Detailed MS Automaton   (b) Simplified MS Automaton $\mathcal{H}_M$

Fig. 4. The four-mode hybrid automaton for the MS model. The primed version of variables is used to indicate the reset map of a given transition. Variables not shown in the reset map are not updated during the jump.

To handle this type of stimulus signal in the MS model, we split the voltage dynamics into two separate modes: a stimulus mode and a non-stimulus mode.

Since the dynamics of variable h is also separable into two modes, we can represent the MS cardiac-cell model as a four-mode hybrid automaton (HA) whose schematic is shown in Fig. 4(a). We add an additional state variable τ that serves as a local clock for time-triggered events; for example, the transition from a stimulus to a non-stimulus mode or the transition from the current AP cycle to the next.

Due to the following observations, we can simplify this HA by removing certain edges:

- $v < V_g$ will not occur in "Stimulation Mode 1", as the value of v always increases in this mode
- $v \geq V_g$ will not occur in "Non-stimulation Mode 2", as v always decreases in this mode

- $v \geq V_g$ occurs before $\tau \geq \tau_s$ in "Stimulus Mode 2"
- For a chosen BCL range, $v < V_g$ occurs before $\tau \geq BCL$ in "Non-stimulus Mode 1"

3.2   Encoding Alternans and Non-Alternans as Hybrid Automata

We now encode a modified definition of alternans that incorporates transient cycles and a tolerance threshold r_{th}, $0 \leq r_{th} \leq 1$, which establishes the relative difference between APDs. Transients are important since, when starting from an initial state and a set of parameters that are known to produce alternans, the voltage signal only settles into period-doubling after the transient phase is over. Failure to incorporate transient cycles can result in unwanted effects on the alternans calculation. We add the tolerance threshold r_{th} to take into account noise and measurement errors in the clinical data that is used to calculate alternans.

Definition 1. *Let σ be a (possibly infinite) voltage signal that begins with N_{trans} AP cycles, followed by at least two AP cycles, where N_{trans} is the number of transient cycles. Let $\tau_1 > 0$ and $\tau_2 > 0$ be the APDs of any two consecutive AP cycles after the initial N_{trans} cycles in σ. Further, let $r = \frac{\tau_2}{\tau_1}$. We say that σ exhibits alternans with respect to a given r_{th} when $|r - 1| > r_{th}$ is an invariant. Likewise, we say that σ exhibits non-alternans with respect to r_{th} when $|r - 1| \leq r_{th}$ is an invariant.*

As opposed to using the absolute value of the difference of consecutive APDs ($|APD_1 - APD_2|$) for the definition of alternans, Definition 1 yields a *normalized* (between 0 and 1) basis for comparison. Note that as r_{th} is increased, the estimated bifurcation point is moved away from the exact value and farther into the alternans region. In the limit as r_{th} approaches zero, the estimated bifurcation point approaches the exact value, as shown in Fig. 5.

We first explain the steps used to encode alternans as an HA based on $\mathcal{H}_M$, and then follow similar steps to encode non-alternans as another HA. We consider alternans as a safety property and characterize it using a so-called *safety automaton* [2]. For our purposes, a safety automaton is an HA with modes additionally marked as *accepting* or *non-accepting*, and with the property that no

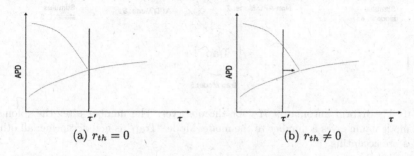

(a) $r_{th} = 0$ (b) $r_{th} \neq 0$

Fig. 5. Effect of r_{th} on bifurcation point.

accepting mode can be reaching from a non-accepting mode. After first deter-mining that $\mathcal{H}_M$ has completed N_{trans} transient cycles, our safety, or *observer*, automaton $\mathcal{H}_O$ repeatedly computes two successive APDs τ_1 and τ_2, and checks if the condition for alternans (Definition 1) is violated. If so, the automaton enters a trap (i.e. non-accepting) state, from which it never exits. If no such violation is detected, then the observed sequence of cycles is accepted. Thus, in $\mathcal{H}_O$, there is a single non-accepting mode named "Trap"; all other modes are accepting. Note that $\mathcal{H}_O$ uses the v and τ values from $\mathcal{H}_M$ to determine when a cycle has completed and to compute APD values.

Figure 6 presents observer HA $\mathcal{H}_O$ for the alternans problem. As, by defini-tion, *APD* is the time period in each AP cycle during which $v \geq V_T$, an APD event can occur only in "Stimulus Mode: 1" and "Non-stimulus Mode: 1" in $\mathcal{H}_M$. So to compute *APD*, the observer splits "Non-stimulus Mode: 1" into two modes: "APD Mode" (when $v \geq V_T$) and "Non-APD Mode" (when $v < V_T$). As the "Stimulus Mode: 1" is at most τ_s and τ_s (typically 1 ms) is negligible compared to the duration of "Non-stimulus Mode: 1" (> 200 ms), we ignore the event $v \geq V_T$ inside "Stimulus Mode: 1" for the *APD* computation. This helps us avoid splitting "Stimulus Mode: 1" and thus reduces the number of modes in the observer HA.

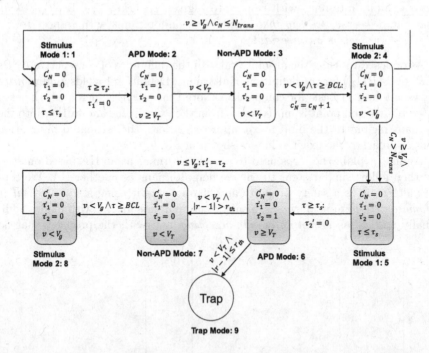

Fig. 6. The hybrid automaton $\mathcal{H}_O$ for the observer. The number after the colon in each mode name gives a number to the mode. Mode "Trap" is non-accepting; all other modes are accepting.

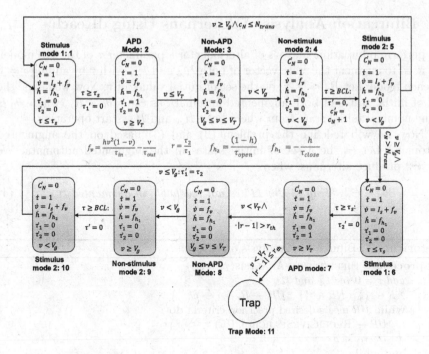

Fig. 7. The 11-mode hybrid automaton $\mathcal{H}_A$ for alternans.

To determine whether $\mathcal{H}_M$ completes N_{trans} transient cycles, we add a counter C_N in $\mathcal{H}_O$ which is increased by 1 during the jump from "Non-APD Mode: 3" to "Stimulus Mode 2: ". In $(N_{trans} + 1)$ cycle, $\mathcal{H}_O$ computes τ_1 in "APD Mode: 2" and then compute τ_2 in the consequent cycle in "APD Mode: 6". When $v < V_T \wedge |r - 1| > r_{th}$ does not hold, a transition from "APD Mode: 6" to "Trap Mode: 9" occurs, i.e., the alternans property is violated. All the other modes are the accepting states for this safety (Buechi) automaton.

To check the alternans property, we combine $\mathcal{H}_M$ and $\mathcal{H}_O$ into a single automaton $\mathcal{H}_A$ as shown in Fig. 7. This approach is known as shared-variable composition [4].

Let Θ_0 be a set of initial states in $\mathcal{H}_O$. We say Θ_0 produce alternans when:

$$\exists \theta_0 \in \Theta_0. \text{"Trap Mode: 11" is not reachable in } \mathcal{H}_A. \tag{2}$$

Similar to Fig. 7, we can encode the dual behavior, non-alternans, as an HA $\mathcal{H}_N$ by inter-changing guard conditions of the outgoing transitions in "APD Mode: 7". We then say that that Θ_0, a set of initial states in $\mathcal{H}_N$, produces non-alternans when:

$$\exists \theta_0 \in \Theta_0. \text{"Trap Mode: 11" is not reachable in } \mathcal{H}_N. \tag{3}$$

4 Bifurcation Analysis of Alternans Using dReach

To perform bifurcation analysis of alternans for a parameter τ of the MS model, we need to augment the state vector of both $\mathcal{H}_A$ and $\mathcal{H}_N$ with τ by adding $\dot{\tau} = 0$ in each mode. Let $R_\tau = [\underline{\tau}, \overline{\tau}]$ be the set of initial values of τ. Now we define the set of initial states of both augmented automata as $\Theta_0^a = \theta_0 \times R_\tau$, where θ_0 is some nominal initial state from where both $\mathcal{H}_A$ and $\mathcal{H}_N$ start operating.

Now we will redefine the problem (2) and (3) based on the augmented automata. Let Θ_0^a be a set of initial states in the augmented automata. We say Θ_0^a produce alternans, when

$$\exists \theta_0^a \in \Theta_0^a. \text{``Trap Mode: 11'' is not reachable in augmented } \mathcal{H}_A. \qquad (4)$$

Algorithm 1. Bifurcation Analysis on dReach

1: **procedure** BIFURCATION-ANALYSIS(τ, R_τ, δ_0)
2:     add $\dot{\tau} = 0$ *in* $\mathcal{H}_A$ *and* $\mathcal{H}_N$
3:     $AR \leftarrow \{\}$ $NR \leftarrow \{\}$ $UR \leftarrow R_\tau$ $\delta \leftarrow \delta_0$
4:     **while** *UR* meets desired precision criteria **do**
5:         $UR \leftarrow$ RECURSIVESEARCH(δ, UR)
6:         *Decrease* δ
7:     **end while**
8: **end procedure**

Similarly, we say Θ_0^a produce non-alternans, when

$$\exists \theta_0^a \in \Theta_0^a. \text{``Trap Mode: 11'' is not reachable in augmented } \mathcal{H}_N. \qquad (5)$$

Algorithm 1 serves as an outline of our bifurcation analysis of alternans, for τ varying in range R_τ, using dReach-based reachability analysis on problems (4) and (5). The algorithm will partition R_τ into three regions: 1) *Alternans Region* (AR), 2) *Non-alternans Region* (NR) and 3) *Uncertainty Region* (UR) which contains the bifurcation point (BP).

Algorithm 1 starts by augmenting $\mathcal{H}_A$ and $\mathcal{H}_N$ with τ and initializing AR, NR, UR and δ. In the while-loop, it then calls a recursive search procedure to reduce the size of the UR, while concomitantly computing AR and NR. The algorithm terminate when size of the UR meets the desired precision criteria (i.e., the UR is small enough).

In the recursive search procedure, we first initialize Θ_0^a, which we will use for both automata. We then run dReach on problem (4). If dReach returns *unsat* for this problem, we add UR to NR and return the empty set for the new UR. If it returns δ-*sat*, however, we run dReach on the dual problem as shown on line 8. If dReach returns *unsat* for the dual problem, we add UR to AR and return the empty set for the new UR.

In both cases, when dReach returns δ-*sat* and the size of UR becomes less than or equal to current δ, we return UR as the new uncertainty region as

```
1:  procedure RECURSIVESEARCH(UR,δ)
2:      Θ₀ᵃ = θ₀ × UR
3:      α ← dReach(ℋₐ, Θ₀ᵃ, δ)
4:      if α = unsat then
5:          NR ← NR ∪ UR
6:          return {}
7:      end if
8:      β ← dReach(ℋₙ, Θ₀ᵃ, δ)
9:      if β = unsat then
10:         AR ← AR ∪ UR
11:         return {}
12:     end if
13:     if |UR| ≤ δ then
14:         return UR
15:     end if
16:     (URˡ, URʳ) ← BISECT(UR)
17:     return RECURSIVESEARCH(URˡ, δ) ∪ RECURSIVESEARCH(URʳ, δ)
18: end procedure
```

shown on line 14. If, however, the size of UR is greater than δ, we bisect UR and recursively call the search method on both branches, returning their union as the new UR.

Figure 8 provides an example of our bifurcation analysis of alternans. Figure 8(a) shows the exact bifurcation analysis that we wish to achieve using δ-decidability over the reals. Figure 8(b) shows the bifurcation analysis using Algorithm 1. Initially, the entire range is considered as an UR in Algorithm 1. The algorithm then iteratively reduces UR and increases AR and NR. Figure 8(c), shows how the recursive search procedure, in a binary-search-tree-like fashion, computes AR and NR and reduces UR.

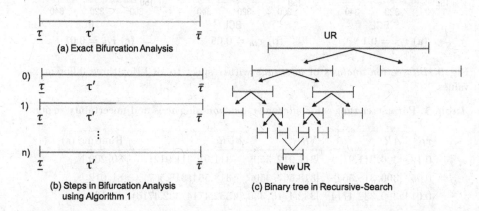

Fig. 8. Bifurcation analysis of alternans. Red: AR, Green: NR, Gray: UR. (Color figure online)

5   Results

In this section, we present the results of performing *bifurcation analysis* of alternans over five parameters in the MS model using Algorithm 1. When we perform bifurcation analysis for a parameter, we fix the other parameter as follows:
$[V_g, r_{th}, N_{trans}, BCL, \tau_{in}, \tau_{out}, \tau_{open}, \tau_{close}]$ are set to $[0.1, 0.2, 2, 300, 0.3, 6, 20, 150]$ unless specified otherwise. The fixed initial condition θ_0 for $\mathcal{H}_A$ and $\mathcal{H}_N$ were taken as $v(0) = 0.2$, $h(0) = 1$ with $C_N(0)$, $\tau(0)$, $\tau_1(0)$ and $\tau_2(0)$ all set to zero. In all cases, we consider voltage signal that contains $N_{trans} + 2$ AP cycles.

For the *bifurcation analysis* of *alternans* for BCL, we consider the range as $[300, 330]$, $\delta_0 = 0.5$. We perform the bifurcation analysis for three different r_{th} values. Figure 9, for three different r_{th}, illustrates the partitioning of the range of BCL into three regions: AR, NR and UR and Table 1 shows the corresponding subranges computed by Algorithm 1. We also overlay the simulation-based *bifurcation diagram* to help visualizing the position of the *bifurcation point*. The sequence of figures illustrate how the bifurcation region returned by dReach approaches the exact bifurcation point as r_{th} approaches zero.

We summarize the *bifurcation analysis* for other parameters in Table 2 for $r_{th} = 0.01$ and Fig. 10 shows their bifurcation diagrams. Note that we are not able to find any BP for τ_{open}. All computation is performed using Intel Core i7-4770 CPU @ 3.40 GHz × 8 on Linux platform.

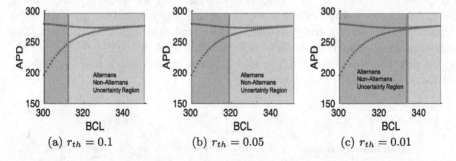

(a) $r_{th} = 0.1$ (b) $r_{th} = 0.05$ (c) $r_{th} = 0.01$

Fig. 9. *Bifurcation analysis* of *alternans* with respect to BCL for three different r_{th} values.

Table 1. Parameter ranges for *alternans* and *non-alternans* and uncertainty region.

r_{th}	AR	NR	UR	Runtime (s)
0.1	$[300, 311.91]$	$[311.912, 350]$	$(311.91, 311.912)$	$80, 209$
0.05	$[300, 318.564]$	$[318.567, 350]$	$(318.564, 318.567)$	$81, 012$
0.01	$[300, 332.4714]$	$[332.4716, 330]$	$(332.4714, 332.4716)$	$81, 162$

Table 2. Parameter ranges for *alternans* and *non-alternans* and uncertainty region.

Parameter	AR	NR	UR	Runtime (s)
τ_{in}	$[0.3, 0.3729]$	$[0.3730, 0.4]$	$(0.3729, 0.3720)$	176010
τ_{out}	$[4.9995, 6]$	$[3, 4.9990]$	$(4.9990, 4.9995)$	66000
τ_{open}	$[7.5, 20]$	–	–	110231
τ_{close}	$[131.8586, 150]$	$[130, 131.8584]$	$(131.8584, 131.8586)$	84938

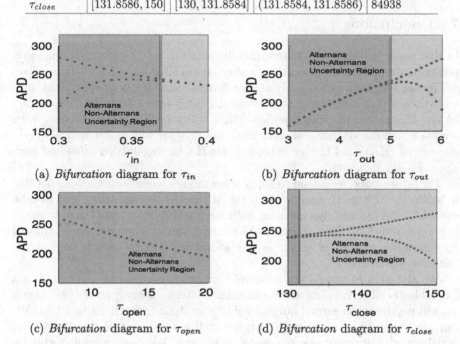

(a) *Bifurcation* diagram for τ_{in} (b) *Bifurcation* diagram for τ_{out}

(c) *Bifurcation* diagram for τ_{open} (d) *Bifurcation* diagram for τ_{close}

Fig. 10. *Bifurcation analysis* of *alternans* with respect to four parameters of the MS model with $r_{th} = 0.01$.

6   Related Work

Reachability analysis has emerged as a promising solution for many biological systems [6,11,17,31]. SMT-based verification using dReal [14] has been applied in various problems [5,8,18,20,25,26]. Liu et al. successfully applied SMT-based reachability analysis using dReach in identifying patient-specific androgen ablation therapy schedules for postponing the potential cancer relapse in [22].

Brim et al. present a bifurcation analysis technique to analyze stability of genetic regulatory networks in [7]. They first express various stability-related properties by a temporal logic language extended by directional propositions and then verify those properties by varying the model parameters. Even though they apply their method only on piece-wise affine dynamics, the authors claim that it can be extended for piece-wise multiaffine dynamics. The method, however, is not applicable for general nonlinear dynamical systems.

In [16], Huang et al. presents a reachability analysis technique for a hybrid model of cardiac dynamics for a 1-d cable of cells and show the presence of alternans based on computed reachtube. The authors, however, neither define nor verify the alternans property formally. They just do reachability analysis for two *BCL* values and show, by visual inspection, that one *BCL* value produces alternans and another does not.

7   Conclusions

In this paper, we have applied reachabilty analysis to identify the bifurcation points that represent the transition to alternans in the Mitchell-Schaefer cardiac-cell model. Our bifurcation analysis is performed using the bounded-reachability tool dReach [21], and uses a sophisticated guided-search strategy to "zoom in" on the bifurcation point in question. Since this tool is designed to work with nonlinear hybrid systems, we converted the original MS model into a hybrid automaton (HA), and further extended this HA to encode alternans- and non-alternans-like behavior.

For future work, we intend to study other models where alternans are not due to solely the voltage dynamics, as in the MS model. Rather, they may also be caused by the calcium dynamics, as both mechanisms have been found to occur in cardiac cells [28]. Such models can have multiple BPs and our algorithm will automatically find all of them, as it searches for BPs in each branch of the recursive search tree.

We also plan to extend the cell-level bifurcation analysis we conducted to a 1-d cable of cells. Traveling waves can exhibit alternans along cables [30]. Doing so, will require us to extend our reachability analysis from ODEs to PDEs. We can also extend our analysis by varying multiple parameters simultaneously; currently, we only vary one parameter at a time. We can accomplish this by augmenting the state vector with each of these parameters.

Acknowledgments. We would like to thank the anonymous reviewers for their helpful comments. Research supported in part by the following grants: NSF IIS-1447549, NSF CPS-1446832, NSF CPS-1446725, NSF CNS-1446665, NSF CPS 1446365, NSF CAR 1054247, AFOSR FA9550-14-1-0261, AFOSR YIP FA9550-12-1-0336, CCF-0926190, ONR N00014-13-1-0090, and NASA NNX12AN15H.

References

1. Shrier, A., Dubarsky, H., Rosengarten, M., Guevara, M.R., Nattel, S., Glass, L.: Prediction of complex atrioventricular conduction rhythms in humans with use of the atrioventricular nodal recovery curve. Circulation **76**(6), 1196–1205 (1987)
2. Alpern, B., Schneider, F.B.: Recognizing safety and liveness. Distrib. Comput. **2**(3), 117–126 (1987)
3. Alur, R., Courcoubetis, C., Henzinger, T.A., Ho, P.: Hybrid automata: an algorithmic approach to the specification and verification of hybrid systems. In: Hybrid Systems, pp. 209–229 (1992)

4. Alur, R., Henzinger, T.A., Ho, P.-H.: Automatic symbolic verification of embedded systems. IEEE Trans. Softw. Eng. **22**(3), 181–201 (1996)
5. Bae, K., Ölveczky, P.C., Kong, S., Gao, S., Clarke, E.M.: SMT-based analysis of virtually synchronous distributed hybrid systems. In: Proceedings of the 19th International Conference on Hybrid Systems: Computation and Control, pp. 145–154. ACM (2016)
6. Batt, G., De Jong, H., Page, M., Geiselmann, J.: Symbolic reachability analysis of genetic regulatory networks using discrete abstractions. Automatica **44**(4), 982–989 (2008)
7. Brim, L., Demko, M., Pastva, S., Šafránek, D.: High-performance discrete bifurcation analysis for piecewise-affine dynamical systems. In: Abate, A., et al. (eds.) HSB 2015. LNCS, vol. 9271, pp. 58–74. Springer, Heidelberg (2015). doi:10.1007/978-3-319-26916-0_4
8. Bryce, D., Gao, S., Musliner, D.J., Goldman, R.P.: SMT-based nonlinear PDDL+ planning. In: 29th AAAI Conference on Artificial Intelligence, p. 3247
9. Fenton, F.H., Cherry, E.M., Hastings, H.M., Harold, M., Evans, S.J.: Multiple mechanisms of spiral wave breakup in a model of cardiac electrical activity. chaos: an interdisciplinary. J. Nonlinear Sci. **12**(3), 852 (2002)
10. Fenton, F.H., Cherry, E.M.: Models of cardiac cell. Scholarpedia **3**(8), 1868 (2008)
11. Feret, J.: Reachability analysis of biological signalling pathways by abstract interpretation. In: Computation in Modern Science and Engineering, vol. 2, Part A (AIP Conference Proceedings vol. 963), pp. 619–622 (2007)
12. Gao, S., Avigad, J., Clarke, E.M.: δ-complete decision procedures for satisfiability over the reals. In: Gramlich, B., Miller, D., Sattler, U. (eds.) IJCAR 2012. LNCS, vol. 7364, pp. 286–300. Springer, Heidelberg (2012)
13. Gao, S., Kong, S., Chen, W., Clarke, E.M.: Delta-complete analysis for bounded reachability of hybrid systems. CoRR, abs/1404.7171 (2014)
14. Gao, S., Kong, S., Clarke, E.M.: dReal: an smt solver for nonlinear theories over the reals. In: Bonacina, M.P. (ed.) CADE 2013. LNCS, vol. 7898, pp. 208–214. Springer, Heidelberg (2013)
15. Gizzi, A., Cherry, E.M., Gilmour, R.F., Luther, S., Filippi, S., Fenton, F.H.: Effects of pacing site and stimulation history on alternans dynamics and the development of complex spatiotemporal patterns in cardiac tissue. Front. Physiol. **4**, 71 (2013)
16. Huang, Z., Fan, C., Mereacre, A., Mitra, S., Kwiatkowska, M.: Invariant verification of nonlinear hybrid automata networks of cardiac cells. In: Biere, A., Bloem, R. (eds.) CAV 2014. LNCS, vol. 8559, pp. 373–390. Springer, Heidelberg (2014)
17. Islam, M.A., De Francisco, R., Fan, C., Grosu, R., Mitra, S., Smolka, S.A.: Model checking tap withdrawal in C. Elegans. In: Hybrid Systems Biology, p. 195
18. Islam, M.A., Murthy, A., Girard, A., Smolka, S.A., Grosu, R.: Compositionality results for cardiac cell dynamics. In: Proceedings of the 17th international conference on Hybrid systems: computation and control, pp. 243–252. ACM (2014)
19. Weiss, J.N., Alain, S., Shiferaw, Y., Chen, P., Garfinkel, A., Qu, Z.: From pulsus to pulseless the saga of cardiac alternans. Circ. Res. **98**(10), 1244–1253 (2006). WOS: 000237812200006
20. Kapinski, J., Deshmukh, J.V., Sankaranarayanan, S., Arechiga, N.: Simulation-guided Lyapunov analysis for hybrid dynamical systems. In: Hybrid Systems: Computation and Control (HSCC), pp. 133–142. ACM Press (2014)
21. Kong, S., Gao, S., Chen, W., Clarke, E.M.: dReach: δ-reachability analysis for hybrid systems. In: Proceedings of Tools and Algorithms for the Construction and Analysis of Systems - 21st International Conference, TACAS, London, UK, April 11–18, 2015, pp. 200–205 (2015)

22. Liu, B., Kong, S., Gao, S., Zuliani, P., Clarke, E.M.: Towards personalized prostate cancer therapy using delta-reachability analysis. In: Proceedings of the 18th International Conference on Hybrid Systems: Computation and Control, pp. 227–232. ACM (2015)
23. Guevara, M., Glass, L., Shrier, A.: Phase locking, period-doubling bifurcations, and irregular dynamics in periodically stimulated cardiac cells. Science **214**(4527), 1350–1353 (1981)
24. Mitchell, C.C., Schaeffer, D.G.: A two-current model for the dynamics of cardiac membrane. Bull. Math. Biol. **65**(5), 767–793 (2003)
25. Murthy, A., Islam, M.A., Smolka, S.A., Grosu, R.: Computing bisimulation functions using SOS optimization and δ-decidability over the reals. In: Proceedings of the 18th International Conference on Hybrid Systems: Computation and Control, pp. 78–87. ACM (2015)
26. Murthy, A., Islam, M.A., Smolka, S.A., Grosu, R.: Computing compositional proofs of input-to-output stability using SOS optimization and δ-decidability. Hybrid Systems, Nonlinear Analysis (2016)
27. Gilmour, R.F., Chialvo, D.R.: Electrical restitution, critical mass, and the riddle of fibrillation. J. Cardiovas. Electrophysiol. **10**(8), 1087–1089 (1999)
28. Shiferaw, Y., Sato, D., Karma, A.: Coupled dynamics of voltage and calcium in paced cardiac cells. Phys. Rev. E **71**(2), 021903 (2005)
29. Quail, T., Shrier, A., Glass, L.: Predicting the onset of period-doubling bifurcations in noisy cardiac systems. Proc. Nat. Acad. Sci. **112**(30), 9358–9363 (2015)
30. Watanabe, M.A., Fenton, F.H., Evans, S.J., Hastings, H.M., Karma, A.: Mechanisms for discordant alternans. J. Cardiovasc. Electrophysiol. **12**(2), 196–206 (2001)
31. Yang, Y., Lin, H.: Reachability analysis based model validation in systems biology. In: 2010 IEEE Conference on Cybernetics and Intelligent Systems (CIS), pp. 14–19. IEEE (2010)

A Stochastic Hybrid Approximation
for Chemical Kinetics Based on the Linear
Noise Approximation

Luca Cardelli[1,2], Marta Kwiatkowska[2], and Luca Laurenti[2(✉)]

[1] Microsoft Research, Cambridge, UK
[2] Department of Computer Science, University of Oxford, Oxford, UK
luca.laurenti@cs.ox.ac.uk

Abstract. The Linear Noise Approximation (LNA) is a continuous approximation of the CME, which improves scalability and is accurate for those reactions satisfying the leap conditions. We formulate a novel stochastic hybrid approximation method for chemical reaction networks based on adaptive partitioning of the species and reactions according to leap conditions into two classes, one solved numerically via the CME and the other using the LNA. The leap criteria are more general than partitioning based on population thresholds, and the method can be combined with any numerical solution of the CME. We then use the hybrid model to derive a fast approximate model checking algorithm for Stochastic Evolution Logic (SEL). Experimental evaluation on several case studies demonstrates that the techniques are able to provide an accurate stochastic characterisation for a large class of systems, especially those presenting dynamical stiffness, resulting in significant improvement of computation time while still maintaining scalability.

1 Introduction

Biochemical systems are inherently stochastic: the time for the next reaction to occur and which reaction fires next are both random variables. When the reactant molecules are in low number the resulting dynamic behaviour can be highly stochastic and deterministic models are unable to correctly approximate it [4,23]. Thus, an accurate characterisation of stochastic fluctuations in biological systems is essential [30]. It is well known that a biochemical system evolving in a spatially homogeneous environment, at constant volume and temperature, can be described as a continuous-time Markov chain (CTMC) [10] Transient analysis is generally performed through solving the Chemical Master Equation (CME) [30] or with the Stochastic Simulation Algorithm (SSA) [12]. The SSA produces a single realization of the stochastic process, whereas the CME gives the probability distribution of each species over time. The CME can be solved numerically through solving differential equations or methods based on uniformisation, both requiring exploration of the reachable state space and thus infeasible for systems with large or infinite state spaces. On the other hand, the SSA is generally

© Springer International Publishing AG 2016
E. Bartocci et al. (Eds.): CMSB 2016, LNBI 9859, pp. 147–167, 2016.
DOI: 10.1007/978-3-319-45177-0_10

faster, although obtaining good accuracy necessitates potentially large numbers of simulations and can be time consuming.

An alternative is to approximate the CME as a *continuous-state stochastic* process. The *Linear Noise Approximation (LNA)* is a Gaussian process which has been derived as an approximation of the CME [30]. Thus, the LNA is inherently unimodal and not accurate for multimodal dynamics. Its solution involves a number of differential equations that is quadratic in the number of species and independent of the molecular populations. As a consequence, the LNA is generally much more scalable than a discrete stochastic representation. For these reasons, the LNA has recently been used for model checking of large biochemical systems [5,8]. The solution given by the LNA is accurate if conditions on species and reactions known as the *leap conditions* are satisfied, which holds in the limit of high populations, but typically only for a subset of species and reactions (i.e. stiff systems). As a result, a *discrete stochastic* representation is necessary for the remaining species. A natural approach is thus to consider a *stochastic hybrid* semantics that combines a continuous approximation based on the LNA for species respecting the leap conditions and maintains a discrete stochastic representation for the remaining species. Fortunately, for a large class of biological systems the species that respect the leap conditions are in high number [31], which necessitates solving the CME only for a significantly reduced state space.

Contributions. We present a *stochastic hybrid* model for biochemical systems, where a subset of species and reactions is treated with a *continuous* state-space stochastic process, the LNA, while the remaining species are treated as a *discrete* state-space stochastic process. A key advantage is that transient analysis of a discrete stochastic process is needed only for a substantially reduced set of species, ameliorating state-space explosion. The main novelty of our approach is that we partition species and reactions using the leap conditions. This allows us to dynamically and automatically update the partitions, which is necessary since the satisfaction of the leap conditions may change with time. We derive equations for the joint and marginal probability distributions of the partitioned system. Continuous species are modelled as a mixture of Gaussian distributions, enabling us to treat multimodality. We present a numerical method for solving the CME, which adaptively and automatically decides for which species a discrete characterization is needed, and which species can be approximated with the LNA, thus resulting in significant improvement of computation time while still maintaining scalability. We then employ the presented hybrid semantics to build a fast and scalable probabilistic model checking algorithm for Stochastic Evolution Logic (SEL), a temporal logic presented in [8]. We implement the techniques and demonstrate on several case studies their ability to provide an accurate stochastic characterization of systems for which the LNA is imprecise, but full solution of the CME, even using advanced numerical techniques, is not feasible because of scalability issues. We emphasise that our method can be used in conjunction with any existing numerical solution of the CME.

Related Work. The work of Henzinger et al. [18], where a hybrid method is presented with a subset of species treated as a continuous approximation and the remaining species by solving the CME, differs from ours in at least two key aspects. Firstly, their continuous approximation is *deterministic*, whereas ours is continuous stochastic. Secondly, they partition the species based on a *threshold* on the molecular population, rather than the leap conditions, which may lead to inaccuracies, since the error of the deterministic model depends not only on the molecular population but also on model parameters [10]. Our use of the leap conditions guarantees the accuracy of the stochastic approximation. Thomas et al. [29] develop a conditional LNA method and apply it to gene expression networks. They approximate the probability distribution of gene expression products with the conditional LNA, while still treating promoters with the CME. Our approach is similar in the sense that we also consider the LNA for a subset of the species and a discrete-time Markov process for the remaining ones. However, it is not clear in [29] how to partition the species. Instead, we provide criteria based on the leap conditions to automatically decide for which species the LNA is accurate, and which species instead need a discrete characterization.

In [17], the authors present the method of conditional moments for approximating the moments of the solution of the CME, where small populations are treated via a discrete process and high using approximate moment closure. However, how to automatically partition the species is left as an open problem.

Partitioning of species and reactions of a reaction network for the purpose of speeding up the SSA in multi-scale systems has been explored in [15,25,26]. For instance, Yao et al. introduced the slow-scale stochastic simulation algorithm [6], where they distinguish between fast and slow species. Fast species are then treated assuming they reach equilibrium much faster than the slow ones. Adaptive partitioning of the species has been considered in [11,19]. However, in both cases, the authors consider continuous models that differ from the LNA. In particular, in [11] the authors use a jump diffusion Markov process to approximate the original CTMC and derive error bounds to decide the species partitioning.

2   Background

Chemical Reaction Networks. A *chemical reaction network (CRN) $C =$ (Λ, R)* is a pair of finite sets, where Λ is the set of *chemical species*, $|\Lambda|$ denotes its size, and R is a set of reactions. Species in Λ interact according to the reactions in R. A *reaction* $\tau \in R$ is a triple $\tau = (r_\tau, p_\tau, k_\tau)$, where $r_\tau \in \mathbb{N}^{|\Lambda|}$ is the *reactant complex*, $p_\tau \in \mathbb{N}^{|\Lambda|}$ is the *product complex* and $k_\tau \in \mathbb{R}_{>0}$ is the coefficient associated to the rate of the reaction. r_τ and p_τ represent the stoichiometry of reactants and products. Given a reaction $\tau_1 = ([0,1,1],[0,0,2],k_1)$ we often refer to it as $\tau_1 : \lambda_1 + \lambda_2 \to^{k_1} 2\lambda_3$. The *state change* associated to a reaction τ is defined by $v_\tau = p_\tau - r_\tau$. Assuming well mixed environment, constant volume V and temperature, a *configuration* or *state* $x \in \mathbb{N}^{|\Lambda|}$ of the system is given by a vector of the number of molecules of each species. Given a configuration x then $x(\lambda_i)$ represents the number of molecules of λ_i in the configuration and $\frac{x(\lambda_i)}{N}$ is the concentration of λ_i in the same configuration, where $N = V \cdot N_A$ is the volumetric factor, V is the volume and N_A Avogadro's number. The *deterministic* semantics approximates the concentrations of species over time as the solution $\Phi(t)$ of a set of differential equations of the form:

$$\frac{d\Phi(t)}{dt} = F(\Phi(t)) = \sum_{\tau \in R} \upsilon_\tau \cdot (k_\tau \prod_{i=1}^{|\Lambda|} \Phi_i^{r_{i,\tau}}(t)) \tag{1}$$

where $\Phi_i^{r_{i,\tau}}(t)$ is the ith component of vector $\Phi(t)$ raised to the power of $r_{i,\tau}$, the ith component of vector r_τ. The initial condition is $\Phi(0) = \frac{x_0}{N}$. It is known that Eq. (1) is accurate in the limit of high populations [10].

Stochastic Semantics. The propensity rate α_τ of a reaction τ is a function of the current configuration x of the system such that $\alpha_\tau(x)dt$ is the probability that a reaction event occurs in the next infinitesimal interval dt. We assume mass action kinetics, therefore $\alpha_\tau(x) = k_\tau \frac{\prod_{i=1}^{|\Lambda|} r_{i,\tau}!}{N^{|r_\tau|-1}} \prod_{i=1}^{|\Lambda|} \binom{x(\lambda_i)}{r_{i,\tau}}$, where $r_{i,\tau}$ is the ith component of the vector r_τ, $r_{i,\tau}!$ is its factorial, and $|r_\tau| = \sum_{i=1}^{|\Lambda|} r_{i,\tau}$ [3]. To simplify the notation, N is considered embedded inside the coefficient k_τ for any τ. The stochastic semantics of the CRN $C = (\Lambda, R)$ is represented by a *time-homogeneous continuous-time Markov chain* (CTMC) [10] $(X(t), t \in \mathbb{R}_{\geq 0})$ with state space $S \subseteq \mathbb{N}^{|\Lambda|}$. $X(t)$ is a random vector describing the molecular population of each species at time t. Let $x_0 \in \mathbb{N}^{|\Lambda|}$ be the initial condition of X then $P(X(0) = x_0) = 1$. For $x \in S$, we define $P(x,t) = P(X(t) = x \mid X(0) = x_0)$. The transient evolution of X is described by the Chemical Master Equation (CME), a set of differential equations

$$\frac{d}{dt}(P(x,t)) = \sum_{\tau \in R} \{\alpha_\tau(x - \upsilon_\tau)P(x - \upsilon_\tau, t) - \alpha_\tau(x)P(x,t)\}. \tag{2}$$

Solving Eq. (2) requires computing the solution of a differential equation for each reachable state. The size of the reachable states depends on the number of species and molecular populations and can be huge or even infinite. As a consequence, solving the CME is generally feasible only for CRNs with very few species and small molecular populations.

Linear Noise Approximation. A promising line of research is to consider continuous state-space approximations of $X(t)$. The *Linear Noise Approximation* (LNA) [30] is a continuous approximation of the CME, which permits a *stochastic* characterization of the evolution of a CRN, while still maintaining scalability comparable to that of deterministic models. The LNA is accurate for processes satisfying the *leap conditions* [31]. Given a CRN $C = (\Lambda, R)$, we say that the Markov process $X(t)$ induced by C satisfies the leap conditions at time t if, for any $\tau \in R$, there exists a finite time interval dt such that:

$$\alpha_\tau(X(t)) \text{ constant in } [t, t + dt] \tag{3}$$

$$\alpha_\tau(X(t)) \cdot dt \gg 1. \tag{4}$$

In [13], Gillespie shows that if these conditions are satisfied then the solution of the CME can be approximated by a *Chemical Langevin Equation (CLE)*. Then, under the assumption that stochastic fluctuations are of the order of $N^{\frac{1}{2}}$ [10,30], we can assume that $X(t)$ admits a solution of the form

$$X(t) = N\Phi(t) + N^{\frac{1}{2}}G(t) \tag{5}$$

where $G(t) = (G_1(t), G_2(t), ..., G_{|\Lambda|})$ is a random vector, independent of N, representing the stochastic fluctuations at time t and $\Phi(t)$ is the solution of Eq. (1). It is possible to show that the probability distribution of $G(t)$ can be modelled by a linear Fokker-Planck equation [31]. For every $t \in \mathbb{R}_{\geq 0}$, $G(t)$ has a multivariate normal distribution whose expected value $E[G(t)]$ and covariance matrix $C[G(t)]$ are the solution of the following differential equations:

$$\frac{\mathrm{d}E[G(t)]}{\mathrm{d}t} = J_F(\Phi(t))E[G(t)] \tag{6}$$

$$\frac{\mathrm{d}C[G(t)]}{\mathrm{d}t} = J_F(\Phi(t))C[G(t)] + C[G(t)]J_F^T(\Phi(t)) + W(\Phi(t)) \tag{7}$$

where $J_F(\Phi(t))$ is the Jacobian of $F(\Phi(t))$, $J_F^T(\Phi(t))$ its transpose, $W(\Phi(t)) = \sum_{\tau \in R} \upsilon_\tau \upsilon_\tau^T \alpha_{c,\tau}(\Phi(t))$ and $F_j(\Phi(t))$ the jth component of $F(\Phi(t))$. We assume $X(0) = x_0$ with probability 1; as a consequence $E[G(0)] = 0$ and $C[G(0)] = 0$, which implies $E[G(t)] = 0$ for every t. The following theorem illustrates the nature of the approximation using the LNA.

Theorem 1 [10]. *Let $C = (\Lambda, R)$ be a CRN and X the CTMC induced by C. Let $\Phi(t)$ be the solution of Eq. (1) with initial condition $\Phi(0) = \frac{x_0}{N}$ and G be the Gaussian process with expected value and variance given by Eqs. (6) and (7). Then, for $t \in \mathbb{R}_{\geq 0}$*

$$N^{\frac{1}{2}}|\frac{X(t)}{N} - \Phi(t)| \Rightarrow_N G(t)$$

In the above $\Rightarrow_N$ indicates convergence in distribution [10]. Theorem 1 shows that $G(t)$ models the stochastic fluctuations around the rate equations and guarantees that the leap conditions are always verified in the limit of high populations. However, they could be satisfied even for relatively small numbers of molecules [31]. To compute the LNA it is necessary to solve $O(|\Lambda|^2)$ first order differential equations, and the complexity is independent of the initial number of molecules of each species. Therefore, one can avoid the exploration of the state space that methods based on uniformization rely upon.

3 Stochastic Hybrid Approximation

The key idea behind our approximation is to partition the species into two classes, those that satisfy the leap conditions, which we approximate by a continuous process using the LNA, and the remaining species, for which we need a discrete model. The stochastic process $X(t)$ induced by the CRN can then be approximated by a *hybrid* combination of the continuous and discrete processes describing the evolution of the partitions. The set of reactions satisfying the leap conditions may change with time and, as a consequence, the partitions of species and reactions need to adapt with time.

Partitioning of Species and Reactions. Given a CRN $C = (\Lambda, R)$, condition (3) is satisfied for reaction $\tau \in R$ at time t and during the interval dt if $\alpha_\tau(X(t))$ is approximately constant during dt. Reaction $\tau \in R$, at time t, satisfies condition (4) if it fires many times during dt. Given $\sigma_1, \sigma_2 \in \mathbb{R}_{\geq 0}$, it can be equivalently stated that a CRN $C = (\Lambda, R)$ satisfies the leap conditions at time t for an interval dt and reaction $\tau \in R$ if:

$$X_{\lambda_i}(t) \geq \sigma_1 \cdot |v_\tau^{\lambda_i}| \quad \text{for } \lambda_i \text{ such that } v_\tau^{\lambda_i} \neq 0 \text{ and } r_\tau^{\lambda_i} \neq 0 \tag{8}$$

$$\alpha_\tau(X(t)) \geq \sigma_2 \tag{9}$$

where $v_\tau^{\lambda_i}$ represents the state change induced by the occurrence of reaction τ with respect to species λ_i, and $r_\tau^{\lambda_i}$ is the component of the reactant complex relative to species λ_i. A method for choosing $\sigma_1, \sigma_2 \in \mathbb{R}_{\geq 0}$ is given in [26] for SSA (see also below). These criteria induce a partition $R = (R^f, R^s)$ over reactions, where R^f includes reactions for which the leap conditions are satisfied and R^s the remaining reactions, respectively called *continuous* (or fast) reactions and *discrete* (or slow). This induces a partition $\Lambda = (\Lambda^f, \Lambda^s)$ over the species of the CRN, where Λ^f and Λ^s are respectively called *continuous* and *discrete* species. $\lambda \in \Lambda$ is in Λ^f if and only if it is changed by at least one reaction in R^f and it is not changed by reactions in R^s whose propensity is of the same order of magnitude as the reactions in R^f that change it, and otherwise it is in Λ^s. For some systems these criteria may result in species with large populations treated with a discrete stochastic process. This happens for systems where the LNA is not accurate. We illustrate partitioning with the following example.

Example 1. We consider the gene expression model described in [28]. There are two species, $mRNA$ and the protein P, and the following set of reactions

$$\tau_1 : \emptyset \rightarrow^{0.5} mRNA; \quad \tau_2 : mRNA \rightarrow^{0.0058} mRNA + P;$$

$$\tau_3 : mRNA \rightarrow^{0.0029} \emptyset; \quad \tau_4 : P \rightarrow^{0.0001} \emptyset.$$

All species are initialized with 0 molecules. We consider $\sigma_1 = 30$ and $\sigma_2 = 0.05$. At time $t = 0$, the initial partition is $\Lambda^f = \{mRNA\}$ and $R^f = \{\tau_1\}$, meaning that the continuous subsystem is given by the only reaction τ_1. In fact, in τ_1 $mRNA$ is not a reagent but only a product. Note that, using a simple threshold on the molecular population of each species to decide if it has a discrete or continuous characterization, as done in [18], would not consider $mRNA$ as a continuous species. After the first molecule of $mRNA$ is produced, the propensity rate of τ_3 increases and its influence needs to be considered. The new species partition becomes $\Lambda^f = \{\}$ and $\Lambda^s = \{mRNA, P\}$. Under our initial conditions, there exists t' such that $mRNA(t') > 30$ with probability 1. As a consequence, in t' τ_3 is a continuous reaction and the continuous subsystem is:

$$\tau_1 : \emptyset \to^{0.5} mRNA; \quad \tau_3 : mRNA \to^{0.0029} \emptyset.$$

Thus, P is considered a discrete species until both τ_2 and τ_4 become continuous reactions, and thus partitions change over time.

Derivation of the Transient Probability in the Hybrid Model. Based on the partitioning described above, the stochastic process $X(t)$ induced by a CRN can be written as $X(t) = (X^f(t), X^s(t))$, where X^f and X^s respectively describe the evolution of species in Λ^f and species in Λ^s. $X(t)$ is a Markov process, but $X^f(t)$ and $X^s(t)$, if taken separately, are *not* Markovian because they depend on each other. To tackle this issue, following Cao et al. [6], we consider the *virtual* process $\bar{X}^f(t)$ that describes the same species as X^f, but with all the discrete reactions turned off. Therefore, $\bar{X}^f$ is Markovian because it is independent of X^s, and species in Λ^s are now only parameters. Note that $\bar{X}^f$ is only an approximation of the real stochastic process X^f. This approximation is accurate when continuous and discrete species evolve in different time scales. Generally, partitioning using the leap conditions guarantees that. However, it may happen that some reactions satisfy the second leap condition (Eq. 4), but not the first one (Eq. 3). This particular scenario requires attention because these reactions would be classified as discrete, and, in this case, the introduction of the virtual process may introduce some inaccuracies.

Now, we derive equations to study the transient evolution of the continuous and discrete species. Given $x^s \in S^s$ and $x^f \in S^f$, where S^s and S^f are the state spaces of discrete and continuous species, then $P(X^s(t) = x^s, \bar{X}^f(t) = x^f)$, the joint distribution of $X^s(t)$ and $\bar{X}^f(t)$, can be described by the CME (Eq. (2)). However, this would lead to state space explosion. As a consequence, in what follows, we first separate the evolution of continuous and discrete species, and then approximate the continuous subsystem using the LNA. This enables analysis of the transient evolution of the resulting hybrid process.

We denote $P(X^s(t) = x^s, \bar{X}^f(t) = x^f | X^s(0) = x^s_0, \bar{X}^f(0) = x^f_0) = P(x^s, x^f, t)$, $P(X^s(t) = x^s | X^s(0) = x^s_0, \bar{X}^f(0) = x^f_0) = P(x^s, t)$ and $P(\bar{X}^f(t) = x^f | X^s(t) = x^s, \bar{X}^f(0) = x^f_0) = P(x^f | x^s, t)$. Then, as illustrated in [25], by using the axioms of probability we have the following equivalent representation for the CME.

Lemma 1. *Let $x^s \in S^s$ and $x^f \in S^f$. Then, for $t \in \mathbb{R}_{\geq 0}$*

$$\frac{dP(x^f, x^s, t)}{dt} = \frac{dP(x^f | x^s, t)}{dt} P(x^s, t) + P(x^f | x^s, t) \frac{dP(x^s, t)}{dt}$$

So, to solve the CME in this form it is necessary to calculate $P(x^f | x^s, t)$ and $P(x^s, t)$. The first term is Markovian because of the assumption that in the virtual continuous subsystem the continuous species are independent of the discrete species, which are only parameters. Thus, it can be described by the following master equation for continuous species

$$\frac{dP(x^f|x^s,t)}{dt} = \sum_{\tau \in R^f} \alpha_\tau(x^f - \upsilon_\tau, x^s)P(x^f - \upsilon_\tau|x^s,t) - \alpha_\tau(x^f, x^s)P(x^f|x^s,t) \quad (10)$$

where υ_τ is considered restricted to the components relative to continuous species in $x^f - \upsilon_\tau$. Since the criteria for applicability of the LNA are ensured by partitioning, Eq. (10) can be approximated by the LNA.

On the other hand, $P(x^s,t)$ is not Markovian. However, Proposition 1, whose proof is in the Appendix, guarantees that $P(x^s,t)$ can be derived by solving a set of equations which have the same form as a master equation, and so numerical techniques developed for the CME can still be employed

Proposition 1. *Let $x^s \in S^s$ and $x^f \in S^f$. Then, for $t \in \mathbb{R}_{\geq 0}$ we have*

$$\frac{dP(x^s|t)}{dt} = \sum_{\tau \in R} \beta_\tau(x^s - \upsilon_\tau, t)P(x^s - \upsilon_\tau, t) - \beta_\tau(x^s, t)P(x^s, t) \quad (11)$$

where $\beta_\tau(x^s,t) = \sum_{x^f \in S^f} \alpha_\tau(x^f, x^s)P(x^f|x^s,t)$.

$\beta_\tau(x^s,t)$ is the conditional expectation of the propensity rate of τ at time t given $X^s(t) = x^s$. Reactions of higher order than bi-molecular are not likely [7], and they can always be simulated as a sequence of bi-molecular reactions. As a consequence, we can assume we are limited to at most bi-molecular reactions. Given $\lambda_i^s, \lambda_j^s \in \Lambda^s$ and $\lambda_i^f, \lambda_j^f \in \Lambda^f$, if $\alpha_\tau = k_\tau \cdot \lambda_i^f \cdot \lambda_j^s$ then $\beta_\tau(x^s,t) = k_\tau \cdot E[\bar{X}_{\lambda_i}^f(t)|x^s,t] \cdot x^s(\lambda_j)$. Similarly, if $\alpha_\tau = k_\tau \cdot \lambda_i^f \cdot \lambda_j^f$ then $\beta_\tau(x^s,t) = k_\tau \cdot E[\bar{X}_{\lambda_i}^f(t) \cdot \bar{X}_{\lambda_j}^f(t)|x^s,t]$. If $\alpha_\tau = k_\tau \cdot \lambda_i^s \cdot \lambda_j^s$ then $\beta_\tau(x^s) = k_\tau \cdot x^s(\lambda_i) \cdot x^s(\lambda_j)$. The unimolecular case follows in a straightforward way. Therefore, to fully characterize $P(x^s,t)$ only the first two moments of the conditional distribution of $\bar{X}^f(t)$ given x^s are needed. In general, this would require solving the entire CME (Eq. (2)). However, thanks to our partitioning criteria, we can safely approximate Eq. (10) by using the LNA and calculating coefficients β using Eqs. (6) and (7).

Example 2. Consider the following CRN, taken from [9]:

$$\lambda_z \rightarrow^{k_1} \lambda_1; \quad \lambda_z \rightarrow^{k_2} \emptyset; \quad \lambda_1 \rightarrow^1 \lambda_1 + \lambda_{out}$$

with $k_1, k_2 \in \mathbb{R}_{\geq 0}$ and initial condition x_0 such that $x_0(\lambda_z) = 1$ and $x_0(\lambda_1) = x_0(\lambda_{out}) = 0$. According to the partitioning criteria, for $\sigma_1 > 1$ and $\sigma_2 < \frac{k_1}{k_1+k_2}$ there exists $t' > 0$ such that for $t > t'$ the set of discrete species is $\Lambda^s = \{\lambda_z, \lambda_1\}$ and the set of continuous species is $\Lambda^f = \{\lambda_{out}\}$ and the partition remains constant over time. A state of the discrete state space is a vector $x^s = (x^s(\lambda_z), x^s(\lambda_1))$. It is easy to verify that the discrete state space S^s is composed of only 3 states: $S^s = \{x_0^s = (1,0), x_1^s = (0,0), x_2^s = (0,1)\}$. According to Eq. (10), and using the law of total probability, the distribution of λ_{out} for $t > t'$ is given by

$$P(\bar{X}^f_{\lambda_{out}}(t) = k) = P(\bar{X}^f_{\lambda_{out}}(t) = k|x^s_0, t)P(x^s_0, t) +$$
$$P(\bar{X}^f_{\lambda_{out}}(t) = k|x^s_1, t)P(x^s_1, t) + P(\bar{X}^f_{\lambda_{out}}(t) = k|x^s_2, t)P(x^s_2, t)$$

and $P(\bar{X}^f_{\lambda_{out}}(t) = k|x^s_0, t) = P(\bar{X}^f_{\lambda_{out}}(t) = k|x^s_1, t) = \begin{cases} 1 & \text{if } k = 0 \\ 0 & \text{if } otherwise \end{cases}$. As

explained in [9], for $t \to \infty$ we have $P(X^s(t) = x^s_0) = 0$ and $P(X^s(t) = x^s_1) = \frac{k_2}{k_2 + k_1}$. As a consequence, our partitioned system correctly predicts that, for $t \to \infty$, λ_{out} has a bimodal distribution that is 0 with probability $\frac{k_2}{k_2 + k_1}$.

As shown in Example 2, the distribution of the continuous species can be derived using the law of total probability as $P(x^f, t) = \sum_{x^s \in S^s} P(x^f|x^s, t)P(x^s, t)$. Since each $P(x^f|x^s, t)$ is approximated with the Gaussian distribution given by the LNA, $P(x^f, t)$ is given by a mixture of Gaussian distributions weighted by the probability of being in a particular state of the discrete state space. This enables stochastic characterisation of multimodal distributions for continuous species. Note that the simple LNA, because of its unimodal nature, is unable to represent multimodal behaviours. The following remark shows that, if some assumptions are verified, we can further reduce the computational effort.

Remark 1. Equation (11) requires solving the LNA once for each $x^s \in S^s$. This can be expensive. However, for a large class of systems, especially those where continuous species have a unimodal distribution, we can consider a reasonable approximation. We can assume $\beta_\tau(x^s, t) \approx \sum_{x^f \in S^f} \alpha_\tau(x^f, E[X^s(t)])$ and $P(x^f, t) = \sum_{x^s \in S^s} P(x^f|x^s, t)P(x^s, t) \approx P(x^f|E[X^s(t)], t)$. So, instead of solving the LNA many times, this requires solving the LNA only once and conditioned on the expectation of the discrete population.

Ensuring Satisfaction of the Leap Conditions. We now explain how to choose constants σ_1 and σ_2 introduced in Eqs. (8) and (9). Given a CRN $C = (\Lambda, R)$ and an infinitesimal time interval dt, then $\tau \in R$ satisfies the first leap condition at time t if $\alpha_\tau(X(t))$ is approximately constant during the next dt. This is verified if the relative state change of each reactant species of τ is small enough during dt, that is, if

$$|X_{\lambda_i}(t + dt) - X_{\lambda_i}(t)| \leq max(\epsilon X_{\lambda_i}(t), 1) \text{ for } \lambda_i \in \Lambda \text{ such that } r^{\lambda_i}_\tau \neq 0$$

where $0 \geq \epsilon \geq 1$ is a parameter which quantifies the maximum relative change admitted in reactant species, extensively discussed in [14] for SSA. Rearranging the terms, it is easy to verify that the condition holds if

$$X_{\lambda_i}(t) \geq \frac{|X_{\lambda_i}(t + dt) - X_{\lambda_i}(t)|}{\epsilon} \text{ for } \lambda_i \text{ such that } r^{\lambda_i}_\tau \neq 0 \text{ and } \upsilon^{\lambda_i}_\tau \neq 0.$$

Thus, for a given CRN, σ_1 in Eq. (8) quantifies the minimum number of molecules for which we can assume the inequality is satisfied. This is reasonable, as dt is

---

Algorithm 1. Compute Transient Probabilities at Time t_{fin}

Require: A CRN $C = (\Lambda, R)$ with initial condition $x_0 = (x_0^f, x_0^s)$, a finite time interval $[t_0, t_{fin}]$, and parameters for leap conditions σ_1, δ_2.
1: **function** COMPUTEPROB(C, $x_0, \sigma_1, \delta_2, [t_0, t_{fin}]$)
2:     Compute partitions $\Lambda = (\Lambda^f, \Lambda^s)$, $R = (R^f, R^s)$ at time t_0
3:     $(S^s(t_0), X^f(t_0), t) \leftarrow ((x_0^s, 1), x_0^f, t_0)$
4:     **while** $t < t_{fin}$ **do**
5:         Compute Δt and solve discrete master equation for $[t, t + \Delta t]$
6:         **for each** $(x^s, p) \in S^s(t + \Delta t)$ **do**
7:             Solve the LNA to compute $P(X^f(t + \Delta t)|X^s(t) = x^s)$
8:         $t \leftarrow t + \Delta t$
9:         Compute new partitions $\Lambda = (\bar{\Lambda}^f, \bar{\Lambda}^s)$, $R = (\bar{R}^f, \bar{R}^s)$ at time t
10:         **for each** $\lambda_i \in \Lambda$ **do**
11:             **if** $\lambda_i \in \bar{\Lambda}^f \wedge \lambda_i \in \Lambda^s$ **then**
12:                 Move λ_i from $S^s(t)$ to $X^f(t)$
13:             **if** $\lambda_i \in \bar{\Lambda}^s \wedge \lambda_i \in \Lambda^f$ **then**
14:                 Move λ_i from $X^f(t)$ to $S^s(t)$
15:         $(\Lambda^f, \Lambda^s, R^f, R^s) \leftarrow (\bar{\Lambda}^f, \bar{\Lambda}^s, \bar{R}^f, \bar{R}^s)$
16:         $P(X^f(t)) \leftarrow \sum_{(x^s, p) \in S^s(t)} P(X^f(t)|X^s(t) = x^s) \cdot p$
17:         Compute $P(X^s(t))$ by exploration of $S^s(t)$
18:     **return** $(P(X^f(t)), P(X^s(t)))$

---

considered to be small, and we assume there are no reactions with unbounded propensity rate. $\tau \in R$ satisfies Eq. (9) if it fires many times during dt, that is, if $\alpha_\tau(X(t)) > \frac{\delta_2}{dt} = \sigma_2$, where δ_2 quantifies the number of times that τ must fire during dt in order to assume the condition satisfied. As a consequence, in order to choose σ_1 and σ_2, we need to tune three parameters: σ_1, δ_2 and dt. Empirical values for σ_1 and δ_2 are given in [26]; dt can be computed as for tau-leaping (see Sect. 3 of [14]). A possible strategy is to compute dt only once, at time t_0. Then, we can consider dt constant for any $t > t_0$ and make use of Eqs. (8) and (9). Fixing dt does not affect the correctness of the algorithm, but simply means that, for $t > 0$, there could be a better choice of dt' for which more reactions would be considered continuous.

4   Numerical Implementation

In this section, we present an algorithm to calculate the marginal probability of discrete and continuous species. We first present the general method, where continuous species are modelled as a mixture of Gaussian distributions, and then show how it can be simplified if Remark 1 applies. Algorithm 1 presents the pseudo-code for our routine. In Line 2, we partition species and reactions according to the leap conditions (Eqs. (8) and (9)). In Line 3, we initialize discrete and continuous stochastic processes as follows. The discrete process $X^s(t)$ at time t is represented by its state space, $S^s(t)$, given by a set of pairs (x^s, p), where $x^s \in \mathbb{N}^{|\Lambda^s|}$ and p is such that $P(X^s(t) = x^s) = p$. The continuous process $X^f(t)$

at time 0 is equal to x_0^f with probability 1. From Line 4 to 19, the algorithm iteratively updates the partitions. Δt is determined as the integration step of the numerical method used for characterizing discrete species; we use an explicit 4-th order Runge-Kutta algorithm with fixed time step, as in [18]. Alternatively, methods such as uniformisation [20,22] or aggregation-based techniques [1] could also be used. In Line 5, Eq. (11) is solved numerically for the next Δt. In Lines 6 − 7, for any $(x^s, p) \in S^s(t + \Delta t)$, the algorithm solves the LNA to compute Eq. (10). In Line 9, the partitions are computed at time t according to the leap conditions (Eqs. (8) and (9)) at that time. In general, the probability mass at time t is distributed over a set of states. In some cases the leap conditions can be checked deterministically based on the expected values $E[X^f(t)]$ and $E[X^s(t)]$. In a more general scenario, it may be necessary to compute the probability that the leap conditions are verified for any $\tau \in R$ and then partition according to these probabilities, which can be approximated as, at time t, we know the approximate solution of the CME [3]. In Lines 11 − 15, the species are reclassified and the partitions, $S^s(t)$ and $X^f(t)$, are modified accordingly. If λ_i was previously a discrete species and is now assigned to the continuous set, then all states in $S^s(t)$ that are equal except for the number of molecules of λ_i can now be merged. Then, for any state x^s of the updated discrete state space, we compute $P(X_{\lambda_i}^f(t)|X^s(t) = x^s)$, which is Gaussian. In Line 14, for any $(x^s, p) \in S^s(t)$ we discretize the Gaussian distribution $P(X_{\lambda_i}^f(t)|X^s(t) = x^s)$, where $X_{\lambda_i}^f$ is the component of $X^f(t)$ relative to λ_i. Finally, for $t \geq t_{fin}$, in Lines 16 − 17, the probability distributions of interest are computed.

A Faster Algorithm. If, for a particular CRN, Remark 1 applies then we can assume that $P(X^f(t)) \approx P(X^f(t)|X^s(t) = E[X^s(t)])$. Then we need to compute the LNA only once, and conditioned on the expectation of the discrete stochastic process. The remaining computation can be simplified as well because the virtual continuous process is modelled with a Gaussian distribution and not with a mixture of Gaussians.

Complexity and Error Analysis. The solution of Eq. (11) at time t, using our particular implementation, has a time cost linear in $|S^s(t)|$. We work with the numerical method of [18], which, for each $(x^s, p) \in S^s(t)$, propagates the probability retaining only the x^s such that $P(X^s(t) = x^s) = p > \zeta$. We fix $\zeta = 10^{-14}$. Solving the LNA requires solving a number of differential equations quadratic in the number of continuous species, and independent of the molecular population of such species. In the general case, at time t, we need to solve the LNA during the next Δt a number of times that is of the same order as the dimension of the discrete state space ($O(|S^s(t)||\Lambda^f|)$ differential equations). If Remark 1 is applicable, then the LNA needs to be solved only once.

If all species are partitioned as discrete/continuous, then the solution of Algorithm 1 reduces to that of the CME/LNA. The accuracy depends on the choice of σ_1, σ_2, where it can be shown [14] that, as $\sigma_1, \sigma_2 \to \infty$, then our algorithm guarantees an error equal to the error guaranteed by the numerical method used to solve the discrete master equation. If, instead, both σ_1, σ_2 equal 0, then the error of our hybrid algorithm reduces to the error in computing the LNA, which is model dependent and does not depend only on the molecular counts [10], but also on the validity of assumption (5), which needs to be verified a posteriori [16]. Error bounds would be a viable companion to estimate the committed error, but we are not aware of any explicit formulation of them for the convergence of the LNA. As a result, simulations may be used to validate the results.

5   Model Checking of Stochastic Evolution Logic (SEL)

Employing the hybrid semantics developed here, we present a fast probabilistic model checking algorithms for Stochastic Evolution Logic (SEL) [8]. SEL is a probabilistic logic for analysis of linear combinations of the species of a CRN.

Let $C = (\Lambda, R)$ be a CRN with initial state x_0, then SEL enables evaluation of the probability, variance and expectation of linear combinations of populations of the species of C. The syntax of SEL is given by

$$\eta := P_{\sim p}[B, I]_{[t_1, t_2]} \quad | \quad Q_{\sim v}[B]_{[t_1, t_2]} \quad | \quad \eta_1 \wedge \eta_2 \quad | \quad \eta_1 \vee \eta_2$$

where $Q = \{supV, infV, supE, infE\}$, $\sim = \{<, >\}$, $p \in [0, 1]$, $v \in \mathbb{R}$, $B \in \mathbb{Z}^{|\Lambda|}$, I is a finite set of disjoint intervals and $[t_1, t_2] \subseteq \mathbb{R}_{\geq 0}$. If $t_1 = t_2$ the interval reduces to a singleton.

Formulae η describe global properties of the stochastic evolution of the system. (B, I) specifies a linear combination of the species, where $B \in \mathbb{Z}^{|\Lambda|}$ is a vector defining the linear combination and I represents a set of disjoint closed real intervals. $P_{\sim p}[B, I]_{[t_1, t_2]}$ is the probabilistic operator, which specifies the average value of the probability that the linear combination defined by B falls within the range I over the time interval $[t_1, t_2]$ (we stress that this is not equivlent to reachability). The operators $supE, infE, infV, supV$, see [8], respectively, yield the supremum and infimum of expected value and variance of the random variables associated to B within the specified time interval. The quantitative value associated to a formula can be computed by writing $=?$ instead of $\sim p$ or $\sim v$. For instance, $P_{=?}[B, I]_{[t_1, t_2]}$ gives the probability value associated to the probabilistic property.

Model Checking Algorithm. Given $Z(t) = B \cdot X(t)$, where B is a linear combination of the species of C, then, according to the semantics of SEL [8], in order to perform model checking, we need to compute $P(Z(t) = z|X(0) = x_0), E[Z(t)|X(0) = x_0]$ and $E[Z(t) \cdot Z(t)|X(0) = x_0]$ (transient probability, expected value and variance of Z), where $z \in \mathbb{Z}$, and $x_0 \in \mathbb{N}^{|\Lambda|}$. In general, this requires solving the CME, which leads to state space explosion ot the LNA, which is fast but not always accurate. However, we can use our hybrid approximation in order to derive a fast and approximate model checking algorithm of SEL. We approximate Z with Z^h, which is the linear combination of the hybrid approximation of $X = (X^f, X^s)$. The following theorems, whose proofs are in the Appendix, show that model checking SEL just requires computing the hybrid approximation of the CME. In fact, uni-dimensional Gaussian integrals can be computed numerically in constant time. We denote Λ_t^s as the set of discrete species at time t.

Theorem 2. *Assume Λ_t^s is non-empty and S^s is the state space of $X^s(t)$. Then, the stochastic process $Z^h : \Omega \times \mathcal{R}_{\geq 0} \to \mathcal{S}$, with Ω its sample space and $(\mathcal{S}, \mathcal{B})$ a measurable space, is such that for $A \in \mathcal{B}$ and $t \in \mathbb{R}_{\geq 0}$*

$$P(Z^h(t) \in A | X(0) = x_0) = \sum_{x^s \in S^s} P(Z_{x^s}(t) \in A)P(X^s(t) = x_s)$$

where $Z_{x^s}(t)$ is a Gaussian random variable with expected value and variance

$$E[Z_{x^s}(t)] = B \cdot \begin{pmatrix} E[\bar{X}^f(t)] \\ x^s \end{pmatrix} \quad C[Z_{x^s}(t)] = B \cdot \begin{pmatrix} C[\bar{X}^f(t)] & 0 \\ 0 & 0 \end{pmatrix} \cdot B^T$$

where $\bar{X}^f$ is the virtual fast process introduced in Sect. 3.

Note that if the linear combination, at time t, involves only slow species, then $Z_{x_0}(t)$ is distributed according to a delta-Dirac function. This theorem guarantees that the transient probabilities of Z^h can be computed by solving a set of Guassian integrals, one for each reachable discrete state. The following theorem illustrates that expected value and variance of Z^h can be computed by considering Gaussian properties, even if Z^h is not Gaussian in general.

Theorem 3. *Assume Λ_t^s is non-empty. Then, for $t \in \mathbb{R}_{\geq 0}$*

$$E[Z^h(t) | X(0) = x_0] = \sum_{x^s \in S^s} E[Z_{x^s}(t) \in A]P(X^s(t) = x_s)$$

$$C[Z^h(t) | X(0) = x_0] = \sum_{x^s \in S^s} C[Z_{x^s}(t) \in A]P(X^s(t) = x_s)$$

The basic tools used in the proofs are the law of total expectation and the fact that jointly Gaussian random variables are closed with respect to a linear combination, which is Gaussian [2]. Theorems 2 and 3 assume that, at time t, the set of discrete species is not empty. In fact, if this is the case, all species are treated with the LNA and model checking algorithms for this scenario are given in [8]. We stress that the presented model checking algorithms are accurate only for finite time. In fact, for unbounded time, events that can be neglected in a finite time horizon scenario may fire with probability one. In the next section, SEL is employed in a set of case studies.

6 Experimental Results

We present three case studies showing how our approach significantly improves stochastic analysis of biochemical systems. We implemented Algorithm 1 in Matlab. All the experiments were run on an Intel Dual Core i7 machine with 8 GB of RAM. The first example is a CRN where we need to adaptively partition the species. The second example shows that our hybrid approach can be accurate in cases where the LNA is not, still maintaining comparable time complexity. The third is a system for which advanced numerical techniques for solving the CME such as fast adaptive uniformisation (FAU) [22], as implemented in PRISM [21], fail (out of memory) and using simulations would be too time consuming for comparable accuracy. However, we show that our approach still permits an accurate stochastic characterization.

Gene Expression. We consider the CRN of Example 1. All species in this example follow a unimodal distribution. As a consequence, we employ Remark 1. To ensure a fair comparison, we use the same numerical method for solving the CME and for solving the discrete part of our hybrid model: an explicit 4th order Runge-Kutta algorithm [18]. Even though the stochastic semantics is an infinite CTMC, there are only 2 species in the system with relatively small variance, and thus numerical solution of the CME is feasible. In Fig. 2, in the Appendix, we compare $supE_{=?}[mRNA]_{[T,T]}$ and $supV_{=?}[mRNA]_{[T,T]]}$ for $T \in [0, 200]$, the transient evolution of the expected value and variance of the $mRNA$, as calculated by direct solution of the CME and by our hybrid algorithm. Our algorithm decides to use the LNA for around 70 % of the time points. Moreover, we need to adaptively recompute the partitions, as shown in Example 1. In the table below we compare the performance of the same properties for different methods. We consider the following metrics: $||\epsilon||_\infty$ and $||\epsilon||_1$, respectively, average point-wise error and maximum point-wise error of LNA or hybrid approach with respect to the CME solution. *ProbLost* is the probability lost by the numerical solution of the CME due to the truncation of states with probability mass smaller than 10^{-14}.

Semantics	Time	$\|\epsilon\|_1$	$\|\epsilon\|_\infty$	ProbLost
CME	205 s	-	-	$< 10^{-7}$
Hybrid	35 s	$< 10^{-7}$	$< 10^{-7}$	-
LNA	5 s	$9 \cdot 10^{-5}$	0.0112	-

The LNA yields good accuracy. However, our hybrid algorithm achieves accuracy comparable to that for CME and is faster by one order of magnitude.

Bimodal Switch. We consider the CRN presented in Example 2 for $k_1 = 0.7$ and $k_2 = 0.3$. We are interested in analysing the probability distribution of λ_{out} over time, more specifically the SEL property $P_{=?}[\lambda_{out} = K]_{[100,100]}$, for $K \in [0, 200]$. Because of the bimodal nature of such a distribution, Remark 1 is not applicable and the LNA alone is not able to correctly estimate such a distribution. However, our hybrid model, as described in Eq. (2), correctly characterizes the distribution of λ_{out}. Figure 1 compares the distribution of λ_{out} at

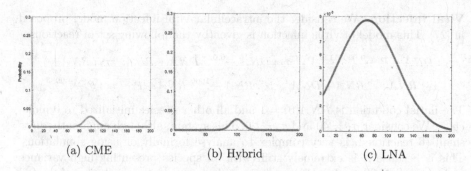

(a) CME   (b) Hybrid   (c) LNA

Fig. 1. Comparison of the probability distribution of λ_{out} at time $t = 100$, as estimated by a numerical solution of the CME (Fig. 1a), by our hybrid semantics for $\sigma_1 = 2$, $\sigma_2 = 0.5$ (Fig. 1b) and by the LNA (Fig. 1c). Note that in Fig. 1a and b there is non-zero probability of having exactly zero molecules.

time $t = 100$ as estimated by our hybrid approach against the LNA and a full solution of the CME. The following table compares our hybrid approach with the other semantics for different values of σ_1 and σ_2. We consider the average point-wise error, $||\epsilon||_1$, and the maximum point-wise error, $||\epsilon||_\infty$, with respect the a numerical solution of the CME, whose error is due to state space truncation (*ProbLost*). For a fair comparison, both for the solution of the master equations of discrete species and for the CME, we use the same numerical method, an explicit 4th order Runge Kutta algorithm with fixed time step [18].

Semantics	σ_1	σ_2	Time	$||\epsilon||_1$	$||\epsilon||_\infty$	ProbLost
CME	-	-	100 s	-	-	$< 10^{-6}$
LNA	-	-	2.3 s	0.081	0.2971	-
Hybrid	2	0.5	2.5 s	$3.284 \cdot 10^{-4}$	0.0024	-
Hybrid	0.5	0.5	2.2 s	0.081	0.2971	-
Hybrid	2	2	96 s	$< 10^{-6}$	$< 10^{-6}$	-

For $\sigma_1 > 1$ and $\sigma_2 < 0.7$, the hybrid approach improves the accuracy of the LNA by around two orders of magnitude, while still maintaining comparable execution time. Note that, for this choice of σ_1 and σ_2, the virtual continuous subsystem ignores the delay induced by the firing of the first reaction, which explains why the accuracy of the hybrid method is worse than CME. For $\sigma_2 > 0.7$, all species are considered as discrete and the hybrid approach reduces to the solution of the CME. For $\sigma_1 = \sigma_2 = 0.5$, all species are continuous and the accuracy of the hybrid approach is identical to that of the LNA.

Viral Infection. We consider the intracellular viral infection model proposed in [27]. This model of virus infection is given by the following set of reactions:

$$\tau_1 : DNA + P \rightarrow^{0.00001125} V; \quad \tau_2 : DNA \rightarrow^{0.025} DNA + RNA; \quad \tau_3 : RNA \rightarrow^{0.25}$$

$$\tau_4 : RNA \rightarrow^1 RNA + DNA; \quad \tau_5 : RNA \rightarrow^{1000} RNA + P; \quad \tau_6 : P \rightarrow^{1.9985}$$

The initial condition is $RNA(0) = 1$ and all other species initialized to 0 molecules. We consider $\sigma_1 = 40$ and $\sigma_2 = 20$. This system, although apparently quite small (6 reactions), is very complex to analyse formally or using simulations. This is because it is extremely stiff, with all species presenting high variance and some also high molecular populations. As a consequence, solution of the full CME, even using advanced techniques such as FAU or finite state projection (FSP) [24], is prohibitive due to state-space explosion. For all the properties we consider, FAU is out of memory on our hardware. Because of the stiffness of the system, simulations are time consuming and ensuring good accuracy is not feasible. Our hybrid approach, by considering P as a continuous species for any time instant, enables an effective and efficient stochastic characterization of such a system. Note that, for this system, the LNA is clearly not accurate because of its multimodality.

In Fig. 3, in the Appendix, we compare the distribution of the RNA at time $t = 200$ as estimated by our hybrid approach and the distribution of the same species with only the LNA. Results show that the LNA is not able to accurately characterize the distribution of interest, while our hybrid approach correctly predicts multimodality and confirms values obtained by Goutsias in [15] (Fig. 5) by using 4000 simulations.

Note that, although the original model is stiff, after species separation the resulting model is much less stiff. This remains true for a large class of systems, and it is a consequence of how we separate the species of a CRN. As a result, for such systems, we need to solve a discrete master equation only for less stiff systems in a reduced state space. As we see in the following table, this results in a marked improvement.

Property (SEL)	Time (Hy)	Time (LNA)	Time (FAU)	RelErr (Hy-LNA)
$P_{=?}[RNA = 0]_{[200,200]}$	4300 s	28	OutOfMem	0.215
$P_{=?}[RNA = 0]_{[50,50]}$	1500 s	20	OutOfMem	0.215

$Time(\cdot)$ represents the execution time of different algorithms. $RelErr(Hy\text{-}LNA)$ is the distance between the quantitative value of the property as computed by our hybrid algorithm (and validated by simulations) and by the LNA.

7   Conclusion

We presented a stochastic hybrid approximation of the CME based on automatically partitioning the species and reactions of a CRN according to the leap conditions, and treating the discrete species as a discrete stochastic process, while treating the continuous species as a mixture of Gaussian distributions. The use of the leap conditions justifies the hybrid approximation compared to simple threshold conditions on molecular populations. Our method can be integrated with any numerical method to solve the CME, such as FAU [22], FSP [24] or aggregation based techniques [1]. We demonstrated through case studies that our method is efficient, scales well and can handle multimodality. The algorithm works particularly well for systems where species evolve on different time scales (i.e. stiff systems), which are common in biology. It also works well when there are no reactions that satisfy the second leap condition, but not the first one. In this case, our hybrid model can introduce some inaccuracies due to the assumptions in partitioning of the species. As future work, we plan to handle this problem by dealing directly with the non-Markovian aspect of the process related to continuous species, without introducing any virtual process. Finally, we plan to implement an algorithm to automatically tune the parameters for species partitioning using stochastic simulations.

A   Proofs

Proposition 1. *Let $x^s \in S^s$ and $x^f \in S^f$. Then, for $t \in \mathbb{R}_{\geq 0}$*

$$\frac{dP(x^s|t)}{dt} = \sum_{\tau \in R} \beta_\tau(x^s - \upsilon_\tau, t)P(x^s - \upsilon_\tau, t) - \beta_\tau(x^s, t)P(x^s, t)$$

where $\beta_\tau(x^s, t) = \sum_{x^f \in S^f} \alpha_\tau(x^f, x^s)P(x^f|x^s, t)$.

Proof. By using the law of total probability we have

$$\frac{dP(x^s|t)}{dt} = \sum_{x^f \in S^f} \frac{dP(x^s, x^f, t)}{dt}$$

Then, using Eq. (2), and rearranging terms we have

$$\sum_{x^f \in S^f} \frac{dP(x^s, x^f, t)}{dt} =$$

$$\sum_{x^f \in S^f} \sum_{\tau \in R^f} \alpha_\tau(x^f - \upsilon_\tau, x^s - \upsilon_\tau)P(x^f - \upsilon_\tau, x^s - \upsilon_\tau, t) - \alpha_\tau(x^f, x^s)P(x^f, x^s, t) =$$

$$\sum_{\tau \in R} \beta_\tau(x^s - \upsilon_\tau, t)P(x^s - \upsilon_\tau, t) - \beta_\tau(x^s, t)P(x^s, t)$$

where $\beta_\tau(x^s, t) = \sum_{x^f \in S^f} \alpha_\tau(x^f, x^s)P(x^f|x^s, t)$, that is, the conditional expectation of the propensity rate of τ at time t given $X^s(t) = x^s$.

Theorem 2. *Assume Λ_t^s is non-empty and S^s is the state space of $X^s(t)$. Then, the stochastic process $Z^h : \Omega \times \mathcal{R}_{\geq 0} \to S$, with Ω its sample space and $(S, \mathcal{B})$ a measurable space, is such that for $A \in \mathcal{B}$ and $t \in \mathbb{R}_{\geq 0}$*

$$P(Z^h(t) \in A | X(0) = x_0) = \sum_{x^s \in S^s} P(Z_{x^s}(t) \in A) P(X^s(t) = x_s)$$

where $Z_{x^s}(t)$ is a Gaussian random variable with expected value and variance

$$E[Z_{x^s}(t)] = B \cdot \begin{pmatrix} E[\bar{X}^f(t)] \\ x^s \end{pmatrix} \quad C[Z_{x^s}(t)] = B \cdot \begin{pmatrix} C[\bar{X}^f(t)] & 0 \\ 0 & 0 \end{pmatrix} \cdot B^T$$

where $\bar{X}^f$ is the virtual fast process introduced in Sect. 3.

Proof. By the law of total probability we have

$$P(Z(t) \in A | X(0) = x_0) = \sum_{x^s \in S^s} P(Z(t) \in A | X^s(t) = x^s, X(0) = x_0) P(X^s(t) = x^s | X(0) = x_0).$$

By application of the LNA it follows that $X^f(t)$ conditioned on the event $X^s(t) = x^s$ is a Gaussian random variable with expected value and variance

$$E[X^f(t) | X^s(t) = x^s] = \begin{pmatrix} E[\bar{X}^f(t)] \\ x^s \end{pmatrix}$$

and covariance matrix

$$C[X^f(t) | X^s(t) = x^s] = \begin{pmatrix} C[\bar{X}^f(t)] & 0 \\ 0 & 0 \end{pmatrix}$$

Given a multidimensional Gaussian distribution, each linear combination of its components is still Gaussian. As a consequence, $E[Z^h(t) | X^s(t) = x^s] = B \cdot E[X^f(t) | X^s(t) = x^s]$ and $C[Z^h(t) | X^s(t) = x^s] = B \cdot C[X^f(t) | X^s(t) = x^s] \cdot B^T$.

Theorem 3. *Assume Λ_t^s is non-empty. Then, for $t \in \mathbb{R}_{\geq 0}$*

$$E[Z^h(t) | X(0) = x_0] = \sum_{x^s \in S^s} E[Z_{x^s}(t) \in A] P(X^s(t) = x_s)$$

$$C[Z^h(t) | X(0) = x_0] = \sum_{x^s \in S^s} C[Z_{x^s}(t) \in A] P(X^s(t) = x_s)$$

Proof. The proof follows from the application of the law of total expectation for random variables with mutually exclusive and exhaustive events.

B   Figures

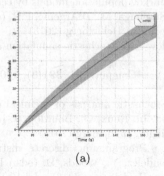

(a)

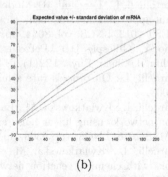

(b)

Fig. 2. Comparison of expected value and variance of $mRNA$ in Example 2 in interval $[0, 200]$ as calculated by direct solution of the CME (Fig. 2a) and by our algorithm (Fig. 2b).

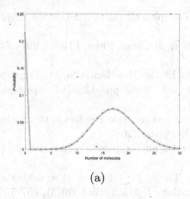

(a)

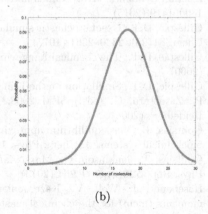

(b)

Fig. 3. Comparison of the probability distribution of RNA at time $t = 200$ as calculated by numerical hybrid algorithm (Fig. 3a) and by the LNA (Fig. 3b).

References

1. Abate, A., Brim, L., Češka, M., Kwiatkowska, M.: Adaptive aggregation of markov chains: quantitative analysis of chemical reaction networks. In: Kroening, D., Păsăreanu, C.S. (eds.) CAV 2015. LNCS, vol. 9206, pp. 195–213. Springer, Heidelberg (2015)
2. Adler, R.J.: An introduction to continuity, extrema, related topics for general Gaussian processes. Lect. Notes-Monogr. Ser. **12**, i-155 (1990)
3. Anderson, D.F., Kurtz, T.G.: Stochastic Analysis of Biochemical Systems. Springer, Heidelberg (2015)

 4. Arkin, A., Ross, J., McAdams, H.H.: Stochastic kinetic analysis of developmental pathway bifurcation in phage λ-infected escherichia coli cells. Genetics **149**(4), 1633–1648 (1998)
 5. Bortolussi, L., Lanciani, R.: Model checking Markov population models by central limit approximation. In: Joshi, K., Siegle, M., Stoelinga, M., D'Argenio, P.R. (eds.) QEST 2013. LNCS, vol. 8054, pp. 123–138. Springer, Heidelberg (2013)
 6. Cao, Y., Gillespie, D.T., Petzold, L.R.: The slow-scale stochastic simulation algorithm. J. Chem. Phys. **122**(1), 014116 (2005)
 7. Cardelli, L.: On process rate semantics. Theoret. Comput. Sci. **391**(3), 190–215 (2008)
 8. Cardelli, L., Kwiatkowska, M., Laurenti, L.: Stochastic analysis of chemical reaction networks using linear noise approximation. In: Roux, O., Bourdon, J. (eds.) CMSB 2015. LNCS, vol. 9308, pp. 64–76. Springer, Heidelberg (2015)
 9. Cardelli, L., Kwiatkowska, M., Laurenti, L.: Programming discrete distributions with chemical reaction networks. In: Rondelez, Y., Woods, D. (eds.) DNA 2016. LNCS, vol. 9818, pp. 35–51. Springer, Heidelberg (2016). doi:10.1007/978-3-319-43994-5_3
10. Ethier, S.N., Kurtz, T.G.: Markov Processes: Characterization and Convergence, vol. 282. Wiley, Hoboken (2009)
11. Ganguly, A., Altintan, D., Koeppl, H.: Jump-diffusion approximation of stochastic reaction dynamics: error bounds and algorithms. Multiscale Model. Simul. **13**(4), 1390–1419 (2015)
12. Gillespie, D.T.: Exact stochastic simulation of coupled chemical reactions. J. Phys. Chem. **81**(25), 2340–2361 (1977)
13. Gillespie, D.T.: The chemical Langevin equation. J. Chem. Phys. **113**(1), 297–306 (2000)
14. Gillespie, D.T.: Simulation methods in systems biology. In: Bernardo, M., Degano, P., Zavattaro, G. (eds.) SFM 2008. LNCS, vol. 5016, pp. 125–167. Springer, Heidelberg (2008)
15. Goutsias, J.: Quasiequilibrium approximation of fast reaction kinetics in stochastic biochemical systems. J. Chem. Phys. **122**(18), 184102 (2005)
16. Goutsias, J., Jenkinson, G.: Markovian dynamics on complex reaction networks. Phys. Rep. **529**(2), 199–264 (2013)
17. Hasenauer, J., Wolf, V., Kazeroonian, A., Theis, F.: Method of conditional moments (mcm) for the chemical master equation. J. Math. Biol. **69**(3), 687–735 (2014)
18. Henzinger, T.A., Mikeev, L., Mateescu, M., Wolf, V.: Hybrid numerical solution of the chemical master equation. In: Proceedings of the 8th International Conference on Computational Methods in Systems Biology, pp. 55–65. ACM (2010)
19. Hepp, B., Gupta, A., Khammash, M.: Adaptive hybrid simulations for multiscale stochastic reaction networks. J. Chem. Phys. **142**(3), 034118 (2015)
20. Kwiatkowska, M., Norman, G., Parker, D.: Stochastic model checking, pp. 220–270 (2007)
21. Kwiatkowska, M., Norman, G., Parker, D.: PRISM 4.0: verification of probabilistic real-time systems. In: Qadeer, S., Gopalakrishnan, G. (eds.) CAV 2011. LNCS, vol. 6806, pp. 585–591. Springer, Heidelberg (2011)
22. Mateescu, M., Wolf, V., Didier, F., Henzinger, T., et al.: Fast adaptive uniformisation of the chemical master equation. IET **4**, 441–452 (2010)
23. McAdams, H.H., Arkin, A.: Stochastic mechanisms in gene expression. Proc. Nat. Acad. Sci. **94**(3), 814–819 (1997)

24. Munsky, B., Khammash, M.: The finite state projection algorithm for the solution of the chemical master equation. J. Chem. Phys. **124**(4), 044104 (2006)
25. Rao, C.V., Arkin, A.P.: Stochastic chemical kinetics and the quasi-steady-state assumption: application to the gillespie algorithm. J. Chem. Phys. **118**(11), 4999–5010 (2003)
26. Salis, H., Kaznessis, Y.: Accurate hybrid stochastic simulation of a system of coupled chemical or biochemical reactions. J. Chem. Phys. **122**(5), 054103 (2005)
27. Srivastava, R., You, L., Summers, J., Yin, J.: Stochastic vs. deterministic modeling of intracellular viral kinetics. J. Theoret. Biol. **218**(3), 309–321 (2002)
28. Thattai, M., Van Oudenaarden, A.: Intrinsic noise in gene regulatory networks. Proc. Nat. Acad. Sci. **98**(15), 8614–8619 (2001)
29. Thomas, P., Popović, N., Grima, R.: Phenotypic switching in gene regulatory networks. Proc. Nat. Acad. Sci. **111**(19), 6994–6999 (2014)
30. Van Kampen, N.G.: Stochastic Processes in Physics and Chemistry, vol. 1. Elsevier, Amsterdam (1992)
31. Wallace, E., Gillespie, D., Sanft, K., Petzold, L.: Linear noise approximation is valid over limited times for any chemical system that is sufficiently large. IET Syst. Biol. **6**(4), 102–115 (2012)

Autonomous and Adaptive Control of Populations of Bacteria Through Environment Regulation

Chieh Lo[(✉)] and Radu Marculescu

Carnegie Mellon University, Pittsburgh, PA, USA
chiehl@andrew.cmu.edu, radum@cmu.edu

Abstract. The proliferation of antibiotic-resistant bacteria poses a significant threat to humans health and welfare. To reduce the bacterial pathogenesis and growth, we propose an autonomous biological controller that can adaptively generate quorum sensing inhibitors and control the iron availability in the environment. As the main theoretical contribution, we provide a detailed analysis of our proposed controller that includes model calibration, system response, and inhibitor effectiveness. We also formulate a constrained optimization problem to choose the values of the biological parameters of the proposed controller under given environment constraints. Finally, we validate our results via detailed population-level simulations and demonstrate that bacteria virulence can be significantly reduced without developing drug resistance or inducing selective pressure among bacteria wild type and mutants. This work represents a first step towards a paradigm change in reducing bacterial pathogenesis via controlling the dynamics of the cell-cell communication through environment regulation.

Keywords: Quorum sensing · Biological controller · Pathogen · Environment regulation · Cell-cell communication

1 Introduction

The fight against bacterial virulence represents one of the big challenges of modern medicine. Indeed, due to the large-scale proliferation and inappropriate use of antibiotics, new strains antibiotic-resistant bacteria begin to emerge. These new, stronger bacteria pose a significant threat to humans health and welfare. To fight antibiotic-resistant bacteria, we propose to engineer synthetic cells, insert them in a population of bacteria, and then control the dynamics and virulence of the entire population [9]. We note that while previous work [6,26] proposed to engineer cells to kill the antibiotic-resistant bacteria, this kind of approaches may actually select strains that can survive under such therapies. In contrast, in this paper, we design an autonomous controller that can not only regulate the cell-cell communication, but also manipulate the environment signals in order

© Springer International Publishing AG 2016
E. Bartocci et al. (Eds.): CMSB 2016, LNBI 9859, pp. 168–185, 2016.
DOI: 10.1007/978-3-319-45177-0_11

to reduce bacterial virulence and prevent selective pressure among antibiotic-resistant strains.

Getting now into details, bacteria can form biofilms, express virulence, and become resistant to antibiotics after reaching a quorum through cell-cell communication. Quorum sensing (QS) is a fundamental cell-cell communication that is used by bacteria to obtain cell density information and hence, alter their genes expression [4]. In particular, the QS system used by Gram-negative bacteria is mediated by diffusible signaling molecules, termed "autoinducers"[1] [4]. For instance, the opportunistic human pathogen *Pseudomonas aeruginosa* (PA) possess a complex QS system that regulates genes and operons which constitute over 6 % of the genome. Those genes coordinate the biofilm formation and produce large amounts of virulence factors, such as elastase, rhamnolipids, and pyocyanin [10].

QS regulation can be strongly affected by various environmental factors [11]; for example, in PA, the nutrient availability have been shown to affect the expression of QS genes [24]. Several other studies have demonstrated that high iron concentrations favor the formation of biofilms and higher growth rates, but restrict the expression of QS signals [12,22]. On the other hand, QS also regulates bacteria access to nutrients and environmental niches that favor their growth and defense.

The intertwined regulation between QS and environmental signals enable bacteria to thrive in a stringent environment [2,15]; indeed, under such conditions, bacteria must coordinate the expression of related genes in order to successfully form and maintain biofilms [16]. For example, a shortage of iron availability in the environment leads to the increased expression of iron acquisition system [3,5] and decreased activity of pathways that rely on relatively large amounts of iron [10]. However, a rigorous mathematical model that can precisely capture the complex relationship between the QS system and bacteria growth has not yet been explored. Additionally, most studies published so far focus on observing the qualitative behaviors of bacteria and lack the ability to predict long term evolution dynamics under different environmental conditions [18].

We argue that having a quantitative model of QS behavior available can not only capture the important dynamics of bacteria growth, but also give credible predictions for the long term behaviors of bacteria virulence; this analytical QS model is the first major contribution of this work. We also raise another important question: Given such an analytical (i.e., quantitative) model, what are the strategies to control bacteria virulence and growth rate, while lowering the chances of developing drug resistance or inducing selective pressure among bacteria wild type and mutants? To address this second question, we propose an autonomous biological controller that can dynamically generate different types of inhibitors; this controller is based on genetic parts used to design genetic circuits [1].

To shed light on the complex relationship between QS and environment signals, we use the opportunistic pathogen PA as a canonical example (Fig. 1(a)).

---

[1] Denoted as AI in this paper.

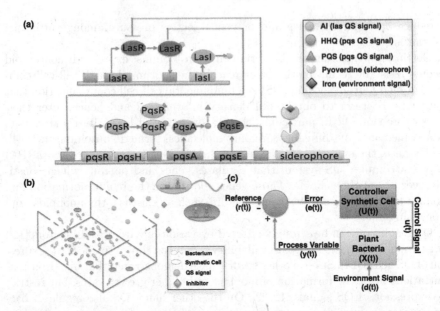

Fig. 1. (a) Interconnection of the *las* and *pqs* QS system of PA. The small orange circles represent autoinducer molecules (AIs) that move freely through the cell membrane; the purple circles and triangles represent the QS signals of *pqs* QS system; the green particles represent the pyoverdine molecules. The red, orange, blue, and purple ovals represent LasR, LasI, PqsR, PqsA, and PqsE, respectively. (b) The simulation environment, where bacteria (green cells) flow in space and release the QS signals (orange circles). By placing the synthetic (red) cells in the environment, they first react to the QS signals and then express the inhibitors (purple diamonds) that can quench the communication among bacteria. (c) The diagram of the proposed control system. The environment signal (d) (e.g., nutrient availability) can be viewed as the input to the intracellular bacterial regulations system. The control variables (u) are the QSI inhibitors which can control the dynamics of bacteria. The process variable (y) can be detected by the synthetic controller. (Color figure online)

PA requires an abundance of iron to produce and sustain infections. Hence, iron depletion prevents bacterial growth and affects their metabolism [19]. By expressing siderophores, PA can sequester iron from environment and regain the ability to form biofilm [25]. Two major genetic components of QS, namely, the *las* and *pqs* QS systems have been identified in PA [5]. As shown in Fig. 1(a), the *las* QS system sits at the upstream of *pqs* QS system and positively regulates the operons of the *pqs* QS system [2]. The *pqs* QS system produces molecules that mediate the expression of the siderophores [5,7]. To enhance the expression of the siderophores, the upstream *las* QS system needs to highly express proteins in order to induce the downstream *pqs* QS system. Hence, as the iron concentration is relatively low, the *las* QS system is highly expressed and vice versa. However, bacteria can become more virulent when the *las* QS system strongly expresses proteins as this can regulate the virulence genes.

In summary, in an iron depletion environment, the growth of bacteria can be delayed but the virulence can actually increase [12]. To control both the virulence and the growth rate of bacteria simultaneously, we use two different kinds of inhibitors that target the *las* QS system and the iron availability in the environment. Different types of inhibitors can have different effects on virulence and growth rate, hence, multi-inhibitor schemes can be more effective. To synthesize inhibitors and control the iron concentration, a few simple genetic circuits can serve as the basic control units which automatically detect and react to environment changes. For example, we can construct the genetic circuits by cloning the genes in the plasmid, such as the *aiiA* gene which expresses the enzyme that hydrolyzes the AI [17] (Fig. 1(b)). However, synthesizing excessive amounts of inhibitors in the environment can have toxic effects on the host.

Therefore, in this paper, we propose a dynamic optimization problem that incorporates bacteria QS, growth, and control dynamics. Solving this optimization problem allows us to choose the biological parameters that can be further used to design controllers that can generate the optimal amount of inhibitors adaptively. By placing the biological controller into the bacterial environment, it becomes possible to detect the concentration of the signaling molecules in the environment and then generate the right amount of inhibitors in real-time. Consequently, the proposed system aims at a paradigm shift from manual to autonomous control of bacteria population dynamics (Fig. 1(c)). Taken together, our contribution is threefold:

- First, we develop a new (cellular-level) ordinary differential equations (ODEs) based model of *pqs* QS system and propose new synthetic circuitry to control bacteria virulence and growth. To the best of our knowledge, this is the first design that formally considers the autonomous control of the QS and environmental signals in populations of bacteria.
- Second, we formulate a constrained optimization problem based on the QS and control dynamics. We illustrate the design procedures for the biological controller by solving this optimization problem; this provides the theoretical basis for synthesizing the controller.
- Third, we verify our proposed controller via detailed simulations at population-level. As such, the design procedure we provide can serve as a general guideline towards *in vitro* construction of synthetic cellular controllers.

The remainder of this paper is organized as follows. Section 2 focuses on the mathematical modeling of the QS regulation system (i.e., *las* and *pqs*) of PA, bacteria growth, and QSI model. Section 3 analyzes the QS system response and bacteria growth model via simulation. Section 4 formulates the constrained optimization problem for designing the biological controller and provides a design example based on the proposed design guidelines. Section 5 utilizes the bacteria simulator proposed in [21] to validate the model under various scenarios that mimic realistic settings. Finally, conclusions are drawn in Sect. 6.

2   Cellular-Level Mathematical Modeling

In this section, we model the dynamics of bacteria QS, growth, and inhibition systems based on ODEs. To uncover the complex interaction between the QS and environment signals (i.e., iron), we first model the *las* and the *pqs* QS systems and the growth of PA explicitly. Next, we calibrate our models with the reported experimental data [12]; that is, at different iron concentration levels, we calibrate the relative concentration change of the LasR protein (main receptor in *las* QS system); this provides the basis for examining the QS system response and subsequently designing the biological controller.

2.1   QS Model of PA

The QS regulatory network of PA consists of two main systems: *las* and *pqs*. The *las* and the *pqs* QS systems are linked by (1) LasR-AI complex which directly *up* regulates the expression of the PqsR and the PqsH proteins and (2) iron-chelated complex which *down* regulates the expression of the LasR protein and then reduces the expression of the siderophore (a negative feedback loop). The entire QS system is modeled as follows:

***las* QS Model.** The regulatory network of the *las* QS system has two feedback loops. As shown in Fig. 1(a), the LasR-AI complex up regulates the expression of both *lasR* and the *lasI* genes. Based on the ODE models proposed in [13,23], we have the following equations for the *las* QS system:

$$\frac{d[A]}{dt} = c_A + \frac{V_A[C]}{K_A + [C]} - \alpha_{RA}[R][A] + \delta_{RA}[RA] - b_A[A] - \frac{d_A}{\rho}([A_{EX}] - [A]) \quad (1)$$

$$\frac{d[A_{EX}]}{dt} = -b_{A_{EX}}[A_{EX}] - \frac{d_A}{1-\rho}([A_{EX}] - [A]) \quad (2)$$

$$\frac{d[R]}{dt} = c_R + \frac{V_R[C]}{K_R + [C]} - \alpha_{RA}[R][A] + \delta_{RA}[RA] - b_R[R] \quad (3)$$

$$\frac{d[RA]}{dt} = \alpha_{RA}[R][A] - 2\alpha_{RA^2}[RA]^2 - \delta_{RA}[RA] + 2\delta_{RA^2}[C] \quad (4)$$

$$\frac{d[C]}{dt} = \alpha_{RA^2}[RA]^2 - \delta_{RA^2}[C] \quad (5)$$

where $[X]$ denotes the concentration of a particular molecular species X. In our formulation, A stands for AI, A_{EX} is the extracellular AI, R is LasR, RA is the LasR-AI complex and C is the dimerized complex. The meaning of biological constants are listed in Table 1 while their numerical values are listed in Tables 2 and 3 in the **Appendix**.

***pqs* QS Model.** The *pqs* QS system consists of two kinds of signaling molecules, PQS (2-heptyl-3,4-dihydroxyquinoline) and HHQ (4-hydroxy-2-heptylquinoline); in addition we have one receptor regulator PqsR. The PqsR protein can bind to the HHQ and the PQS molecules and up regulate the *pqsABCDE* operon; this forms a positive feedback since the PqsA protein directly up regulates the synthesis of the HHQ molecules. Another signaling molecule, PQS, is converted

Table 1. Table with model parameters

Symbol	Parameter	Source
K	Half saturation concentration	[8,13]
U	Utilization coefficient	introduce in this paper
V	Maximum production rate	[8,13]
b	Molecule decay rate	[8,13]
c	Basal production rate	[8,13]
d	Membrane diffusion rate	[8]
α	Binding rate	[8,13]
β	Enzyme production rate	[8,13]
δ	Unbinding rate	[8,13]
ρ	Cell density	[8]
P	Promoter strength	[20]
r	Basal production rate	[20]

from HHQ via PqsH protein. Therefore, pqs QS system forms a second positive feedback loop. By explicitly capturing regulations among proteins and molecules based on molecular transcription and translation, we propose the following new ODEs to describe the pqs QS system:

$$\frac{d[pR]}{dt} = c_{pR} + \frac{V_{pR}[C]}{K_{pR} + [C]} - \alpha_{pR}([pR][A_1] + [pR][A_2]) + \delta_{pR}([C_1] + [C_2]) - b_{pR}[pR] \quad (6)$$

$$\frac{d[pH]}{dt} = c_{pH} + \frac{V_{pH}[C]}{K_{pH} + [C]} - b_{pH}[pH] \quad (7)$$

$$\frac{d[pA]}{dt} = c_{pA} + \frac{V_{pA,1}[C_1]}{K_{pA,1} + [C_1]} \frac{V_{pA,2}[C_2]}{K_{pA,2} + [C_2]} - b_{pA}[pA] \quad (8)$$

$$\frac{d[pE]}{dt} = c_{pE} + \frac{V_{pE,1}[C_1]}{K_{pE,1} + [C_1]} \frac{V_{pE,1}[C_2]}{K_{pE,1} + [C_2]} - b_{pE}[pE] \quad (9)$$

$$\frac{d[C_1]}{dt} = \alpha_{pR}[pR][A_1] - \delta_{pR}[C_1] \quad (10)$$

$$\frac{d[C_2]}{dt} = \alpha_{pR}[R_2][A_2] - \delta_{pR}[C_2] \quad (11)$$

$$\frac{d[A_1]}{dt} = \beta_{pA}[pA]\frac{K_{A_1}}{K_{A_1} + [pE]} - \alpha_{pR}[pR][A_1] + \delta_{pR}[C_1] - b_{A_1}[A_1] + \frac{d_{A_1}}{\rho}([A_{1EX}] - [A_1]) \quad (12)$$

$$\frac{d[A_{1EX}]}{dt} = -b_{A_1}[A_{1EX}] - \frac{d_{A_1}}{1 - \rho}([A_{1EX}] - [A_1]) \quad (13)$$

$$\frac{d[A_2]}{dt} = \beta_{pH}[pH][A_1] - \alpha_{pR}[pR][A_2] + \delta_{pR}[C_2] - b_{A_2}[A_2] + \frac{d_{A_2}}{\rho}([A_{2EX}] - [A_2]) \quad (14)$$

$$\frac{d[A_{2EX}]}{dt} = -b_{A_2}[A_{2EX}] - \frac{d_{A_2}}{1 - \rho}([A_{2EX}] - [A_2]) \quad (15)$$

$$\frac{d[Pyo]}{dt} = c_{Pyo} + \frac{V_{Pyo}[pE]}{K_{Pyo} + [pE]} - b_{Pyo}[Pyo] + \frac{d_{Pyo}}{\rho}([Pyo_{EX}] - [Pyo]) \quad (16)$$

$$\frac{d[Pyo_{EX}]}{dt} = -\alpha_I[Pyo_{EX}][I] - b_{Pyo_{EX}}[Pyo_{EX}] + \frac{d_{Pyo}}{1 - \rho}([Pyo_{EX}] - [Pyo]) \quad (17)$$

$$\frac{d[Q]}{dt} = -b_Q[Q] + \frac{d_Q}{\rho}([Q_{EX}] - [Q]) \tag{18}$$

$$\frac{d[Q_{EX}]}{dt} = \alpha_I[Pyo_{EX}][I] + \frac{d_Q}{1-\rho}([Q_{EX}] - [Q]) \tag{19}$$

where $[X]$ denotes the concentration of a particular molecular species X. In our formulation, pA, pE, pH, and pR stand for PqsA, PqsE, PqsH and PqsR, respectively. A_1, A_2, C_1 and C_2 represent HHQ, PQS, PqsR-HHQ and PqsR-PQS, respectively. Pyo and I represent pyoverdine and iron, respectively. Q is the iron-chelated complex.

Given the pqs QS system, we modify the expression of the LasR protein in (3) as follows:

$$\frac{d[R]}{dt} = c_R + \frac{V_R[C]}{K_R + [C]}\frac{V_Q K_Q}{K_Q + [Q]} - \alpha_{RA}[R][A] + \delta_{RA}[RA] - b_R[R] \tag{20}$$

where we add a new term to account for the effect of iron-chelated complex Q. In this equation, the parameters V_Q and K_Q represent the maximum production rate and Michaelis-Menten constant, respectively.

2.2 Bacteria Growth Model and Virulence Measure

To describe bacteria growth, Monod introduced the concept of single nutrient controlled kinetics [14], which relates the specific growth rate (μ_X) of a bacterium cell mass (X) to the substrate concentration (S). The kinetic parameters, i.e., maximum specific growth rate (k_X) and substrate affinity (K_S), are assumed to be constant and dependent on strain, medium, and growth conditions (e.g. temperature, pH). In our model, however, we need to consider a second nutrient source and add a new term Q to describe it. However, when cells are metabolically active, but not growing or dividing, they may still take up substrate.

To address bacteria size reduction, a maintenance rate (m) is generally used; consequently, we improve Monod's model as follows:

$$\mu_X = k_X \cdot \frac{S+Q}{S+Q+K_g} \tag{21}$$

$$\frac{dX}{dt} = (\mu_X - m) \cdot X \tag{22}$$

We also define the virulence (V) as the concentration of LasR-AI complex as it controls the downstream virulence expressions; therefore, the total virulence (TV) of the bacteria population is defined as the product of the virulence and the number of bacteria (N)[2]:

$$TV = V \times N \tag{23}$$

We note that, as discussed later in Sect. 4, both V and N are variables that depend on time (t) and the set of biological parameters (**p**).

---

[2] Since the number of bacteria is proportional to the biomass, we use biomass and the number of bacteria to account for the total virulence interchangeably.

2.3   Inhibition Model

We target the bacterial iron acquisition as a strategy to control the virulence of bacteria. From our previous discussion, the QS signaling pathways are the primary target. More precisely, to control the iron uptake rate of PA, we propose two strategies that can either modulate the iron uptake or inhibit the upstream *las* QS system:

AI Inhibitors. The AI inhibitor hydrolyzes the extracellular AI molecules which can be viewed as a degradation source and assumed to follow the Michaelis-Menten kinetics. Accordingly, (2) should be modified as:

$$\frac{d[A_{EX}]}{dt} = -b_{A_{EX}}[A_{EX}] - \frac{d_A}{1-\rho}([A_{EX}] - [A]) - \frac{V_E[A_{I_{EX}}][A_{EX}]}{K_{A_{EX}} + [A_{EX}]} \tag{24}$$

where $A_{I_{EX}}$ denotes the extracellular AI inhibitor.

Iron Inhibitors. Different species of bacteria can produce different kinds of siderophores that trap the iron from environment, e.g. *Enterobactin* produced by *E. coli* cannot be up-taken by PA. If the amount of iron is limited, bacteria compete with each other in order to retain the essential resources. Therefore, we consider the siderophores produced by other bacteria as iron inhibitors that can limit the availability of iron in the environment. The dynamics of the available iron in the environment can be simply modeled as:

$$[I_{ava}] = [I](\frac{[Pyo_{EX}]}{[I_I] + [Pyo_{EX}]}) \tag{25}$$

where I_{ava} denotes the available iron in the environment and I_I stands for the iron inhibitor. By replacing I with I_{ava} in (17) and (19), we can incorporate the iron inhibitor dynamics to the QS model.

2.4   Biological Controller

To dynamically and autonomously generate either AI or iron inhibitors, we propose to use synthetic methods to construct the genetic circuitry. To obtain variable combinations of the inhibitors with optimal expression levels, we build two circuits separately. More precisely, to generate AI inhibitor, we can assemble the *aiiA* genes with the *lux* promoter to sense the concentration of LasR-AI (C). Similarly, the iron inhibitor circuit is built with genes that can express the competing siderophores and sense the concentration of iron-chelated complexes (Q). Based on the general modeling of genetic circuitry [20], we model the dynamics of the new biological controller with the following ODEs:

$$\frac{d[A_I]}{dt} = P_{A_I}(\frac{1}{r_{A_I}} + \frac{[C]^2}{K_{A_I}^2 + [C]^2}) - b_{A_I}[A_I] + \frac{d_{A_I}}{\rho_s}([A_{I_{EX}}] - [A_I]) \tag{26}$$

$$\frac{d[A_{I_{EX}}]}{dt} = -\frac{d_{A_I}}{1-\rho_s}([A_{I_{EX}}] - [A_I]) - b_{A_{I_{EX}}}[A_{I_{EX}}] \tag{27}$$

$$\frac{d[I_I]}{dt} = P_{I_I}(\frac{1}{r_{I_I}} + \frac{[Q]^2}{K_{I_I}^2 + [Q]^2}) - b_{I_I}[I_I] + \frac{d_{I_I}}{\rho_s}([I_{I_{EX}}] - [I_I]) \tag{28}$$

$$\frac{d[I_{I_{EX}}]}{dt} = -\frac{d_{I_I}}{1-\rho_s}([I_{I_{EX}}] - [I_I]) - b_{I_{I_{EX}}}[I_{I_{EX}}] \tag{29}$$

where A_I (I_I) and $A_{I_{EX}}$ ($I_{I_{EX}}$) denote the intracellular and extracellular concentration of the AI (iron) inhibitors, respectively. The inhibitors production rate (second term) in (26) and (28) can be characterized by the binding of the LasR-AI and iron-chelated complex, respectively. The product of the promoter strength (P_{A_I} and P_{I_I}) and the basal production rate (r_{A_1} and r_{I_I}) characterize the minimal expression rate when there is no LasR-AI and iron-chelated complex present, respectively[3].

3   QS System Analysis

In this section, we first examine the QS system responses to different chemical substances. Next, we examine the effects of substrate utilization constant on bacteria growth. Finally, we examine the effectiveness of AI and iron inhibitors.

3.1   QS System Responses

We first examine the responses of the *las* and *pqs* QS systems by varying the concentration of available iron in the environment. Figure 2 shows the QS system responses to several chemical substances. At first, the concentration of iron is 0.01 (arbitrary unit (a.u.)); at $t = 5000$ (a.u.), the concentration of iron is changed to 1.00 (a.u.) (i.e., one hundred fold increase). We observe that the LasR protein concentration decreases due to the increase of the iron concentration; this is discussed in [12] and illustrated with the negative feedback (see also Fig. 1(a)). The other chemical substances show similar patterns except the iron-chelated complex which directly increases the growth rate. This way, the system responses confirm that our model can precisely describe the changes of chemical substance concentrations when the concentration of iron changes; this confirms the experiments in [12].

3.2   Growth Model: Utilization Constant

In Monod's bacterial growth model, bacteria consume the substrate for their growth. We assume the utilization of substrate (U_S) is constant under different iron concentrations. However, the exact values of the utilization constant are hard to measure and estimate experimentally. To determine the U_S value, we examine the changes of total virulence and biomass under different iron concentrations.

---

[3] We discuss a design example for the biological parameters in subsequent sections.

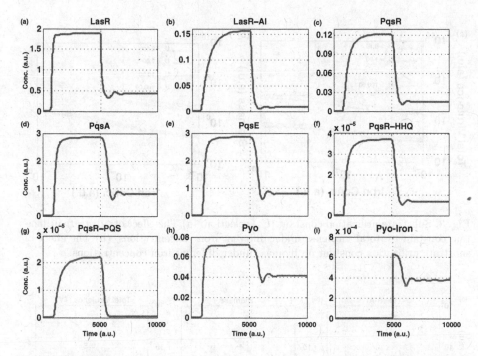

Fig. 2. Simulation results of the PA QS system responses to different iron concentrations. At first, the iron concentration is 0.01 (a.u.); later at time t = 5000 (a.u.), it is changed to 1 (a.u.). (a) As iron concentration increases, the expression of LasR proteins is repressed due to the negative feedback of iron-chelated complexes (see Fig. 1(a)). (b) The concentration of the LasR-AI complex decreases accordingly. The downstream proteins (i.e., (c) PqsR, (d) PqsA, and (e) PqsE) are all positively regulated by LasR. Hence, they change in accordance with LasR protein. The (f) PqsR-HHQ and (g) PqsR-PQS concentrations also decrease due to the decrease of PqsR. (h) Pyoverdine (Pyo) concentration shows similar profile since it is positively regulated by PqsE. (i) The concentration of iron-chelated (Pyo-Iron) complex increases due to the high affinity of pyoverdine and iron.

As shown in Fig. 3(a), once a certain concentration of iron is reached, the larger the U_S, the greater the total virulence; this is because a consumption rate of substrate that is low results in a nutrient abundant environment that favors bacteria growth. We can observe that the biomass and the total virulence are almost identical if U_S is greater than 10. Hence, in the following analysis, we set U_S to 10.

3.3 Inhibitors Effectiveness

From Fig. 2(a), we note that when the iron concentration is high, the expression of the LasR is repressed (Fig. 4(a)). On the other hand, the biomass increases due to the higher growth rate (Fig. 4(b)). By using (23), the TV increases as the concentration of iron increases as shown in Fig. 4(c).

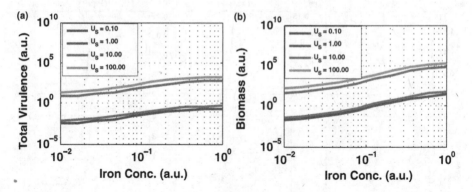

Fig. 3. Substrate utilization constant (U_s) selection. (a) The effect of substrate utilization constant on total virulence under different iron concentrations. (b) The effect of substrate utilization constant on biomass under different iron concentrations.

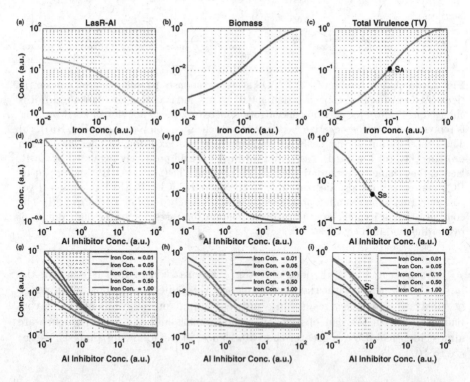

Fig. 4. The concentration changes of LasR-AI, biomass, and the TV due to the effect of different inhibitors. (a), (b), and (c) show the effect of iron concentration alone. (d), (e), and (f) show the effect of AI inhibitors alone. (g), (h), and (i) show the combined effect of iron and AI inhibitors. The set points in (c), (f), and (i) are denoted as S_A, S_B, and S_C, respectively; they are used to derive results in Fig. 5

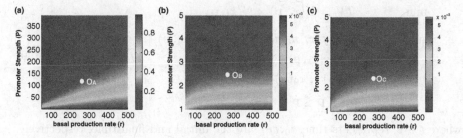

Fig. 5. Operation points for different types of inhibitors. Different color indicates the TV of a certain combination of biological parameters (promoter strength (P) and basal production rate (r) described in (26)–(29)). (a) and (b) show the operation points O_A and O_B for iron and AI inhibitor alone; they can achieve TV around 0.1 and 0.001 where we choose the setting points S_A and S_B in Fig. 4(c) and (f), respectively. Based on the operation points we choose, we can solve the optimization problem and obtain the corresponding biological parameters where $P = 100, r = 250$ and $P = 2, r = 250$, respectively. (c) The operation point O_C for multi-inhibitors. In this case, $P = 100$ and $r = 250$.

Figures 4(d)–(f) show the effect of adding the AI inhibitor into the environment, both LasR-AI complex and biomass decrease (Fig. 4(d), (e)). Hence, the TV decreases as the amount of inhibitors increases.

Our most important observation shows that if we vary both the iron concentration and AI inhibitors, we may decrease the TV. Indeed, Fig. 4(i) shows that TV decreases as we increase the concentration of AI inhibitors and decrease the iron concentration. The AI inhibitor and iron concentration have opposite effects on the LasR-AI complex and the biomass. More precisely, lower concentrations of iron result in higher concentrations of the LasR-AI complex (Fig. 4(g)), but a decrease in the biomass production (Fig. 4(h)).

4   Autonomous Biological Control System

The autonomous biological controller we propose can automatically detect signals, react to environment, and adaptively release chemical substances for intended objectives. To control the TV, the objective is to find a set of biological parameters **p** that minimize (23). However, this objective function is subject to various biological constraints including the bacteria QS, growth and QSI, as well as control dynamics. Given the mathematical model in Sect. 2, we formulate a constrained dynamic optimization problem and solve it through numerical methods.

4.1   Problem Formulation

Based on the general constrained dynamic optimization formulation and control dynamics (see **Appendix**), we can formulate our problem as follows:

$$\min_{\mathbf{p}} \qquad TV_t = V(x_t, \mathbf{p}) \times N(x_t, \mathbf{p})$$

$$\text{subject to} \quad \dot{x}_t = f(x_t, u_t, d_t, \mathbf{p}) \; \dot{u}_t = h(u_t, e_t, \mathbf{p})$$

$$y_t = g(x_t, \mathbf{p}) \qquad\quad e_t = y_t - r_t \qquad \forall t \in [t_0, t_{FL}] \qquad (30)$$

$$x_{t_0}(\mathbf{p}) = x_0(\mathbf{p})$$

$$\mathbf{p}^L \le \mathbf{p} \le \mathbf{p}^U$$

where $t \in \mathbb{R}$ represents time, t_0, t_{FL} are the initial and final time, respectively, $t_i \in [t_0, t_{FL}]$, $\mathbf{x}$ and $\dot{\mathbf{x}} \in \mathbb{R}^n$ are the state variables and their time derivatives, respectively, and $\mathbf{p} \in \mathbb{R}^r$ capture the time-invariant biological parameters that can vary within $[\mathbf{p}^L, \mathbf{p}^U]$. The functions f and h are QS and QSI models, respectively. The function g selects the process variables (y) (i.e., LasR-AI and iron-chelated complexes in our case). The state variable $\mathbf{x}$ represents the set of concentrations of chemical substances described by (1)–(19); the environment input (d) describes the environment conditions such as the nutrient availability. The control variables (u) are the inhibitors which target the AI and iron availability.

The genetic circuit can be thought of as an integral controller which reacts to the concentration of LasR-AI and iron-chelated complexes, respectively. The error signal (e) is computed as the difference between the process variable and the reference signal (r); this then feeds back to the controller, which forms a closed loop (see Fig. 1(c)).

4.2 Biological Parameters Design

To design the biological parameters for our controller, we can numerically solve the above optimization problem by sampling biological parameters within the given constraints. From our analyses in Sect. 3, we notice that TV is a monotonically decreasing function (Fig. 4(i)). Consequently, by setting (23) to a desired value, we can solve (30) for biological parameters to fulfill the design specifications.

We now provide a design example for the control circuitry that can effectively achieve the setting objective value. As shown in Fig. 4(c), (f), (i), we first choose the setting points S_A, S_B, and S_C for three different strategies that can achieve desired TVs (0.1, 0.001 and 0.001 in this design examples). The biological parameters we choose to engineer are the promoter strength (P) and the basal production rate (r) in (26)–(29) since we can tune their values through the evolution method [1]. Next, by solving (30) through varying the value of a set of biological parameters within the given constraints ($[\mathbf{p}^L, \mathbf{p}^U]$), we can obtain the most suitable combination of biological parameters that express the minimal amount of inhibitors. Figure 5 shows the operation points O_A, O_B and O_C for three strategies that can achieve the setting values (S_A, S_B, and S_C), respectively. Based on the operation points, we obtain the set of desired biological parameters.

5 Population-Level Simulation Results

In this section, we validate the proposed control system by using a 3D microfluidic environment agent-based simulator [21]. First, we explicitly apply the cellular-level model to each agent (bacterium). Next, we consider several physical and stochastic effects (physical interactions between bacteria, variation in the QS systems, growth model, etc.) and examine the growth and virulence of populations of bacteria.

The environmental configurations used in these simulations are presented in the **Appendix**. As shown in Fig. 6(a), for the case without inhibitors, the values of TV surpass the other strategies after 70 h of cultivation (i.e., the time we grow bacteria in wet-lab); this is because the bacteria growth rate (μ_X in (21)) without inhibitors is larger compared to inhibitor schemes (Fig. 6(c)). If we use AI inhibitors alone, the concentration of LasR-AI complex is reduced, but this can not repress the growth of bacteria. On the contrary, the iron inhibitor alone can inhibit the bacteria growth but the LasR-AI concentration increases (Fig. 6(b)). The multi-inhibitor strategy shows the best results; indeed it can lower the concentration of LasR-AI and bacteria growth simultaneously.

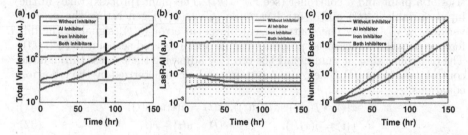

Fig. 6. The simulation results for (a) TV (b) concentration of LasR-AI complex (c) number of bacteria for four different scenarios. Note that TV is the product of the concentration of LasR-AI complex and the number of bacteria as shown in (23). We observe that the multi-inhibition strategy is the most effective in reducing TV.

6 Conclusion

In this work, we have proposed an autonomous optimal controller that incorporates the bacteria QS regulation and growth models and operates within a synthetic cell. By analyzing the system characteristics through numerical methods and simulations, we have shown that such synthetic cells can control the expression level of QS signals and cells growth.

We have also formulated a dynamic optimization problem to design the biological parameters of the proposed controller; this provides general guidelines to synthesize such optimal controllers *in vitro*. The proposed autonomous controlled system represents a first step towards a paradigm change in controlling the dynamics of communicating bacteria.

Appendix

General form of a dynamic constrained optimization problem. A general dynamic optimization problem can be formulated as follows:

$$\min_{p} \quad J(x_t, p)$$

$$\text{subject to} \quad \dot{x}_t = f(x_t, p) \; \forall t \in [t_0, t_{FL}]$$

$$x_{t_0}(p) = x_0(p)$$

$$p^L \leq p \leq p^U$$

where $t \in R$ is time, t_0, t_f are the initial and final time, respectively, $t_i \in [t_0, t_{FL}]$, x and $\dot{x} \in R^n$ are the state variables and their time derivatives, respectively, and $p \in R^r$ are the time-invariant parameters and is subjected to the lower constraints p^L and upper constraints p^U. The function J is the objective that we want to minimize. f describes the system dynamics. x_0 is the initial conditions of the state variables.

Control problem formulation. Consider a general control system which consists of a plant and a controller (see Fig. 1(c)). The plant (process) takes in the input variable $(d(t))$ and control variable (CV) $(u(t))$ generating the process variable (PV) $(y(t))$. The controller calculates an error $(e(t))$ signal as the difference between a measured process variable and a desired setpoint (SP) $(r(t))$. The controller aims at minimizing the error by adjusting the process through the control variable $(u(t))$. The control system can be characterized by the following equations:

$$\dot{x}(t) = f(x(t), u(t), d(t)), \quad \dot{u}(t) = h(e(t), u(t)) \tag{31}$$

$$y(t) = g(x(t)), \qquad e(t) = y(t) - r(t) \tag{32}$$

where f, g and h are arbitrary functions. The controller, in this case, can be viewed as an integral controller since the control signal is proportional to the integral of the error signal.

Simulation Environment Configuration. We model bacterial growth in a 3D microfluidic environment ($100\mu m$ x $100\mu m$ x $100\mu m$) that is initialized and inoculated with 1000 wild-type cells, all of which are non-overlapping and randomly attached to the substrate. We set up the simulation time up to 150 h in order to observe the evolution dynamics of bacteria growth.

Model Calibration. We calibrate the model parameters of the pqs QS system shown in Table 3. We first use similar values from [8,13] as our initial values. Next, we tune the model parameters to capture the behavior of the QS system. More precisely, we tune the model parameters based on the relative concentration change of the LasR protein under different iron concentration levels.

As shown in Fig. 2, when we change the iron concentration from 0.01 (a.u.) to 1 (a.u.), the LasR concentration changes from 2 (a.u.) to 0.5 (a.u.) which preserves the fold changes reported in Fig. 4 of reference [12].

Table 2. Table with numerical values of model parameters from [8,13]

Symbol	Value
c_A, c_R	$1e-4/s$
$\alpha_{RA}, \alpha_{RA^2}$	$1e-1$
$\delta_{RA}, \delta_{RA^2}$	$1e-1$
b_A, b_R	$1e-2$
d_A	$1e-1$
V_A, V_R	$2e-3$
K_A	$1e-6$
K_R	$1e-5$

Table 3. Table with numerical values of model parameters calibrated in this paper as explained below.

Symbol	Value
$c_{pR}, c_{pH}, c_{pA}, c_{pE}, c_{Pyo}$	$1e-7/s$
α_{pR}	$1e-1$
α_I	$1e-2$
δ_{pR}	$1e-1$
$b_{pR}, b_{pH}, b_{pA}, b_{pE}, b_{A_1}, b_{A_2}$	$1e-2$
$b_{A_{1EX}}, b_{A_{2EX}}, b_{Pyo}, b_{Pyo_{EX}}, b_Q$	$1e-1$
$d_{A_1}, d_{A_2}, d_{Pyo}, d_Q$	$1e-1$
$V_{pR}, V_{pH}, V_{pA,1}, V_{pA,2}, V_{pE,1}, V_{pE,2} V_{Pyo}$	$2e-3$
$K_{pA,1}, K_{pA,2}, K_{pE,1}, K_{pE,2}$	$1e-6$
K_{A_1}	$1e-3$
K_{pR}, K_{pH}	$1e-1$
K_{Pyo}	1
β_{pA}	$1e-2$
β_{pH}	$1e-1$

References

1. Arpino, J., et al.: Tuning the dials of synthetic biology. Microbiology **159**(7), 1236–1253 (2013)
2. Balasubramanian, D., et al.: A dynamic and intricate regulatory network determines Pseudomonas aeruginosa virulence. Nucleic Acids Res. **41**(1), 1–20 (2013)
3. Banin, E., et al.: Iron and Pseudomonas aeruginosa biofilm formation. Proc. Natl. Acad. Sci. USA **102**(31), 11076–11081 (2005)
4. Bassler, B.L., Losick, R.: Bacterially speaking. Cell **125**(2), 237–246 (2006)
5. Bredenbruch, F., et al.: The Pseudomonas aeruginosa quinolone signal (PQS) has an iron-chelating activity. Environ. Microbiol. **8**(8), 1318–1329 (2006)

6. Cheng, A.A., et al.: Enhanced killing of antibiotic-resistant bacteria enabled by massively parallel combinatorial genetics. Proc. Natl. Acad. Sci. USA **111**(34), 12462–12467 (2014)

7. Diggle, S.P., et al.: The Pseudomonas aeruginosa 4-quinolone signal molecules HHQ and PQS play multifunctional roles in quorum sensing and iron entrapment. Chem. Biol. **14**(1), 87–96 (2007)

8. Fagerlind, M.G., et al.: Modeling the effect of acylated homoserine lactone antagonists in Pseudomonas aeruginosa. Biosyst. **80**(2), 201–213 (2005)

9. Fischbach, M.A., et al.: Cell-based therapeutics: the next pillar of medicine. Sci. Transl. Med. **5**(179), 179ps7 (2013)

10. Häussler, S., Becker, T.: The pseudomonas quinolone signal (PQS) balances life and death in Pseudomonas aeruginosa populations. PLoS Pathog. **4**(9), e1000166 (2008)

11. Hazan, R., He, J., Xiao, G., Dekimpe, V., Apidianakis, Y., Lesic, B., Astrakas, C., Déziel, E., Lépine, F., Rahme, L.G.: Homeostatic interplay between bacterial cell-cell signaling and iron in virulence. PLoS Pathog. **6**(3), e1000810 (2010)

12. Kim, E.J., Wang, W., Deckwer, W.D., Zeng, A.P.: Expression of the quorum-sensing regulatory protein LasR is strongly affected by iron and oxygen concentrations in cultures of Pseudomonas aeruginosa irrespective of cell density. Microbiology **151**(4), 1127–1138 (2005). (Reading, England)

13. Melke, P., Sahlin, P., Levchenko, A., Jönsson, H.: A cell-based model for quorum sensing in heterogeneous bacterial colonies. PLoS Comput. Biol. **6**(6), e1000819 (2010)

14. Monod, J.: The growth of bacterial cultures. Ann. Rev. Microbiol. **3**, 371–394 (1949)

15. Oglesby, A.G., Farrow, J.M., Lee, J.H., Tomaras, A.P., Greenberg, E.P., Pesci, E.C., Vasil, M.L.: The influence of iron on Pseudomonas aeruginosa physiology: a regulatory link between iron and quorum sensing. J. Biol. Chem. **283**(23), 15558–15567 (2008)

16. Oglesby-Sherrouse, A.G., Djapgne, L., Nguyen, A.T., Vasil, A.I., Vasil, M.L.: The complex interplay of iron, biofilm formation, and mucoidy affecting antimicrobial resistance of Pseudomonas aeruginosa. Pathog. Dis. **70**(3), 307–320 (2014)

17. Park, S., et al.: The role of AiiA, a quorum-quenching enzyme from bacillus thuringiensis on the rhizosphere competence. J. Microbiol. Biotechnol. **18**(9), 1518–1521 (2008)

18. Schaadt, N.S., Steinbach, A., Hartmann, R.W., Helms, V.: Rule-based regulatory and metabolic model for Quorum sensing in P. aeruginosa. BMC Syst. Biol. **7**, 81 (2013)

19. Vasil, M., Ochsner, U.: The response of Pseudomonas aeruginosa to iron: genetics, biochemistry and virulence. Mol. Microbiol. **34**, 399–413 (1999)

20. Voigt, C.A.: Genetic parts to program bacteria. Curr. Opin. Biotechnol. **17**(5), 548–557 (2006)

21. Wei, G., et al.: Efficient modeling and simulation of bacteria-based nanonetworks with BNSim. IEEE J. Sel. Areas Commun. **31**(12), 868–878 (2013)

22. Whiteley, M.: Identification of genes controlled by quorum sensing in Pseudomonas aeruginosa. Proc. Natl. Acad. Sci. **96**(24), 13904–13909 (1999)

23. Williams, J.W., Cui, X., Levchenko, A., Stevens, A.M.: Robust and sensitive control of a quorum-sensing circuit by two interlocked feedback loops. Mol. Syst. Biol. **4**(234), 234 (2008). (Track II)

24. Withers, H., Swift, S., Williams, P.: Quorum sensing as an integral component of gene regulatory networks in gram-negative bacteria. Curr. Opin. Microbiol. **4**, 186–193 (2001)
25. Yang, L., Barken, K.B., Skindersoe, M.E., Christensen, A.B., Givskov, M., Tolker-Nielsen, T.: Effects of iron on DNA release and biofilm development by Pseudomonas aeruginosa. Microbiology **153**(5), 1318–1328 (2007)
26. Yosef, I., Manor, M., Kiro, R., Qimron, U.: Temperate and lytic bacteriophages programmed to sensitize and kill antibiotic-resistant bacteria. Proc. Natl. Acad. Sci. USA **112**(23), 7267–7272 (2015)

Parameter Estimation for Reaction Rate Equation Constrained Mixture Models

Carolin Loos[1,2], Anna Fiedler[1,2], and Jan Hasenauer[1,2(✉)]

[1] Helmholtz Zentrum München - German Research Center for Environmental Health, Institute of Computational Biology, 85764 Neuherberg, Germany
jan.hasenauer@helmholtz-muenchen.de
[2] Center for Mathematics, Chair of Mathematical Modeling of Biological Systems, Technische Universität München, 85748 Garching, Germany

Abstract. The elucidation of sources of heterogeneity in cell populations is crucial to fully understand biological processes. A suitable method to identify causes of heterogeneity is reaction rate equation (RRE) constrained mixture modeling, which enables the analysis of subpopulation structures and dynamics. These mixture models are calibrated using single cell snapshot data to estimate model parameters which are not measured or which cannot be assessed experimentally. In this manuscript, we evaluate different optimization methods for estimating the parameters of RRE constrained mixture models under the normal distribution assumption. We compare gradient-based optimization using sensitivity analysis with two other optimization methods – gradient-based optimization with finite differences and a stochastic optimization method – for simulation examples with artificial data. Furthermore, we compare different numerical schemes for the evaluation of the log-likelihood function. We found that gradient-based optimization using sensitivity analysis outperforms the other optimization methods in terms of convergence and computation time.

Keywords: Parameter estimation · Reaction rate equations · Mixture models · Sensitivity analysis

1 Introduction

In the past years, methods for studying biological processes on a single cell level have been developed and improved. It is possible to quantify the (relative) abundance of molecular species in single cells using, e.g. flow cytometry [2] or single cell microscopy [11]. With these techniques, it is possible to also detect heterogeneity in expression for cells of a same cell population. This heterogeneity has been shown to play an important role for e.g. cancer cells or neurons [10,14]. For homogeneous cell populations, dynamic mathematical models are convenient tools to study biological systems [8]. However, they only capture the dynamic of the mean response in the cell population and cannot account for possible subpopulations. To exploit

© Springer International Publishing AG 2016
E. Bartocci et al. (Eds.): CMSB 2016, LNBI 9859, pp. 186–200, 2016.
DOI: 10.1007/978-3-319-45177-0_12

the information available in single cell data, dynamical models that are able to account for subpopulation structures of the cells are needed.

A suitable method to study subpopulation structures of heterogeneous cell populations is the method of RRE constrained mixture modeling introduced by Hasenauer et al. [5]. These models can in principle be fitted to experimental single cell data to estimate unknown parameters of the biological system, such as kinetic rates, initial conditions or subpopulation weights. Subsequently, hypotheses about mechanistic differences between individual subpopulations can be tested. However, it has not yet been discussed how the parameters of RRE constrained mixture models can be estimated in an efficient and accurate way and there is no comparison of methods available.

In this manuscript, we consider maximum likelihood methods for parameter estimation. For this, a likelihood function which provides a measure of how well the data is explained by the current parametrization of the model is maximized. This maximization can be performed using e.g. local deterministic or global stochastic optimization techniques [3,12,15]. Most deterministic optimizers employ information about the gradient of the likelihood function. This gradient with respect to the parameters can be approximated by finite differences or, if possible, calculated with sensitivity analysis [12,13]. An example of a global stochastic optimizer is particle swarm optimization presented in [15]. This optimizer does not rely on information about the gradient and has been shown to outperform other global optimizers [15].

We describe the concept of RRE constrained mixture models and provide the likelihood function and the sensitivity equations for the calculation of its gradient with respect to the parameters. Additionally, we explain the standard and a robust approach for the evaluation of a mixture likelihood. We compare the deterministic optimization using sensitivities to the deterministic method using finite differences and to the stochastic particle swarm optimization algorithm for artificial single cell snapshot data of a one stage and three stage cascade.

2   Methods

In this section, we outline the method of RRE constrained mixture modeling for single cell snapshot data and the corresponding likelihood formulation for the parameter estimation. We establish the gradient of the likelihood with respect to the model parameters and the sensitivity equations. Further, a numerically robust evaluation of the log-likelihood is presented.

2.1   RRE Constrained Mixture Models

RRE modeling provides the temporal evolution for the mean concentrations $\boldsymbol{x} = (x_1, \ldots, x_{n_x})$ of n_x chemical species involved in a biological process, which is stimulated by an external stimulus u. These RREs can be written as

$$\dot{\boldsymbol{x}} = f(\boldsymbol{x}, \boldsymbol{\psi}, u), \qquad \boldsymbol{x}(0) = \boldsymbol{x}_0(\boldsymbol{\psi}, u), \tag{1}$$

an ODE system with initial conditions $x_0(\psi, u)$ and vector field f. The parameter vector ψ comprises e.g. kinetic rates, initial concentrations or observation parameters. Often, the concentrations x of the species cannot be measured directly or only a subset of them can be observed. In most experiments, only a single property is assessed. Therefore, we considered an observable

$$y = h(x, \psi, u),$$

with h denoting the mapping. The observation process depends on observation parameters included in ψ such as scaling and offset constants.

Mixture models enable the depiction of subpopulations within an overall population. The probability distribution is described by the weighted sum of probability density functions ϕ for individual mixture components, i.e., subpopulations

$$p(y|w_s, \mu_s, \sigma_s) = \sum_{s=1}^{n_s} w_s \phi(y|\mu_s, \sigma_s^2).$$

In this manuscript, we assumed ϕ to be a normal distribution, which is parametrized by its mean μ and variance σ^2.

Combining these, every subpopulation is treated as a mixture component for which the mean concentration is simulated using RREs [5]. This yields the following model for the distribution of an observable y for some given parameters θ at a time point t_k,

$$p(y|\theta, t_k) = \sum_{s=1}^{n_s} w_s(\theta)\, \phi\left(y|\mu_s, \sigma_s^2(\theta, t_k)\right)$$

$$\text{with } \dot{x}_s = f\left(x_s, \psi_s(\theta), u\right),\ x_s(0) = x_0(\psi_s(\theta), u),$$

$$\mu_s = h\left(x_s, \psi_s(\theta), u\right).$$

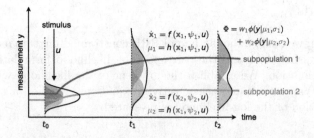

Fig. 1. Illustration of RRE constrained mixture modeling for an example of two subpopulations. The means of measurement y for the individual subpopulations are calculated with RREs and plotted as purple and orange lines for the high and low responsive subpopulation, respectively. The overall cell distribution Φ is plotted as black curve and is calculated by a weighted mixture of the individual distributions for the subpopulations (purple and orange areas). (Color figure online)

The parameter vector can comprise e.g. $\boldsymbol{\theta} = (\{w_s, \sigma_s, \boldsymbol{\xi}_s\}_{s=1}^{n_s}, \boldsymbol{\xi})$, the subpopulation specific mixture weights w_s, standard deviations σ_s and mechanistic parameters $\boldsymbol{\xi}_s$ as well as mechanistic parameters $\boldsymbol{\xi}$ that are shared across subpopulations. The mean of the mixture distribution is linked to the RREs, while the mixture weights and standard deviations do not depend on the RREs. The parameters for the RREs of an individual subpopulation as defined in (1) are thus given by $\boldsymbol{\psi}_s = (\boldsymbol{\xi}_s, \boldsymbol{\xi})$. The concept of RRE constrained mixture models is illustrated in Fig. 1. For a more detailed explanation of these models, we refer to [5].

2.2   Single Cell Snapshot Data

We considered single cell snapshot data

$$\mathcal{D} = \left\{ \{y_j^k\}_{j=1}^{n_c} \right\}_{k=1}^{n_t}.$$

These data contain the measurements y for n_c cells, indexed by j, at n_t time points, indexed by k. In the case considered, the data captures the dynamics of the population on a single cell level after stimulation with some input u.

2.3   Parameter Estimation for RRE Constrained Mixture of Normal Distributions

To obtain the parameters of a RRE constrained mixture model, the model needs to be fitted to experimental data $\mathcal{D}$. This is done by maximum likelihood estimation. A likelihood function $\mathcal{L}(\boldsymbol{\theta})$ describes the probability of observing the data $\mathcal{D}$ given the parameters $\boldsymbol{\theta}$. For the case of RRE constrained mixture models, this function is given by

$$\mathcal{L}(\boldsymbol{\theta}) := \prod_{k,j} \sum_{s=1}^{n_s} w_s(\boldsymbol{\theta}) \, \phi \left(y_j^k | \mu_s, \sigma_s^2(\boldsymbol{\theta}, t_k) \right)$$

$$\text{with } \dot{\boldsymbol{x}}_s = f\left(\boldsymbol{x}_s, \boldsymbol{\psi}_s(\boldsymbol{\theta}), u\right), \ \boldsymbol{x}_s(0) = \boldsymbol{x}_0(\boldsymbol{\psi}_s(\boldsymbol{\theta}), u),$$

$$\mu_s = h\left(\boldsymbol{x}_s, \boldsymbol{\psi}_s(\boldsymbol{\theta}), u\right).$$

The mixture parameters μ_s implicitly depend on the parameter vector $\boldsymbol{\theta}$. A different variance parameter σ_s can be used for every measured time point t_k and subpopulation s. Since the number of parameters increases with the number of measured time points and the number subpopulations, an efficient method for parameter estimation is required. Due to its better numerical properties, we used the negative log-likelihood function

$$J(\boldsymbol{\theta}) = -\log \mathcal{L}(\boldsymbol{\theta})$$

$$= -\sum_{k,j} \log \sum_{s=1}^{n_s} w_s(\boldsymbol{\theta}) \, \phi \left(y_j^k | \mu_s, \sigma_s^2(\boldsymbol{\theta}, t_k) \right)$$

in the optimization, which has the same extrema as the likelihood function. In the following, we derive the gradient of J with respect to $\boldsymbol{\theta}$, which can be employed by deterministic local optimization methods.

Gradient of Negative Log-Likelihood Function. For a simpler notation, we neglect the arguments of w_s and σ_s. The gradient of the log-likelihood with respect to parameters $\boldsymbol{\theta} = (\theta_1, \ldots, \theta_{n_\theta})$, with θ denoting an entry of the vector, is given by

$$
\begin{aligned}
\frac{dJ}{d\theta} &= -\sum_{k,j} \frac{d}{d\theta} \log \left(\sum_{s=1}^{n_s} w_s\, \phi \left(y_j^k | \mu_s, \sigma_s^2 \right) \right) \\
&= -\sum_{k,j} \frac{1}{\sum_{s=1}^{n_s} w_s\, \phi \left(y_j^k | \mu_s, \sigma_s^2 \right)} \frac{d}{d\theta} \sum_{s=1}^{n_s} w_s\, \phi \left(y_j^k | \mu_s, \sigma_s^2 \right) \\
&= -\sum_{k,j} \frac{1}{\sum_{s=1}^{n_s} w_s\, \phi \left(y_j^k | \mu_s, \sigma_s^2 \right)} \sum_{s=1}^{n_s} \left(\frac{dw_s}{d\theta} \phi \left(y_j^k | \mu_s, \sigma_s^2 \right) + w_s \frac{d\phi \left(y_j^k | \mu_s, \sigma_s^2 \right)}{d\theta} \right).
\end{aligned}
$$

Under the assumption that ϕ is a normal distribution, it holds that

$$
\frac{d\phi \left(y_j^k | \mu_s, \sigma_s^2 \right)}{d\theta} = \frac{1}{\sigma_s} \phi \left(y_j^k | \mu_s, \sigma_s^2 \right) \left(\frac{y_j^k - \mu_s}{\sigma_s} \frac{d\mu_s}{d\theta} + \left(\left(\frac{y_j^k - \mu_s}{\sigma_s} \right)^2 - 1 \right) \frac{d\sigma_s}{d\theta} \right),
$$

with

$$
\frac{d\sigma_s^k}{d\theta} = \begin{cases} 1 & \theta = \sigma_s^k \\ 0 & \text{otherwise} \end{cases}, \qquad \frac{dw_s}{d\theta} = \begin{cases} 1 & \theta = w_s \\ 0 & \text{otherwise} \end{cases}.
$$

The gradient of the objective function comprises $\frac{d\mu_s}{d\theta}$, which can be calculated using sensitivity analysis. The sensitivities $z^{\boldsymbol{x}_s} = \left(\frac{\partial x_{s,1}}{\partial \theta}, \ldots, \frac{\partial x_{s,n_x}}{\partial \theta} \right)$ are defined by

$$
\frac{\partial z^{\boldsymbol{x}_s}}{\partial t} = \frac{\partial f}{\partial \boldsymbol{x}_s} z^{\boldsymbol{x}_s} + \frac{\partial \boldsymbol{x}_s}{d\theta}, \quad z^{\boldsymbol{x}_s}(0) = \frac{\partial \boldsymbol{x}_0}{\partial \theta},
$$

$$
z^{\mu_s} = \frac{\partial h}{\partial \boldsymbol{x}_s} z^{\boldsymbol{x}_s} + \frac{\partial h}{\partial \theta},
$$

with $\frac{\partial f}{\partial \boldsymbol{x}_s} = \left(\frac{\partial f_m}{\partial x_{s,l}} \right)_{m,l} \in \mathbb{R}^{n_x \times n_x}$ and $\frac{\partial h}{\partial \boldsymbol{x}_s} = \left(\frac{\partial h_m}{\partial x_{s,l}} \right)_{m,l} \in \mathbb{R}^{n_x \times n_y}$. For the case of RRE constrained mixture models, we obtain μ_s and $\frac{d\mu_s}{d\theta} = z^{\mu_s}$ by simulating an ODE system comprising the RREs and sensitivity equations.

Robust Evaluation of the Log-Likelihood Function and Its Gradient. We explain and tackle the problem occuring when numerically evaluating (log-) likelihood functions of mixture distributions. For this, we formulate the standard and robust approach to evaluate the log-likelihood function following [9]. As already mentioned, rather the log-likelihood than the likelihood function is calculated due to numerical properties. This means, instead of the probability density p, the logarithm $\log(p)$ is evaluated. For the assumption of a normal distribution this circumvents e.g. exponentiation of the difference between measurement and simulation. This is especially advantageous for high differences, since e^{-x} might be numerically evaluated to zero for finite values of x. However, for mixture models, if $n_s > 1$ and $p_s := \phi(y|\mu_s, \sigma_s^2)$, it holds that

$$\log(p) = \log\left(\sum_{s=1}^{n_s} w_s p_s\right) \neq \sum_{s=1}^{n_s} \log(w_s p_s),$$

i.e., for these cases it is not possible to use the logarithm of the probability density of an individual mixture component directly. This problem also occurs in the calculation of the gradient. We refer to this approach of evaluating the likelihood function as standard approach.

A more robust approach for the log-likelihood calculation is given in the following. With $q_s = \log(p_s)$ and $\hat{s} = \operatorname{argmax}_s q_s$, we reformulate

$$\log(p) = \log\left(\sum_{s=1}^{n_s} w_s e^{q_s}\right)$$

$$= \log\left(1 + \sum_{s \neq \hat{s}} \frac{w_s}{w_{\hat{s}}}\left(e^{q_s - q_{\hat{s}}}\right)\right) + \log(w_{\hat{s}}) + q_{\hat{s}}. \tag{2}$$

Considering p_s to be a normal distribution it follows that

$$\log(p_s) = q_s = -\frac{1}{2}\left(\frac{y - \mu_s}{\sigma_s}\right)^2 - \log(\sqrt{2\pi}) - \log(\sigma_s).$$

Regarding the calculation of the gradient it holds that

$$\frac{d\log(p)}{d\theta} = \frac{1}{p}\frac{dp}{d\theta} = \sum_{s=1}^{n_s} \frac{p_s}{\sum_{j=1}^{n_s} w_j p_j} H_s$$

$$= \frac{1}{\sum_{j=1}^{n_s} w_j e^{q_j - q_{\hat{s}}}} \sum_{s=1}^{n_s} e^{q_s - q_{\hat{s}}} H_s, \tag{3}$$

with H_s defined by

$$H_s = \frac{1}{p_s}\frac{dw_s p_s}{d\theta} = \frac{dw_s}{d\theta} + \frac{w_s}{p_s}\frac{dp_s}{d\theta}.$$

Under the assumption that p_s is a normal distribution this is

$$H_s = \frac{dw_s}{d\theta} + \frac{w_s}{\sigma_s}\left(\frac{y - \mu_s}{\sigma_s}\frac{d\mu_s}{d\theta} + \left(\left(\frac{y - \mu_s}{\sigma_s}\right)^2 - 1\right)\frac{d\sigma_s}{d\theta}\right).$$

The proposed reformulations (2) and (3) are used for the robust evaluation of the log-likelihood function and its gradient. For further details we refer to [9].

2.4 Implementation

The RRE constrained mixture models were implemented in MATLAB. The sensitivity equations were derived and simulated using the toolbox CERENA [7]. For parameter estimation with deterministic optimization, we used the toolbox PESTO[1], which employs the MATLAB function fmincon. For stochastic global optimization we employed a toolbox for the algorithm PSwarm [15].

3 Results

We compared the different optimizers in terms of convergence and computation time for artificial data of a one stage and a three stage cascade.

3.1 One Stage Cascade

For a first comparison of the optimizers we considered a small example of a one stage cascade comprising a conversion between two species A and B.

Model and Artificial Data. A conversion process describes a reversible reaction between two species, A and B that have the concentrations $[A]$ and $[B]$, respectively. In our example, we assumed that the conversion from A to B takes place with a basal rate $k_2[A]$ and is additionally increased by external stimulus u. Furthermore, B is converted back to A with kinetic parameter k_3 yielding the reactions

$$R_1: \quad A \rightarrow B, \quad rate = k_1 u[A],$$
$$R_2: \quad A \rightarrow B, \quad rate = k_2[A],$$
$$R_3: \quad B \rightarrow A, \quad rate = k_3[B].$$

We considered that there exist two subpopulations, s_1 and s_2, differing in the stimulus-dependent conversion from A to B. This is described by the kinetic parameter k_1, i.e., the subpopulations share the parameters k_2 and k_3 but have individual parameters k_{1,s_1} and k_{1,s_2} with s_1 and s_2 indicating the kinetic parameters of subpopulation 1 and 2, respectively. The system is in steady state before stimulation ($u = 0$ for $t < 0$). To generate the artificial data we used the parameters $(k_{1,s_1}, k_{1,s_2}, k_2, k_3, w) = (0.1, 0.75, 0.5, 1.5, 0.7)$ and assumed that only the concentration of species B can be measured, yielding the observation model $y = h(\boldsymbol{x}, \boldsymbol{\psi}, u) = x_2$, with $\boldsymbol{x} = (x_1, x_2)^T = ([A], [B])^T$. An illustration of the system including the subpopulations is given in Fig. 2A. This system was simulated using the stochastic simulation algorithm [4], which models random births and deaths of individual molecules. We considered a system size of $\Omega = 1000$ and divided the number of molecules by Ω to obtain the concentration of the species. Moreover, the external stimulus is set to $u = 1$ at $t \geq 0$ and

---

[1] Available at https://github.com/ICB-DCM/PESTO.

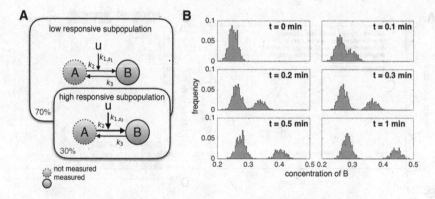

Fig. 2. Artificial data of a conversion process. (A) Illustration of a conversion process between chemical species A and B in a cell population. The conversion from A to B is enhanced by a stimulus u. 30 % of the cells show a higher response to the external stimulus u than the other cells. Only the concentration of B denoted by $[B]$ is measured. **(B)** Artificial data for the conversion process. The system is stimulated with $u = 0$ for $t < 0$ and $u = 1$ for $t \geq 0$.

measurements of the concentration of B are recorded at $t = 0, 0.1, 0.2, 0.3, 0.5, 1$ minutes. The data are shown in Fig. 2B: For $t = 0$, the system is in steady state and no subpopulation structure is visible, since the subpopulations differ only in the response to stimulation. For $t = 0.1$, the subpopulation structure becomes visible, but the subpopulations still highly overlap. However, for later time points the subpopulations are clearly separated.

The mean of the stochastic single cell trajectories can be described by RREs, i.e., the temporal evolution of x_2 can be described by the ODE

$$\dot{x}_2 = k_1 u + k_2 - (k_1 u + k_2 + k_3)\, x_2, \qquad x_2(0) = \frac{k_2}{k_2 + k_3},$$

using mass conservation, $[A] + [B] = 1$. We then assumed the parameters $\theta = (k_{1,s_1}, k_{1,s_2}, k_2, k_3, w, \{\{\sigma_{s(t_k)}\}_{s=1}^2\}_{k=1}^6)$ to be unknown and estimated them from the data. Since the data comprised six time points and we accounted for two subpopulations, 12 parameters for the standard deviation $\sigma_s(t_k)$ need to be estimated.

Convergence of Optimization Methods. To evaluate the optimizers, we compared deterministic gradient-based optimization using sensitivities with deterministic gradient-based optimization using finite differences and a stochastic particle swarm algorithm [15]. For all optimizers, the parameter values for the kinetic rates k_i were restricted to the interval $[10^{-6}, 10^4]$, the mixture weight w to $[0, 1]$ and the parameters for the standard deviation of the normal distributions $\sigma_s(t_k)$ to $[10^{-2.5}, 10^{2.5}]$. Each algorithm was started 100 times and the deterministic optimizers were started from the same randomly drawn start points.

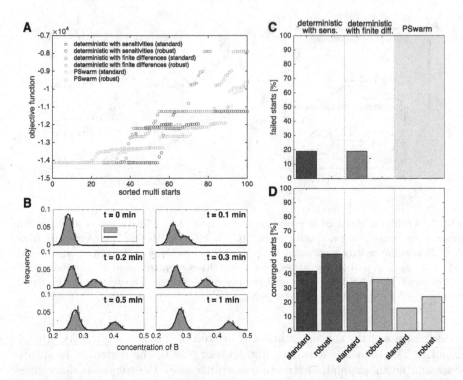

Fig. 3. Comparison of optimization methods. (**A**) Convergence plot for the final negative log-likelihood values for 100 starts. The values are sorted from lowest to highest implying a decreasing goodness of fit. (**B**) Data and fit for the optimal value, which was found by all methods. Percentage of starts for which the initial value was ∞ (**C**) and converged starts (**D**).

The final negative log-likelihood values for every start are sorted with decreasing goodness of fit and shown in Fig. 3A. The data and fit, which correspond to the optimal value found by all methods, are shown in Fig. 3B. The model shows a good agreement with the data. For a detailed comparison of the results obtained by the different optimization methods, we assessed the percentage of failed starts, i.e., the starts for which the log-likelihood function was infinite at the start point (Fig. 3C). For almost 20 % of all drawn start points the log-likelihood has an infinite value when using the standard evaluation of the log-likelihood. However, the log-likelihood can be evaluated for all start points when using the robust calculation approach. Since for PSwarm an initial particle population is used instead of a single initial value, there are no failed starts and it is not possible to compare this property with the deterministic optimizers. We expect the percentage of failed log-likelihood evaluations for the initial particle population to be similar to the percentages found for the failed starts in the deterministic optimization using the standard approach. The likelihood was numerically evaluated to zero for all start points. For the log-likelihood, we counted the number of objective function values that are close to the minimal objective function value found, i.e., below a statistical threshold according to a likelihood ratio test [6]. These starts are then likely to have converged to the global optimum. The percentage of converged starts determined for each optimizer is depicted in Fig. 3D. Clearly, the best convergence is obtained by deterministic local optimization with an analytical gradient that is calculated with sensitivities. For this optimizer, the robust calculation of the log-likelihood and the gradient yielded better convergence compared to the standard approach. For both approaches, standard and robust evaluation of the log-likelihood function, deterministic local optimiziation with finite difference approximation to the gradient shows less convergence than when using sensitivites. The stochastic optimization with PSwarm has even less converged runs than the deterministic optimization with finite differences.

Computation Time of Optimization Methods. We compared the performance of the optimizers in terms of computation time (Fig. 4A). The best computation time was achieved for the deterministic optimization with sensitivities, while the highest computation time is needed for stochastic optimization. Also regarding the number of function evaluations, the stochastic optimization needed most function evaluation and the deterministic optimization with sensitivities performed best (Fig. 4B). Furthermore, regarding the average computation time needed per converged start shown in Fig. 4C, the deterministic optimizer using sensitivities outperforms the other optimizers. However, there were almost no additional computational costs when using the robust approach instead of the standard approach to evaluate the log-likelihood function for all optimizers.

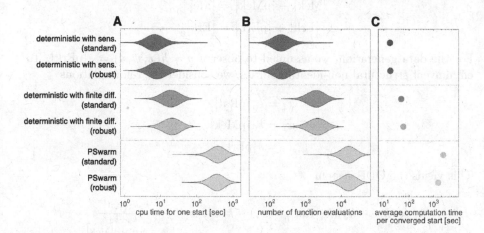

Fig. 4. Performance comparison of optimization methods. (A) Time needed for one optimization start. **(B)** Number of objective function evaluations for one optimization start. **(C)** Average computation time needed per converged start.

3.2   Three Stage Cascade

To validate the results obtained for the simple conversion process, we studied artificial data of a three stage cascade, namely the Raf/Mek/Erk cascade.

Model and Artificial Data. The considered pathway comprises the protein kinases Raf, Mek and Erk and their corresponding phosphorylated/active forms pRaf, pMek and pErk. Raf is activated with a stimulus-dependent rate $k_1 u [\text{Raf}]$ and a basal rate $k_2 [\text{Raf}]$. The activation rate of Mek is proportional to the amount of phosphorylated Raf, while active Mek in turn phosphorylates Erk. These reactions and the dephosphorylation of the active kinases are given by

$$
\begin{aligned}
R_1 : &\quad \text{Raf} \rightarrow \text{pRaf}, &\quad \text{rate} &= k_1 u [\text{Raf}], \\
R_2 : &\quad \text{Raf} \rightarrow \text{pRaf}, &\quad \text{rate} &= k_2 [\text{Raf}], \\
R_3 : &\quad \text{pRaf} \rightarrow \text{Raf}, &\quad \text{rate} &= k_3 [\text{pRaf}], \\
R_4 : &\quad \text{Mek} \rightarrow \text{pMek}, &\quad \text{rate} &= k_4 [\text{pRaf}] [\text{Mek}], \\
R_5 : &\quad \text{pMek} \rightarrow \text{Mek}, &\quad \text{rate} &= k_5 [\text{pMek}], \\
R_6 : &\quad \text{Erk} \rightarrow \text{pErk}, &\quad \text{rate} &= k_6 [\text{pMek}] [\text{Erk}], \\
R_7 : &\quad \text{pErk} \rightarrow \text{Erk}, &\quad \text{rate} &= k_7 [\text{pErk}],
\end{aligned}
$$

with mass conservation

$$
\begin{aligned}
[\text{Raf}] + [\text{pRaf}] &= [\text{Raf}]_0, \\
[\text{Mek}] + [\text{pMek}] &= [\text{Mek}]_0, \\
[\text{Erk}] + [\text{pErk}] &= [\text{Erk}]_0.
\end{aligned}
$$

For the data generation, we assumed to observe $y = h(\boldsymbol{x}, \boldsymbol{\psi}, u) = s [\text{pErk}]$. To circumvent structural non-identifiabilities, we consider the reformulations

$$
\begin{aligned}
x_1 &= k_4 [\text{pRaf}], \\
x_2 &= k_6 [\text{pMek}], \\
x_3 &= s [\text{pErk}].
\end{aligned}
$$

This yields the ODE system

$$
\begin{aligned}
\dot{x}_1 &= (k_1 u + k_2)(k_4 [\text{Raf}]_0 - x_1) - k_3 x_1, &\quad x_1(0) &= \frac{k_2 k_4 [\text{Raf}]_0}{k_3 + k_2}, \\
\dot{x}_2 &= x_1 (k_6 [\text{Mek}]_0 - x_2) - k_5 x_2, &\quad x_2(0) &= \frac{x_1(0) k_6 [\text{Mek}]_0}{x_1(0) + k_5}, \\
\dot{x}_3 &= x_2 (s [\text{Erk}]_0 - x_3) - k_7 x_3, &\quad x_3(0) &= \frac{x_2(0) s [\text{Erk}]_0}{x_2(0) + k_7},
\end{aligned}
$$

with $y = x_3$ and parameters $(k_1, k_2, k_3, k_5, k_7, k_4 [\text{Raf}]_0, k_6 [\text{Mek}]_0, s [\text{Erk}]_0)$. For details regarding the model we refer to [5]. In this example, we considered two

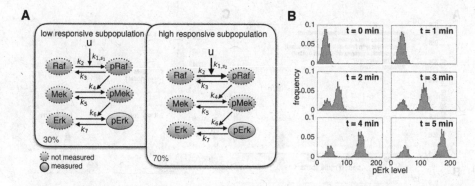

Fig. 5. Artificial data of the Raf/Mek/Erk cascade. (A) Illustration of the considered signaling pathway, which comprises the kinases Raf, Mek and Erk and its corresponding actived forms. The model comprises two subpopulations differing in their response to stimulus u. **(B)** Artificially generated data of the Raf/Mek/Erk cascade for measurements of pErk levels.

subpopulations that differ in their response to stimulus u, captured by parameter k_1 (Fig. 5A). We generated measurements of 1000 cells by simulating the ODE system for $\log_{10}(k_{1,s_1}, k_{1,s_2}, k_2, k_3, k_5, k_7, k_4[\text{Raf}]_0, k_6[\text{Mek}]_0, s[\text{Erk}]_0) = (-2, -1, -2, -0.15, -0.15, -0.15, -2, 2, 3)$, $w = 0.7$ and normally-distributed measurement noise (Fig. 5B). The stimulus u is set to 0 for $t < 0$ and to 1 for $t \geq 0$.

Convergence of Optimization Methods. For parameter estimation, the intervals for the parameters were set to $[10^{-3}, 10^5]$ for the kinetic parameters, to $[0, 1]$ for the mixture weight and to $[10^{-3}, 10^2]$ for $\sigma_s(t_k)$. The resulting objective function values for 100 runs of the optimization procedures are shown in Fig. 6A, and a zoom in of the five best runs in Fig. 6B. The optimization with sensitivities and a robust evaluation of the log-likelihood function converged to the optimal value 44 times. This optimal value yields a good fit to the data (Fig. 6C). Using deterministic optimization with sensitivities and the standard evaluation of the log-likelihood function the same optimal value as with the robust evaluation was found only once. The other optimizers were not able to find the optimal value at all. For the deterministic optimization and the standard evaluation of the log-likelihood function, only three out of 100 initial parameter values yielded a finite log-likelihood value. Consequently, the remaining runs could not be started. These findings indicate that for higher-dimensional estimation problems, the use of sensitivity-based methods and robust log-likelihood evaluation becomes increasingly important.

Performance of Optimization Methods. We compared the computation times and needed function evaluations of the different optimization methods (Fig. 7). Since only the deterministic optimization with sensitivities and robust

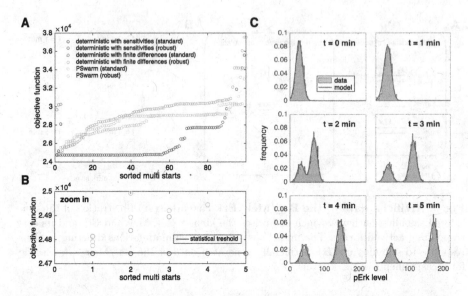

Fig. 6. Comparison of optimization methods. (A) Final negative log-likelihood values for 100 runs, sorted according to a decreasing goodness of fit. **(B)** Zoom for the five best starts. The black line indicates the statistical threshold according to a likelihood ratio test, which was used to obtain the number of converged starts. **(C)** Data and fit for the optimal parameter value found by deterministic optimization with sensitivities and a robust evaluation of the log-likelihood function.

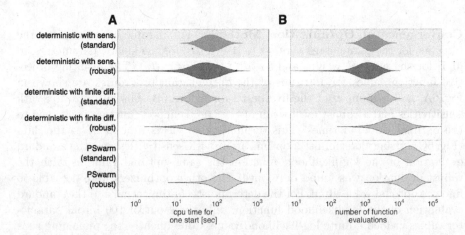

Fig. 7. Performance of optimization methods. (A) CPU time needed for one optimization start. **(B)** Number of objective function evaluations for one optimization start. The representation is based on three starts for deterministic optimization with the standard approach to evaluate the log-likelihood (grey shaded), while it is based on 100 starts for the other optimizers.

evaluation reached a sufficient number of converged starts, we did not compare the optimizers in terms of average computation time per converged starts. The analysis for the deterministic optimization with standard evaluations is only based on three starts that have not failed and is therefore not meaningful for the comparison. Among the optimizers for which 100 starts could be analyzed in terms of their computation time and number of function evaluations, the optimization with sensitivities and the robust evaluation of the log-likelihood function performs best. The proposed approach therefore yields better optimization results and is also more efficient than the other optimizers.

4   Conclusion

In this manuscript, we summarized the concept of RRE constrained mixture modeling and studied the calibration of those models to experimental data under the normal distribution assumption. An often used approach to estimate the parameters of mixture models in general is the Expectation-Maximization (EM) algorithm (see e.g. [1]). This algorithm highly depends on the initialization of the mixture components, which is challenging for RRE constrained mixture models since the components depend on the dynamic parameters of the model. In preliminary studies the EM algorithm showed poor convergence. Therefore, we did not consider the EM algorithm in this manuscript and focused on a maximum likelihood approach.

We derived the log-likelihood function and its gradient, which can be used to perform gradient-based deterministic optimization. Additionally, a robust approach of numerically evaluating these terms has been provided. We compared three optimization schemes, two deterministic gradient-based methods, one using the analytical gradient and one using an approximation of the gradient by finite differences, and a stochastic particle swarm algorithm. For each optimizer, we assessed performance and convergence for the standard and robust approach to evaluate the log-likelihood function. The comparison was carried out for examples of artificial single cell snapshot data of a one stage and a three stage cascade. We found that deterministic gradient-based optimization with sensitivities and robust calculation of the mixture probability outperformed all other methods in terms of robustness and convergence. This is especially important, since the complexity of RRE constrained mixture models increases with the number of measured time points. For the example of the three stage cascade only gradient-based optimization with sensitivites and a robust evaluation of the log-likelihood function yielded a reasonable calibration of RRE constrained mixture models to the data. We expect this also to hold when considering even more complicated systems. Accordingly, the proposed approach facilitates a robust and efficient calibration of RRE constrained mixture models to elucidate the sources of heterogeneity.

References

1. Bishop, C.M.: Pattern recognition and machine learning, vol. 4. Springer, New York (2006)
2. Davey, H.M., Kell, D.B.: Flow cytometry and cell sorting of heterogeneous microbial populations: the importance of single-cell analyses. Microbiolog. Rev. **60**(4), 641–696 (1996)
3. Gábor, A., Banga, J.R.: Robust and efficient parameter estimation in dynamic models of biological systems. BMC Syst. Biol. **9**, 74 (2015)
4. Gillespie, D.T.: Stochastic simulation of chemical kinetics. Ann. Rev. Phys. Chem. **58**, 35–55 (2007)
5. Hasenauer, J., Hasenauer, C., Hucho, T., Theis, F.J.: ODE constrained mixture modelling: a method for unraveling subpopulation structures and dynamics. PLoS Comput. Biol. **10**(7), e1003686 (2014)
6. Hross, S., Hasenauer, J.: Analysis of CFSE time-series data using division-, age- and label-structured population models. Bioinformatics (2016) btw131
7. Kazeroonian, A., Fröhlich, F., Raue, A., Theis, F.J., Hasenauer, J.: CERENA: ChEmical REaction network analyzer - a toolbox for the simulation and analysis of stochastic chemical kinetics. PloS ONE **11**(1), e0146732 (2016)
8. Kitano, H.: Computational systems biology. Nature **420**(6912), 206–210 (2002)
9. Loos, C.: Analysis of single-cell data: ODE-constrained mixture modeling and approximate Bayesian computation. Springer, Best Masters, Heidelberg (2016)
10. Michor, F., Polyak, K.: The origins and implications of intratumor heterogeneity. Cancer Prev. Res. **3**(11), 1361–1364 (2010)
11. Miyashiro, T., Goulian, M.: Single-cell analysis of gene expression by fluorescence microscopy. Methods Enzymol. **423**, 458–475 (2007)
12. Raue, A., Schilling, M., Bachmann, J., Matteson, A., Schelker, M., Kaschek, D., Hug, S., Kreutz, C., Harms, B.D., Theis, F.J., et al.: Lessons learned from quantitative dynamical modeling in systems biology. PLoS ONE **8**(9), e74335 (2013)
13. Sengupta, B., Friston, K., Penny, W.: Efficient gradient computation for dynamical models. NeuroImage **98**, 521–527 (2014)
14. Torres-Padilla, M.-E., Chambers, I.: Transcription factor heterogeneity in pluripotent stem cells: a stochastic advantage. Development **141**(11), 2173–2181 (2014)
15. Vaz, A.I.F., Vicente, L.N.: A particle swarm pattern search method for bound constrained global optimization. J. Glob. Optim. **39**(2), 197–219 (2007)

Normalizing Chemical Reaction Networks by Confluent Structural Simplification

Guillaume Madelaine[1,2]($\boxtimes$), Elisa Tonello[3], Cédric Lhoussaine[1,2],
and Joachim Niehren[2,4]

[1] University of Lille, Lille, France
[2] CRISTAL, CNRS UMR 9189, Lille, France
[3] University of Nottingham, Nottingham, UK
[4] Inria, Lille, France

Abstract. Reaction networks can be simplified by eliminating linear intermediate species in partial steady states. In this paper, we study the question whether this rewrite procedure is confluent, so that for any given reaction network, a unique normal form will be obtained independently of the elimination order. We first contribute a counter example which shows that different normal forms of the same network may indeed have different structures. The problem is that different "dependent reactions" may be introduced in different elimination orders. We then propose a rewrite rule that eliminates such dependent reactions and prove that the extended rewrite system is confluent up to kinetic rates, i.e., all normal forms of the same network will have the same structure. However, their kinetic rates may still not be unique, even modulo the usual axioms of arithmetics. This might seem surprising given that the ODEs of these normal forms are equal modulo these axioms.

1 Introduction

Chemical reaction networks are widely used in systems biology for modeling the dynamics of biochemical molecular systems [1,4,6,11]. A chemical reaction network has a graph structure that can be identified with a Petri net [2]. Beside of this, it assigns to each of its reactions a kinetic rate that models the reaction's speed. Chemical reaction networks can either be given a deterministic semantics in terms of ordinary differential equations (ODEs), which describes the evolution of the average concentrations of the species of the network over time, or a stochastic semantics in terms of continuous time Markov chains, which defines the evolution of molecule distributions of the different species over time. In this paper, we focus on the deterministic semantics.

Reaction networks may become very large when modeling molecular biological systems in sufficient detail, see e.g. the examples in the BioModels database [8]. Therefore much effort has been spent on their simplification (see [18] for an overview). The traditional approach is by reducing the ODEs of the network by symbolic rewriting techniques [9,10]. While clearly beneficial, such approaches have the disadvantage that the simplified ODEs cannot always be translated

© Springer International Publishing AG 2016
E. Bartocci et al. (Eds.): CMSB 2016, LNBI 9859, pp. 201–215, 2016.
DOI: 10.1007/978-3-319-45177-0_13

back to a reaction network [3], so that these simplifications cannot be understood directly as simplifications of biological systems.

Another major problem with large biological reaction networks is that precise kinetic rates are rarely available [14, 16]. In the worst case, no kinetic information is available, so that no ODEs can be derived. The only simplifications that are possible in this case rely purely on the graph structure of the reaction network [12, 17]. In a less extreme setting, the kinetic rates are given by arithmetic expressions with unknown parameters. In this case, the purely structural methods must be lifted so that they can properly account for the kinetic rates.

The common objective of the structural simplification methods is to eliminate intermediate species that are irrelevant to the external behaviour of the system. This can be done in an exact manner – when assuming partial steady states – so that the solutions of the ODEs of reaction networks are preserved [13, 19, 21]. It should be noticed that any structural reduction algorithm preserving the ODE's solutions necessarily induces an exact reduction method on the underlying ODE level. Indeed the above methods are based on the same idea, which is to resolve the partial steady state equation of some intermediate species along its concentration variable, so that this variable can be eliminated from the ODEs. The restriction that makes this possible is that the kinetic rates of the network's reactions are linear in the concentration of the intermediate species.

The structural reduction method for intermediate elimination from [13] removes the intermediates stepwise one by one. The approach of [21] is similar with an extension to rapid equilibrium assumption. The alternative method of [19] removes several intermediates simultaneously. We verified that both methods perform the same reductions when restricted to a single intermediate, even though these are computed by quite differently algorithms. The yet independent method from [17, 18] also performs simultaneous elimination of intermediates, but not necessarily in a unique manner. The intermediates are eliminated from the reaction graph by computing elementary modes in a first step, and in a second, appropriate kinetic rates are assigned to reduced graph. Their method can also be applied in the nonlinear case, but then with some approximations.

In this paper, we study the question of whether the stepwise elimination of linear intermediates is confluent, so that for any given reaction network, a unique normal form will be obtained independently of the elimination order. If confluence would hold, one could compare reaction networks for equivalence, by computing and comparing their normal forms. Furthermore, the unique normal form would be the natural target for simultaneous reduction methods such as [18, 19]. Indeed, a confluence statement was claimed in Sect. 5 of [19] (for the case without conservation laws), but without proof.

We first contribute a counter example which shows that the elimination of linear intermediates on the same network may lead to normal forms with different graph structure. This example contradicts the confluence statement from [19]. The problem is that different "dependent reactions" may be introduced in different elimination orders. We then propose a rewrite rule that eliminates such dependent reactions and prove that the extended rewrite system is confluent up to kinetic rates, so that all normal forms of a same network will have the same structure. This yields a method to eliminate linear intermediates from a reaction graph in a unique manner, while no uniqueness result was stated in [17,18]. However, the kinetic rates may still not be unique, even not modulo the usual axioms of arithmetics. This might seem surprising given that the ODEs of these normal forms are equal modulo these axioms. Finally, we present an example reaction network from systems biology for the failure of confluence with respect to kinetic rates, that we found in the BioModels SBML database [8] with an implementation of our rewrite rules.

Our positive confluence result shows that the graph structure of reaction networks after intermediate and dependency reduction is unique, and thus potentially meaningful biologically. The two negative confluence results show that the situation may be different without dependency reduction, and also for the kinetic rates that can be assigned to the reactions of the reduced network.

All proofs and missing parts are available in the Appendix of the long version.

2   Confluence Notions

We recall confluence notions and their relationships from the literature.

Let $(S, \sim)$ be a set with an equivalence relation and $\rightarrow \ \subseteq S \times S$ a binary relation. We define $\rightarrow^0 \ = \ \sim$ and $\rightarrow^k \ = \ \rightarrow \circ \rightarrow^{k-1}$ for all $k > 0$. The relation $\rightarrow^* \ = \ \cup_{k \geq 0} \rightarrow^k$ is called the reflexive transitive closure of $\rightarrow$. We write $\rightarrow^\epsilon \ = \ \rightarrow^1 \cup \rightarrow^0$, and $\leftarrow \ = \ \{(s, s') \mid s' \rightarrow s\}$.

Definition 1 (Confluence modulo). *We say that a binary relation $\rightarrow$ on $(S, \sim)$ is* confluent *if $\leftarrow^* \circ \rightarrow^* \ \subseteq \ \rightarrow^* \circ \ ^*\leftarrow$,* locally confluent *if $\leftarrow \circ \rightarrow \ \subseteq \ \rightarrow^* \circ \ ^*\leftarrow$,* strongly confluent *if $\leftarrow \circ \rightarrow \ \subseteq \ \rightarrow^\epsilon \circ \sim \circ \ ^\epsilon\leftarrow$, and* uniformly confluent *if $\leftarrow \circ \rightarrow \ \subseteq \ \sim \cup \, (\rightarrow \circ \sim \circ \leftarrow)$.*

Clearly, uniform confluence implies strong confluence, and strong confluence implies local confluence. It is also folklore that there exist locally confluent relations that are not confluent, while strong confluence implies confluence [7]. Uniform confluence implies for any $s \in S$ that all complete reduction sequences starting with s have the same length [15], which may be ∞ though.

In this paper, we will always use binary relations that are terminating, i.e., for any $s \in S$ there exists a $k \geq 0$ such that $\{s' \mid s \rightarrow^k s'\} = \emptyset$, i.e., the length reduction sequences starting with s is bounded. It is well known that locally confluent and terminating relations are confluent (Newman's lemma).

We say that $\sim$ commutes with $\rightarrow$ if $\sim \circ \rightarrow \ \subseteq \ \rightarrow \circ \sim$.

Lemma 1. *If $\rightarrow$ is confluent for $(S, \sim)$ and commutes with $\sim$, then the relation $\sim \circ \rightarrow \circ \sim$ is confluent for $(S, =_S)$.*

3   Simplification of Systems of Equations

In this section, we recall the definition of arithmetic expressions and ordinary differential equations. It is well known that such systems can be inferred from reaction networks with deterministic semantics and partial steady state assumptions. We will then show how to simplify such systems in a confluent manner by eliminating intermediate variables.

Systems of Equations. Let $\mathbb{R}_+$ be the set of non-negative real numbers, and $\mathbb{N}_0 \subseteq \mathbb{R}_+$ the set of natural numbers including 0. Denote by *Vars* a countable set of variables for functions of type $\mathbb{R}_+ \to \mathbb{R}_+$, and by *Param* a set of parameters. We define the set of *arithmetic expression* as the terms $e, e' \in Expr$ with the following abstract syntax:

$$e, e' \in Expr :: = x \mid k \mid n \mid e + e' \mid e * e' \mid 1/e \mid -e$$

where $x \in Vars$, $k \in Param$, $n \in \mathbb{R}$. In the following, the expression $1/e$ is permitted only if e can never become zero, as explained below. For convenience, we will write ee' for $e * e'$; e/e' for $e * (1/e')$, $e - e'$ for $e + (-e')$ and e^n for $e * \ldots * e$ with n repetitions of e.

We map variables to functions on non-negative real numbers, and parameters to positive numbers (different from 0), which are identified with positive constant functions on non-negative real numbers. Given an assignment $\alpha : (Vars \to (\mathbb{R}_+ \to \mathbb{R}_+)) \cup (Param \to \mathbb{R}_+^*)$, any expression $e \in Expr$ can be interpreted as a function $[\![e]\!]_\alpha : \mathbb{R}_+ \to \mathbb{R}_+$ in the usual way.

A *system of equations* S is a combination of equations and constraints, with some existential variables, defined as follows:

$$S:: = dx/dt = e \mid x = e \mid nzero(e) \mid cst(e) \mid S \wedge S' \mid \exists x.\, S.$$

$dx/dt = e$ is an *ordinary differential equation (ODE)*, and $x = e$ an *arithmetic equation*, for the variable x and with an expression $e \in Expr$. The *non-zero constraint* $nzero(e)$ is satisfied by an assignment α if e is never equal to zero, that is $\forall t.\ [\![e]\!]_\alpha(t) \neq 0$. The *positive constant constraint* $cst(e)$ is satisfied by a variable assignment α if $[\![e]\!]_\alpha$ is a positive constant function. And $\exists x.S$ allows us to *existentially quantify* some variables, that we actually want to remove to simplify S. We denote by $Vars(e)$ the set of variables of an expression e and by $Vars(S)$ the set of free variables of a system S. The *set of solutions* of a system of equations S is the set of assignments on the free variables of S that make S true, that is $sol(S) = \{\alpha \mid [\![S]\!]_\alpha = true\}$.

Example 1. The system of equations in Fig. 1 contains 4 ODEs for the variables $\{x_A, x_B, x_C, x_D\}$, and two arithmetic equations and positive constant constraints for the existentially quantified variables $\tilde{x} = \{x_Y, x_Z\}$.

Similar Systems. We now define a syntactic notion of similarity between systems of equations, so that similar systems will have the same solutions. The *similarity relation* $\sim$ on arithmetic expressions is the least congruence that includes the usual arithmetic axioms of a field: commutativity and associativity of $+$ and $*$, removal of neutral elements 0 in sums and 1 in products, uniqueness and laws of inverses for $-$, distributivity, and simplification of real numbers. Similarity is

$$\exists x_Y, x_Z. \begin{cases} \dfrac{dx_A}{dt} = -(k_1 + k_2)x_A \ \wedge \ \dfrac{dx_B}{dt} = k_3 x_Y \ \wedge \ x_Y = \dfrac{k_1}{k_3 + k_5} x_A + \dfrac{k_6}{k_3 + k_5} x_Z \ \wedge \\ \dfrac{dx_C}{dt} = (k_4 + k_5)x_Y \ \wedge \ \dfrac{dx_D}{dt} = k_6 x_Z \ \wedge \ x_Z = \dfrac{k_2}{k_6} x_A + \dfrac{k_5}{k_6} x_Y \ \wedge \\ cst(x_Y) \hspace{4.2cm} \wedge \ cst(x_Z) \end{cases}$$

Fig. 1. The system of equations $S(N_X)$.

decidable, by rewriting expressions to a fraction of polynomials, with the same denominator, and comparing the numerators.

We always identify arithmetic expressions up to similarity (rather than syntactic equality), i.e., we rewrite modulo $\sim$. Given an assignment α, two similar expressions $e \sim e'$ have trivially the same interpretation $[\![e]\!]_\alpha = [\![e']\!]_\alpha$. The similarity relation is lifted to systems of equations in the obvious manner.

Safe Linear Systems. We will consider only *valid systems of equations* in which there is exactly one arithmetic equation per quantified variable and at most one ODE for all others. We also assume that the systems are linear in the existentially quantified variables as defined below, but not necessarily in the others:

Definition 2. *Given a sequence of variables $\tilde{x} = x_1, \ldots, x_n$, an expression e' is called $\tilde{x}$-linear if e' is similar to some expression $e + \sum_{1 \leq i \leq n} x_i e_i$, where e and e_i do not contain any variables from $\tilde{x}$. We call a system $\exists \tilde{x}.S$ linear (in the quantified variables) if for any quantified variable $x \in Vars(\tilde{x})$, the system S is similar to some system $x = e \wedge S'$ where e is an $\tilde{x}$-linear expression.*

In order to always avoid division by zero during the repeated elimination of quantified variables to come (see Lemmas 2 and 3), we introduce the following safety restriction of linear systems, which will be satisfied most of the time in the applications. Without this restriction, the simplification procedure could be shown to be only partially correct, similarly to [19].

Definition 3. *Let S be a system $\exists x_1, \ldots, x_n. S'$ that is linear in the quantified variables, such that S' has the form $\bigwedge_{1 \leq i \leq n} x_i = e^i + \sum_{1 \leq j \leq n} x_j e^i_j \wedge S''$. We define a set expression $L_{S'}$ in which x and y are fresh variables:*

$$L_{S'} =_{df} \{ (x, y) \mid \bigvee_{1 \leq i, j \leq n} x = x_i \wedge y = x_j \wedge nzero(e^j_i) \}.$$

For any assignment of the free variables in the subexpressions e^i_j, the set expression $L_{S'}$ denotes a binary relation, that we call the linking relation of S'. We call the system S safe if S' entails the following formula:

$$S' \models \bigwedge_{i=1}^{n} \bigvee_{k=1}^{n} L^*_{S'}(x_i, x_k) \wedge nzero(e^k) \wedge (e^i \geq 0 \wedge \bigwedge_{j=1}^{n} e^j_i \geq 0).$$

We denote by SafeLin the set of safe linear systems of equations.

$$\frac{x \notin \mathit{Vars}(e)}{\exists x. \ (S \wedge x = e) \quad \Rightarrow \quad S[x := e]_l} \quad \text{(QUANTIFIED VARIABLE)}$$

Fig. 2. Elimination of an existentially quantified variable x in a system of equations.

Simplifying Safe Linear Systems. We want to simplify safe linear systems of equations by removing existentially quantified variables, while preserving the solutions. To do that, given an expression $x = e$ for a quantified variable x, we will substitute x by e, as described in the simplification rule in Fig. 2.

A *substitution* $[x_1 := e]$ is the replacement of any occurrences of x_1 by the expression e. Additionally, we also want to preserve the linearity and safety. Therefore, we define a *linear substitution*, that rewrites arithmetic expressions into linear ones after the substitution. Formally, given a $\tilde{x}$-linear expression $e \sim e^1 + x_2 e_2^1 + \sum_{3 \leq i \leq n} x_i e_i^1$ and an equation $E_2 = (x_2 = e^2 + x_1 e_1^2 + \sum_{3 \leq i \leq n} x_i e_i^2)$, with $\tilde{x} = \{x_1, \ldots, x_n\}$, the linear substitution of x_1 by e in E_2 is:

$$E_2[x_1 := e]_l = (x_2 = \frac{e^1 e_1^2 + e^2}{1 - e_1^2 e_2^1} + \sum_{3 \leq i \leq n} x_i \frac{e_i^1 e_1^2 + e_i^2}{1 - e_1^2 e_2^1}) \wedge \mathit{nzero}(1 - e_1^2 e_2^1)$$

The idea is to a) substitute x_1 by e in the equation of x_2, b) bring the factor $e_1^2 e_2^1 x_2$ from the right to the left, c) factorize the x_2, and d) divide by the factor $1 - e_1^2 e_2^1$ of x_2 we obtained.

Lemma 2. *If S is safe and with the above equations then $S \models \mathit{nzero}(1 - e_1^2 e_2^1)$.*

We define $S[x_1 := e]_l$ by replacing x_1 by e in the ODEs and the constraints of S and by performing the linear substitution as above to all nondifferential equations of S. The relation $S \Rightarrow S'$ defined in Fig. 2 simplifies a safe linear system S to S': a quantified variable is eliminated by applying a linear substitution.

Lemma 3. *The simplification of a safe linear system is a safe linear system.*

Lemma 4. *The simplification preserves the solutions of safe linear systems: if $S \Rightarrow S'$, then $sol(S) = sol(S')$.*

Example 2. For instance, in the system from Example 1, we can substitute the intermediate variable x_Y by $e = \frac{k_1}{k_3 + k_5} x_A + \frac{k_6}{k_3 + k_5} x_Z$. Since we still have the constraint $cst(x_Z)$, the constraint $cst(e)$ can be simplified into $cst(x_A)$. The never-zero constraint $\mathit{nzero}(1 - \frac{k_5 k_6}{(k_3 + k_5) k_6})$ is similar to $\mathit{nzero}((k_3 + k_5) k_6 - k_5 k_6)$ and then $\mathit{nzero}(k_3 k_6)$, and therefore is always true , and can be removed. We obtain the system depicted in Fig. 3 (left). By doing the same with the variable x_Z, we obtain the system in Fig. 3 (right). Note that we used the fact that $k_6/k_6 \sim 1$, that is always true, since parameters are assigned to positive numbers.

For safe linear systems, this simplification modulo similarity is confluent, implying that whatever the order adopted for the elimination of quantified variables, it is always possible to find the same fully simplified system, modulo similarity. We actually establish uniform confluence, implying that any simplification leading to the fully simplified system will have the same number of steps.

$$\exists x_Z. \begin{cases} \dfrac{dx_A}{dt} = -(k_1 + k_2)x_A \qquad \dfrac{dx_D}{dt} = k_6 x_Z \\[2mm] \dfrac{dx_B}{dt} = \dfrac{k_1 k_3}{k_3 + k_5} x_A + \dfrac{k_3 k_6}{k_3 + k_5} x_Z \\[2mm] \dfrac{dx_C}{dt} = \dfrac{k_1(k_4 + k_5)}{k_3 + k_5} x_A + \dfrac{k_6(k_4 + k_5)}{k_3 + k_5} x_Z \\[2mm] x_Z = \dfrac{k_1 k_5 + k_2 k_3 + k_2 k_5}{k_3 k_6} x_A \\[2mm] cst(x_A) \wedge cst(x_Z) \end{cases} \qquad \begin{cases} \dfrac{dx_A}{dt} = -(k_1 + k_2)x_A \\[2mm] \dfrac{dx_B}{dt} = (k_1 + k_2)x_A \\[2mm] \dfrac{dx_C}{dt} = \dfrac{(k_1 + k_2)(k_4 + k_5)}{k_3} x_A \\[2mm] \dfrac{dx_D}{dt} = \dfrac{k_1 k_5 + k_2 k_3 + k_2 k_5}{k_3} x_A \\[2mm] cst(x_A) \end{cases}$$

Fig. 3. Simplifications of $S(N_X)$.

Theorem 1. *The binary relation $\Rightarrow$ on $(SafeLin, \sim)$ is uniformly confluent.*

4   Reaction Networks

In this section, we introduce reaction networks, intermediate species, and the interpretation of a network as a system of equations.

Let *Spec* be a countable set of *molecular species* ranged over by A. We associate to each species A a *concentration variable* x_A, and denote the set of these variables by $Vars = \{x_A \mid A \in Spec\}$. A kinetic expression is a non-negative arithmetic expression on variables $Vars$, i.e. for any non-negative assignment α for the concentrations, $[\![e]\!]_\alpha(t) \geq 0$ for all t.

We define a *(chemical) solution* $s \in Sol : Spec \rightarrow \mathbb{N}_0$ as a multiset of molecular species, i.e. a function from species to natural numbers, with finite support. Given numbers $n_1, \ldots, n_k$, we denote by $n_1 A_1 + \cdots + n_k A_k$ the solution that contains n_i molecules of species A_i for $1 \leq i \leq k$, and 0 molecules of others species. Given $s_1, s_2 \in Solutions$, their intersection is defined for any A by $(s_1 \cap s_2)(A) = min(s_1(A), s_2(A))$. A *kinetic reaction* $r = (s_1 \rightarrow s_2; e)$ is a pair composed of a *reaction* $s_1 \rightarrow s_2$ and a kinetic expression $e \in Expr$. The reaction transforms the solution s_1, called *reactants*, into the solution s_2, called *products*. The reaction vector vr_r of the reaction r is defined for any $A \in Spec$ by $vr_r(A) = s_2(A) - s_1(A)$. We denote by $kin(r) = e$ the kinetic expression of r.

Given a reaction $r = (s_1 \rightarrow s_2; e)$ and the solution $s = s_1 \cap s_2$, the *normalization* of r is the reaction $(s_1 - s \rightarrow s_2 - s; e)$. In the following, we always assume that every reaction is normalized, and normalization is implicitly applied after every simplification. A *reaction network* N is composed of normalized kinetic reactions, constraints, and bound species (that we want to remove):

$$N ::= r \mid cst(e) \mid N \wedge N' \mid \exists X. N$$

We assume the usual structural congruence rules for conjunction and existential quantification. We denote by $C(N)$ the set of constraints of N.

Once again, we need to add some conditions on the bound species, called *intermediate species*, in order to be able to fully remove them in a confluent way. We usually denote by $\mathcal{U}$ the intermediate species, and by $\bar{\mathcal{U}}$ the other species.

Given a set $\mathcal{U}$ of molecules, and a reaction $r = (s_1 \rightarrow s_2; \ e)$, we define the consumption $Cons_{\mathcal{U}}(r) = s_1 \cap \mathcal{U}$ (resp. production $Prod_{\mathcal{U}}(r) = s_2 \cap \mathcal{U}$) of r with respect to $\mathcal{U}$ as the molecules of $\mathcal{U}$ that are consumed (resp. produced) by r.

A molecule $X \in \mathcal{U}$ is *output-connected* (resp. *input-connected*) in N with respect to $\mathcal{U}$ if $\exists r \in N$ with $Cons_{\mathcal{U}}(r) = \{X\}$ (resp. $Prod_{\mathcal{U}}(r) = \{X\}$) and either $Prod_{\mathcal{U}}(r) = \emptyset$ (resp. $Cons_{\mathcal{U}}(r) = \emptyset$), or $Prod_{\mathcal{U}}(r) = \{Y\}$ (resp. $Cons_{\mathcal{U}}(r) = \{Y\}$) with Y output-connected (resp. input-connected). This property will correspond to the safety property of quantified variables in linear systems of equations.

A reaction network $\exists \mathcal{U}. \ N$ is *linear* if the following properties hold:

- connectivity: for any $X \in \mathcal{U}$, X is output and input-connected in N,
- $\mathcal{U}$-stoichiometry: $\forall r \in N$, $|Cons_{\mathcal{U}}(r)| \leq 1$ and $|Prod_{\mathcal{U}}(r)| \leq 1$,
- $\mathcal{U}$-linearity: $\forall r \in N. \ Cons_{\mathcal{U}}(r) = \{X\} \Rightarrow kin(r) = x_X e$, with $\forall Y \in \mathcal{U}.x_Y \notin e$,
- kinetic non-interaction: $\forall r \in N$, $Cons_{\mathcal{U}}(r) = \emptyset$ and $Prod_{\mathcal{U}}(r) \neq \emptyset$ implies $x_X \notin kin(r)$ for any $X \in \mathcal{U}$,
- partial steady-state: $\forall X \in \mathcal{U}$, $cst(x_X) \in C(N)$.

In the following, we will only consider linear networks, and denote by *Nets* the set of linear reaction networks.

Given a linear network $N \in Nets$, we can define the interpretation of N in terms of a system of equations $S(N)$, as described in Fig. 4.

Lemma 5. *For any $N \in Nets$, the interpretation $S(N)$ is a (valid) safe linear system.*

Example 3. We consider the reaction network N_X in Fig. 5, with the reactions on the left and the reaction graph on the right. The set of species is $\{A, B, C, D, Y, Z\}$, where Y and Z are considered intermediates, and the set of

$$\exists \tilde{x}. \ \left[\frac{dx_A}{dt} = \sum_{r \in N} vr_r(A) kin(r) \right]_{A \in \bar{\mathcal{U}}} \wedge \left[x_X = \frac{\sum\limits_{\{r \in N | X \in Prod(r)\}} kin(r)}{\sum\limits_{\{r \in N | X \in Cons(r)\}} kin(r)/x_X} \right]_{X \in \mathcal{U}} \wedge C(N)$$

Fig. 4. Definition of the system of equations $S(N)$, for the network N, with intermediate species $\mathcal{U}$ and with $\tilde{x} = \{x_X \mid X \in \mathcal{U}\}$.

$\exists Y, Z. \ cst(x_Y) \qquad \wedge cst(x_Z)$

$r_1 = (A \rightarrow Y; \ k_1 x_A) \quad r_4 = (0 \rightarrow C; \ k_4 x_Y)$

$r_2 = (A \rightarrow Z; \ k_2 x_A) \quad r_5 = (Y \rightarrow Z + C; \ k_5 x_Y)$

$r_3 = (Y \rightarrow B; \ k_3 x_Y) \quad r_6 = (Z \rightarrow Y + D; \ k_6 x_Z)$

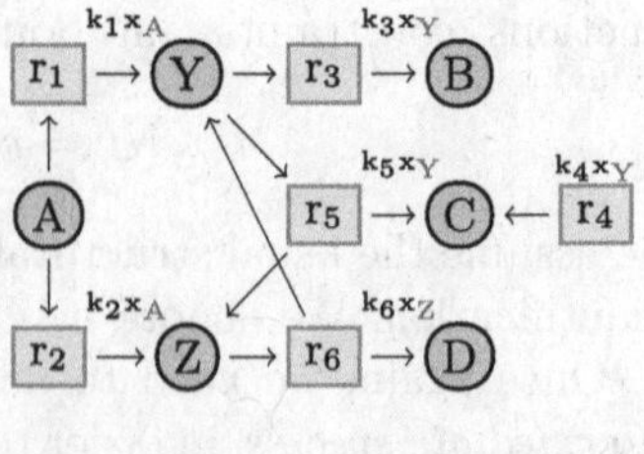

Fig. 5. The reaction network N_X.

$$\dfrac{e = \displaystyle\sum_{\{r \in N | X \in Prod(r)\}} kin(r) \qquad e' = \displaystyle\sum_{\{r \in N | X \in Cons(r)\}} kin(r) \qquad e' = x_X e''}{\bigwedge_{\{r, r' \in N | X \in Prod(r) \cap Cons(r')\}} r \diamond_{e'} r'} \quad (\text{INTERMEDIATE})$$

$$\exists X. \ N \Rightarrow^{\text{Inter}} \bigwedge_{\{r \in N | X \notin Prod(r) \cup Cons(r)\}} r[x_X := \dfrac{e}{e''}]$$
$$\wedge \, C(N)[x_X := \dfrac{e}{e''}]$$

Fig. 6. Intermediate simplification rule, with $C(N)$ the constraints of N.

reactions is $\{r_1, \ldots, r_6\}$. The parameters in the rates are some positive reals $k_1, \ldots, k_6$. All reactions have mass action kinetics, except for reaction r_4 which is activated by Y. Its associated system is $S(N_X)$, described in Example 1.

Given a network N, we can compute its system of equations $S(N)$, and then simplify it in a confluent way, as explained in Sect. 3. But we might sometimes be more interested in the network itself, rather than its system of equations and unfortunately, rebuilding a reaction network from the equations can be difficult, and the network obtained is not unique [3]. It seems then more appropriate to proceed with the simplification directly on the reaction network.

5   Elimination of Intermediate Species

In this section, we introduce the Intermediate simplification rule for reaction networks, and apply it to an example.

The (INTERMEDIATE) rule presented in Fig. 6 aims at removing an intermediate species $X \in \mathcal{U}$: any reaction r_{prod} that produces X is combined with any reaction r_{cons} that consumes X, and x_X is replaced by its value at steady state in the other reactions. This merging operation is achieved by the operator $\diamond_e$:

$$(s_1 \rightarrow s_2; \ e) \diamond_{e''} (s_1' \rightarrow s_2'; \ e') = (s_1 + s_1' \rightarrow s_2 + s_2'; \ \dfrac{ee'}{e''}).$$

Since we only consider normalized reactions, in merged reactions the intermediate molecule is implicitly discarded.

The interpretation $S(N)$ is a simulation from $(Nets, \Rightarrow^{\text{Inter}})$ to $(SafeLin, \Rightarrow)$:

Lemma 6. *Given a network $N \in Nets$, if $N \Rightarrow^{Inter} N'$, then $S(N) \Rightarrow S(N')$.*

This implies as expected that both a network and its simplification have the same deterministic dynamics.

The next example shows that the rewriting system given by the elimination of intermediate species alone is not confluent, given that different dependent reactions may be produced for different elimination orders.

Example 4. Starting from network N_X from Fig. 5, we can either remove Y or Z and obtain the networks depicted in Fig. 7. If we first remove Z, then we obtain

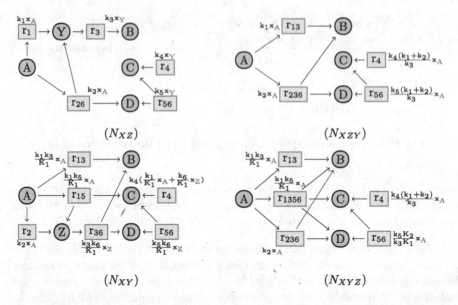

Fig. 7. Two elimination strategies to simplify N_X of Fig. 5: either first eliminate Z to obtain the network N_{XZ} and then Y to obtain N_{XZY}, or swap the elimination order to obtain first N_{XY} and then N_{XYZ}. Simplified networks N_{XZY} and N_{XYZ} are structurally different since the latter has the additional reaction r_{1356}. The new parameters are $K_1 = k_3 + k_5$ and $K_2 = k_1 k_5 + k_2 K_1$.

the reaction network N_{XZ}. From N_{XZ} we can eliminate the intermediate Y and obtain N_{XZY}. This network cannot be simplified any further. Alternatively, we can eliminate Y from N_X in a first step, obtain N_{XY}, and then remove Z and obtain the network N_{XYZ}.

Unfortunately, N_{XYZ} and N_{XZY} do not have the same structure, since N_{XYZ} has an additional reaction r_{1356}, which is a combination of r_{13} and r_{56}. Such dependent reactions can be removed, as we will show in the next section.

6   Elimination of Dependent Reactions

In this section we clarify the notion of dependency between reactions, and introduce an additional simplification rule based on this notion. The addition of this rule is sufficient to establish confluence for the structure of simplified networks. However, we will show that this modification is not enough, in general, to guarantee full confluence.

We formalize the notion of dependency with respect to an initial set of reactions with the notion of *flux*. Flux vectors at steady state are a standard tool for computing elementary modes [5], that correspond to the unique set of reactions in the network normal form that we obtain with the techniques of this paper. Our simplification method, unlike the elementary modes approach, deals with the impact of the simplification on kinetic rates as well as the network structure.

Given an ordered set of m reactions $\mathcal{R} = \{r_1, \ldots, r_m\}$ called *reaction basis*, a flux is a pair $w = (v; e)$ of a *flux vector* $v \in \mathbb{R}^m$ and an expression $e \in Expr$. The function $react_{\mathcal{R}}$ maps fluxes to reactions w.r.t. a reaction basis $\mathcal{R}$ as follows:

$$react_{\mathcal{R}}(v; e) = \left(\sum_{1 \le i \le m} v_i s_i \to \sum_{1 \le i \le m} v_i s_i'; \ e \right).$$

Consequently, the i-th vector u_i of the standard basis is mapped to the i-th reaction r_i of the reaction basis $\mathcal{R}$. Now, instead of simplifying reaction networks, we directly simplify *flux networks* W defined as reaction networks but with fluxes in place of reactions:

$$W ::= w \mid cst(e) \mid W \wedge W' \mid \exists X. \ W.$$

We lift $react_{\mathcal{R}}$ to map flux networks to reaction networks as follows:

$$react_{\mathcal{R}}(cst(e)) = cst(e), \quad react_{\mathcal{R}}(W \wedge W') = react_{\mathcal{R}}(W) \wedge react_{\mathcal{R}}(W'),$$
$$react_{\mathcal{R}}(\exists X. \ W) = \exists X. \ react_{\mathcal{R}}(W).$$

We denote $FNets_{\mathcal{R}}$ the set of flux networks W such that $react_{\mathcal{R}}(W)$ is a linear reaction network for $\mathcal{U}$. The interpretation of $W \in FNets_{\mathcal{R}}$ in terms of system of equations is defined as $S_{\mathcal{R}}(W) = S(react_{\mathcal{R}}(W))$. Finally, we translate some previous definitions to the context of flux networks:

$$Prod_{\mathcal{R}}(w) = Prod(react_{\mathcal{R}}(w)), \qquad Cons_{\mathcal{R}}(w) = Cons(react_{\mathcal{R}}(w)),$$
$$kin(v; e) = e, \qquad (v; e) \diamond_{e''} (v'; e') = (v + v'; \tfrac{ee'}{e''}).$$

We then define two simplification rules for flux networks in Fig. 8. First, (INTERMEDIATE) is simply a reformulation of the one in Fig. 6 but in the terminology of flux networks. The new rule (DEPENDENT) removes a *dependent flux*, that is one whose flux vector can be written as a positive linear combination of the flux vectors of some other fluxes. The rate of the removed flux is added to the rate of the fluxes that it depends on. This guarantees that the system of ordinary differential equations associated to the reaction network is unchanged:

$$\frac{e = \sum_{\{w \in W \mid X \in Prod_{\mathcal{R}}(w)\}} kin(w) \quad e' = \sum_{\{w \in W \mid X \in Cons_{\mathcal{R}}(w)\}} kin(w) \quad e' = x_X e''}{\bigwedge_{\{w, w' \in W \mid X \in Prod_{\mathcal{R}}(w) \cap Cons_{\mathcal{R}}(w')\}} w \diamond_{e'} w'} \text{(INTERMEDIATE)}$$

$$\exists X. \ W \Rightarrow_{\mathcal{R}}^{\text{Inter}} \bigwedge_{\{w \in W \mid X \notin Prod_{\mathcal{R}}(w) \cup Cons_{\mathcal{R}}(w)\}} w[x_X := \frac{e}{e''}]$$
$$\wedge C(W)[x_X := \frac{e}{e''}]$$

$$\frac{}{W \bigwedge_{1 \le i \le k}(v_i, e_i) \wedge \left(\sum_{1 \le i \le k} n_i v_i, e\right) \Rightarrow_{\mathcal{R}}^{\text{Dep}} W \bigwedge_{1 \le i \le k}(v_i, e_i + n_i e)} \text{(DEPENDENT)}$$

Fig. 8. Simplification rules of flux networks.

Lemma 7. *Given $W \in FNets_{\mathcal{R}}$, if $W \Rightarrow_{\mathcal{R}}^{Dep} W'$, then $S_{\mathcal{R}}(W) \sim S_{\mathcal{R}}(W')$.*

Two fluxes are *structurally similar*, denoted $(v, e) \sim^{struc} (v', e')$, if they have the same flux vector, that is $v = v'$. Two vector networks are *structurally similar*, denoted $W \sim^{struc} W'$ if they have structurally similar fluxes.

We can now state the Theorem on the structural confluence for this simplification system. We denote by $\Rightarrow_{\mathcal{R}} = (\Rightarrow_{\mathcal{R}}^{Inter} \cup \Rightarrow_{\mathcal{R}}^{Dep})$ the simplification of vector networks with the rules of Fig. 8.

Theorem 2. *The relation $\Rightarrow_{\mathcal{R}}$ on $(FNets_{\mathcal{R}}, \sim^{struc})$ is confluent.*

Proof (Scetch). The outline of the proof is as follows:

1. the simplification relation $\Rightarrow_{\mathcal{R}}$ preserves the set of intermediate species,
2. the local confluence holds for $\Rightarrow_{\mathcal{R}}$,
3. the binary relation is terminating, so by Newman's lemma, it is confluent.

Note that adding a rule that eliminates reactions whose reaction vectors can be written as sums of the reaction vectors of other reactions in the same network (instead of using a reaction basis) does not guarantee the confluence for the network structure.

Example 5. In Example 4, the elimination of the intermediates Y and Z in two different orders was shown to generate two different networks N_{XZY} and N_{XYZ} the latter having the additional reaction r_{1356}. Let us take $\{r_1, \ldots, r_6\}$ as a reaction basis. If we translate the simplifications to flux networks, the flux vector associated to reaction r_{ij} is $u_i + u_j$. Also, the flux vector associated to r_{1356} is $u_1 + u_3 + u_5 + u_6$, that is the sum of the flux vectors of r_{13} and r_{56}. Thus, the application of the (DEPENDENT) rule to the flux associated to r_{1356} results in a flux network W such that $react_{\mathcal{R}}(W) = N'_{XYZ}$. Since r_{1356} is eliminated, the networks N_{XZY} and N'_{XYZ} have the same structure. The rate of reaction r_{13} in N'_{XYZ} is given by the rate of r_{13} in N_{XYZ}, plus the rate of r_{1356} in N_{XYZ}, and is therefore equal to $\frac{k_1 k_3}{K_1} x_A + \frac{k_1 k_5}{K_1} x_A \sim k_1 x_A$, that is the rate of r_{13} in N_{XZY}. Similarly, one can show that the rates of r_{56} in the two networks also coincide, and both networks have the same kinetics.

The following variation on the same example shows that confluence of the kinetics is not in general guaranteed.

Example 6. Now we shall examine again the simplifications performed in Example 4, but this time we look at the reaction networks as simplifications of the larger network N_ϵ in Fig. 9 from which N_X results after elimination of X. The reaction basis is now $\mathcal{R}' = \{r_{1'}, r_{2'}, r_3, r_{4'}, r_{5'}, r_6\}$ and the reaction r_1 in N_X is obtained from N_ϵ by merging $r_{1'}$ and $r_{2'}$ (that, following our convention, we denote $r_{1'2'}$) and is thus associated to the flux $(u_1 + u_2, k_1 x_A)$ w.r.t. $\mathcal{R}'$. Similarly, $r_2 = r_{1'4'}$ is associated to $(u_1 + u_4, k_2 x_A)$, $r_4 = r_{2'5'}$ to $(u_2 + u_5, k_4 x_Y)$, and $r_5 = r_{4'5'}$ to $(u_4 + u_5, k_5 x_Y)$.

The eliminations of Z first and Y after, represented in Fig. 7, generate the reactions r_{26}, r_{56}, r_{13} and r_{236} (with flux vectors respectively $u_1 + u_4 + u_6$,

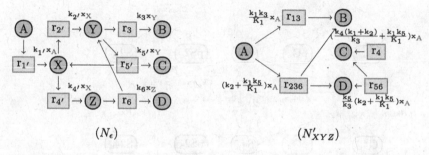

(N_ϵ) (N'_{XYZ})

Fig. 9. Initial network (N_ϵ) and network (N'_{XYZ}) obtained after elimination of X, Y, dependent rule r_{15} and then Z. We have $K_1 = k_3 + k_5$.

$u_4 + u_5 + u_6$, $u_1 + u_2 + u_3$ and $u_1 + u_3 + u_4 + u_6$), with no dependent reactions. Consider now the elimination of Y from N_X. Reaction r_{15} has flux $(u_1 + u_2 + u_4 + u_5, \frac{k_1 k_5}{K_1} x_A)$ in network N_{XY} and is dependent on reactions r_2 and r_4. If we choose to eliminate reaction r_{15} using the (DEPENDENT) rule and apply the (INTERMEDIATE) rule on Z we obtain the network N'_{XYZ} in Fig. 9. No further simplification rule can be applied. Notice that this network is structurally the same as network N_{XZY} in Fig. 7, but all reactions have different kinetic rates.

7   Normalization Modulo Kinetic Rates

We now present the principal result of this paper, about confluence of the simplification system modulo the kinetic rates. In other words, whatever the order of simplification, we can always obtain a fully simplified network with the same structure and with similar system of equations, but the kinetic rates associated to the fluxes can be different, as illustrated before in the Example 6.

Given a fixed set of intermediate species $\mathcal{U}$ and an initial reaction basis $\mathcal{R}$, two networks are *similar*, denoted $W \sim_\mathcal{R} W'$, if they are structurally similar $(W \sim^{struc} W')$, and their systems of equations are similar $(S_\mathcal{R}(W) \sim S_\mathcal{R}(W'))$.

Theorem 3. *The relation $\Rightarrow_\mathcal{R}$ on $(FNets_\mathcal{R}, \sim_\mathcal{R})$ is confluent.*

8   An Example from the BioModels Database

We have shown that the simplification system that we presented can exhibit non-confluence of the rates, even in a simple scenario with a small number of intermediates. To find if such a situation occurs in practice, we investigated the SBML models in the curated BioModels database [8]. For each mass-action model, we created the graph of complexes and searched it for cycles of intermediates, to identify possible candidates for non-confluence. Then, with an implementation

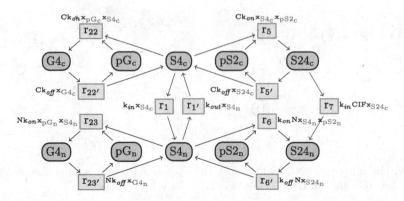

Fig. 10. Subnetwork of the Smad signal transduction network in [20].

of the simplification rules, we considered the elimination of triples or quadruples of intermediates in different orders, and compared the resulting networks.

We were thus able to identify two different reduced networks for model BIOMD0000000173. This is a model of the Smad-based signal transduction mechanisms from the cell membrane to the nucleus, presented in [20]. A subnetwork of this model is represented in Fig. 10. It includes all reactions involving cytoplasmic and nuclear Smad4 and Smad2/Smad4 complexes (abbreviated $S4_c$, $S4_n$, $S24_c$ and $S24_n$): shuttling of Smad4, formation of Smad2/Smad4 complex, import of Smad2/Smad4 into the nucleus, and formation of EGFP-Smad2/Smad4 complex. This network is linear for the four intermediate species $S4_c$, $S4_n$, $S24_c$, $S24_n$. The different orders of elimination yield simplified networks with the same structure but different kinetics. This confirms that the or of simplifying a biological network may indeed affect the result.

Conclusion

We have shown that the elimination of linear intermediate species is not confluent in general. We provided a new simplification rule to remove dependent reactions, and proved that the extended rewrite system is confluent up to kinetic rates, that is, all normal forms of the same network will have the same structure and similar systems of equations, but can have different kinetic rates. Future research efforts is needed to characterize networks that possess a unique normal form.

References

1. Calzone, L., Fages, F., Soliman, S.: BIOCHAM: an environment for modeling biological systems and formalizing experimental knowledge. Bioinformatics **22**(14), 1805–1807 (2006)
2. Chaouiya, C.: Petri net modelling of biological networks. Briefings Bioinf. **8**(4), 210–219 (2007)

3. Fages, F., Gay, S., Soliman, S.: Inferring reaction systems from ordinary differential equations. Theor. Comput. Sci. **599**, 64–78 (2015)
4. Feinberg, M.: Chemical reaction network structure and the stability of complex isothermal reactors-i. The deficiency zero and deficiency one theorems. Chem. Eng. Sci. **42**(10), 2229–2268 (1987)
5. Gagneur, J., Klamt, S.: Computation of elementary modes: a unifying framework and the new binary approach. BMC Bioinf. **5**(1), 175 (2004)
6. Hucka, M., et al.: The systems biology markup language (SBML): a medium for representation and exchange of biochemical network models. Bioinformatics **19**(4), 524–531 (2003)
7. Huet, G.P.: Confluent reductions: abstract properties and applications to term rewriting systems. J. ACM **27**(4), 797–821 (1980)
8. Juty, N., Ali, R., Glont, M., Keating, S., Rodriguez, N., Swat, M.J., Wimalaratne, S.M., Hermjakob, H., Le Novère, N., Laibe, C., Chelliah, V.: BioModels: Content, Features, Functionality and Use. CPT Pharmacometrics Syst. Pharmacol. **4**, 55–68 (2015)
9. King, E.L., Altman, C.: A schematic method of deriving the rate laws for enzyme-catalyzed reactions. J. Phys. Chem. **60**(10), 1375–1378 (1956)
10. Kuo-Chen, C., Forsen, S.: Graphical rules of steady-state reaction systems. Can. J. Chem. **59**(4), 737–755 (1981)
11. Kuttler, C., Lhoussaine, C., Nebut, M.: Rule-based modeling of transcriptional attenuation at the tryptophan operon. Trans. Comput. Syst. Biol. **XII**, 199–228 (2010)
12. Madelaine, G., Lhoussaine, C., Niehren, J.: Attractor equivalence: an observational semantics for reaction networks. In: Fages, F., Piazza, C. (eds.) FMMB 2014. LNCS, vol. 8738, pp. 82–101. Springer, Heidelberg (2014)
13. Madelaine, G., Lhoussaine, C., Niehren, J.: Structural Simplification of Chemical Reaction Networks Preserving Deterministic Semantics. In: Roux, O., Bourdon, J. (eds.) CMSB 2015. LNCS, vol. 9308, pp. 133–144. Springer, Heidelberg (2015)
14. Mäder, U., Schmeisky, A.G., Flórez, L.A., Stülke, J.: Subtiwiki-a comprehensive community resource for the model organism bacillus subtilis. Nucleic Acids Res. **40**, 1278–1287 (2012)
15. Niehren, J.: Uniform confluence in concurrent computation. J. Funct. Program. **10**(5), 453–499 (2000)
16. Niehren, J., John, M., Versari, C., Coutte, F., Jacques, P.: Qualitative reasoning for reaction networks with partial kinetic information. In: Roux, O., Bourdon, J. (eds.) CMSB 2015. LNCS, vol. 9308, pp. 157–169. Springer, Heidelberg (2015)
17. Radulescu, O., Gorban, A., Zinovyev, A., Lilienbaum, A.: Robust simplifications of multiscale biochemical networks. BMC Syst. Biol. **2**(1), 86 (2008)
18. Radulescu, O., Gorban, A.N., Zinovyev, A., Noel, V.: Reduction of dynamical biochemical reactions networks in computational biology. Frontiers in Genetics (2012)
19. Sáez, M., Wiuf, C., Feliu, E.: Graphical reduction of reaction networks by linear elimination of species. arXiv preprint 2015. arXiv:1509.03153
20. Schmierer, B., Tournier, A.L., Bates, P.A., Hill, C.S.: Mathematical modeling identifies Smad nucleocytoplasmic shuttling as a dynamic signal-interpreting system. Proc. Natl. Acad. Sci. **105**(18), 6608–6613 (2008)
21. Tonello, E., Owen, M.R., Farcot, E.: On the elimination of intermediate species in chemical reaction networks (2016, in preparation)

Fast Simulation of Probabilistic Boolean Networks

Andrzej Mizera[1], Jun Pang[1,2], and Qixia Yuan[1(✉)]

[1] Faculty of Science, Technology and Communication, University of Luxembourg,
Luxembourg City, Luxembourg
{andrzej.mizera,jun.pang,qixia.yuan}@uni.lu
[2] Interdisciplinary Centre for Security, Reliability and Trust,
University of Luxembourg, Luxembourg City, Luxembourg

Abstract. As an important mathematical modelling framework, probabilistic Boolean networks (PBNs) are widely used for modelling and analysing biological systems. PBNs are suited for modelling large biological systems, which more and more often arise in systems biology. However, the large system size poses a significant challenge to the analysis of PBNs, in particular, to the crucial analysis of their steady-state behaviour. Numerical methods for performing steady-state analyses suffer from the state-space explosion problem, which makes the utilisation of statistical methods the only viable approach. However, such methods require long simulations of PBNs, rendering the simulation speed a crucial efficiency factor. For large PBNs and high estimation precision requirements, a slow simulation speed becomes an obstacle. In this paper, we propose a structure-based method for fast simulation of PBNs. This method first performs a network reduction operation and then divides nodes into groups for parallel simulation. Experimental results show that our method can lead to an approximately 10 times speedup for computing steady-state probabilities of a real-life biological network.

1 Introduction

Systems biology aims to model and analyse biological systems from a holistic perspective in order to provide a comprehensive, system-level understanding of cellular behaviour. Computational modelling of a biological system plays a key role in systems biology. It connects the field of traditional biology with mathematics and computational science, providing a way to organize and formalize available biological knowledge in a mathematical model and to identify missing biological information using formal means. Together with biochemical techniques, computational modelling promotes the holistic understanding of real-life biological systems, leading to the study of large biological systems. This brings a significant challenge to computational modelling in terms of the state-space size of the system under study. Among the existing modelling frameworks, probabilistic Boolean networks (PBNs) is well-suited for modelling large-size biological systems. It is first introduced by Shmulevich et al. [7,13] as a probabilistic

Q. Yuan—Supported by the National Research Fund, Luxembourg (grant 7814267).

E. Bartocci et al. (Eds.): CMSB 2016, LNBI 9859, pp. 216–231, 2016.
DOI: 10.1007/978-3-319-45177-0_14

generalisation of the standard Boolean networks (BNs) to model gene regulatory networks (GRNs). The framework of PBNs incorporates rule-based dependencies between genes and allows the systematic study of global network dynamics; meanwhile, it is capable of dealing with uncertainty, which naturally occurs at different levels in the study of biological systems.

Focusing on the wiring of a network, PBNs is essentially designed for revealing the long-run (steady-state) behaviour of a biological system. Comprehensive understanding of the long-run behaviour is vital in many contexts. For example, attractors of a GRN are considered to characterise cellular phenotypes [1]. There have been a lot of studies in analysing the long-run behaviour of biological systems for better understanding the influences of genes or molecules in the systems [10]. Moreover, steady-state analyses have been used in gene intervention and external control [9,11], which is of special interest to cancer therapists to predict the potential reaction of a patient to treatment. In the context of PBNs, many efforts have been devoted to computing their steady-state probabilities. In [6,12], efficient numerical methods are provided for computing the steady-state probabilities of small-size PBNs. Those methods utilise an important characteristics of PBNs, i.e., a PBN can be viewed as a discrete-time Markov chain (DTMC) and its dynamics can be studied with the use of the rich theory of DTMCs. The key idea of those methods relies on the computation of the transition matrix of the underlying DTMC of the studied PBN. They perform well for small-size PBNs. However, in the case of large-size PBNs, the state-space size becomes so huge that the numerical methods are not scalable any more.

Many efforts are then spent on addressing the challenge of the huge state-space in large-size PBNs. In fact, the use of statistical methods and Monte Carlo methods remain the only feasible approach to address the problem. In those methods, the simulation speed is an important factor in the performance of these approaches. For large PBNs and long trajectories, a slow simulation speed could render these methods infeasible as well. In our previous work [3], we have considered the two-state Markov chain approach and the Skart method for approximate analysis of large PBNs. Taking special care of efficient simulation, we have implemented these two methods in the tool ASSA-PBN [2] and successfully used it for the analysis of large PBNs with a few thousands of nodes. However, the required time cost is still expected to be reduced. This requirement is of great importance for the construction of a model, e.g., parameter estimation, and for a more precise and deep analysis of the system. In this work, we propose a structure-based method to speed up the simulation process. The method is based on analysing the structure of a PBN and consists of two key ideas: first, it removes the unnecessary nodes in the network to reduce its size; secondly, it divides the nodes into groups and performs simulation for nodes in a group simultaneously. We show with experiments that our structure-based method can significantly reduce the computation time for approximate steady-state analyses of large PBNs. To the best of our knowledge, our proposed method is the first one to apply structure-based analyses for speeding up the simulation of a PBN.

2   Preliminaries

2.1   Probabilistic Boolean Networks (PBNs)

A PBN $G(X, F)$ models elements of a biological system with a set of binary-valued nodes $X = \{x_1, x_2, \ldots, x_n\}$. For each node $x_i \in X$, the update of its value is guided by a set of *predictor functions* $F_i = \{f_1^{(i)}, f_2^{(i)}, \ldots, f_{\ell(i)}^{(i)}\}$, where $\ell(i)$ is the number of predictor functions for node x_i. Each $f_j^{(i)}$ is a Boolean function whose inputs are a subset of nodes, referred to as *parent nodes* of x_i. For each node x_i, one of its predictor functions will be selected to update the value of x_i at each time point t. This selection is in accordance with a probability distribution $C_i = (c_1^{(i)}, c_2^{(i)}, \ldots, c_{\ell(i)}^{(i)})$, where the individual probabilities are the *selection probabilities* for the respective elements of F_i and they sum to 1. The value of node x_i at time point t is denoted as $x_i(t)$ and the state of the PBN at time point t is denoted as $s(t) = (x_1(t), x_2(t), \ldots, x_n(t))$. The state space of the PBN is $S = \{0, 1\}^n$ and it is of size 2^n. There are several variants of PBNs with respect to the selection of predictor functions and the synchronisation of nodes update. In this paper, we consider the *independent synchronous* PBNs, i.e., the choice of predictor functions for each node is made independently and the values of all the nodes are updated synchronously. The transition from state $s(t)$ to state $s(t + 1)$ is performed by randomly selecting a predictor function for each node x_i from F_i and by applying those selected predictor functions to update the values of all the nodes synchronously. We denote $f(t)$ the combination of all the selected predictor functions at time point t. The transition of state $s(t)$ to $s(t + 1)$ can then be denoted as $s(t + 1) = f(t)(s(t))$.

Perturbations of a biological system are introduced by a perturbation rate $p \in (0, 1)$ in a PBN. The dynamics of a PBN is guided with both perturbations and predictor functions: at each time point t, the value of each node x_i is flipped with probability p; and if no flip happens, the value of each node x_i is updated with selected predictor functions synchronously. Let $\gamma(t) = (\gamma_1(t), \gamma_2(t), \ldots, \gamma_n(t))$, where $\gamma_i(t) \in \{0, 1\}$ and $\mathbb{P}(\gamma_i(t) = 1) = p$ for all t and $i \in \{1, 2, \ldots, n\}$. The transition of $s(t)$ to $s(t+1)$ in PBNs with perturbations is given by $s(t + 1) = s(t) \oplus \gamma(t)$ if $\gamma(t) \neq 0$, and $s(t + 1) = f(t)(s(t))$ otherwise, where $\oplus$ is the element-wise *exclusive or* operator for vectors. According to this equation, perturbations allow the system to move from a state to any other state in one transition, hence render the underlying Markov chain irreducible and aperiodic. Thus, the dynamics of a PBN with perturbations can be viewed as an ergodic DTMC [7]. Based on the ergodic theory, the long-run dynamics of a PBN with perturbations is governed by a unique limiting distribution, convergence to which is independent of the choice of the initial state.

The density of a PBN is measured with its predictor function number and parent nodes number. For a PBN G, its density is defined as $\mathcal{D}(G) = \frac{1}{n} \sum_{i=1}^{N_F} \phi(i)$, where n is the number of nodes in G, N_F is the total number of predictor functions in G, and $\phi(i)$ is the number of parent nodes for the ith predictor function.

2.2 Simulating a PBN

A PBN can be simulated via two steps based on its definition. First, perturbation is verified for each individual node and its value is flipped if there is a perturbation. Second, if no perturbation happens for any of the nodes, the network state is updated by selecting predictor functions for all the nodes and applying them. For efficiency reason, the selection of predictor functions for each node x_i is performed with the *alias method* [14], which allows to make a selection among choices in constant time irrespective of the number of choices. The alias method requires the construction of an *alias table* of size proportional to the number of choices, based on the selection probabilities of C_i.

3 Structure-Based Parallelisation

The simulation method described in the above section requires to check perturbations, make a selection and perform updating a node for n times in each step. In the case of large PBNs and huge trajectory (sample) size, the simulation time cost can become prohibitive. Intuitively, the simulation time can be reduced if the n-time operations can be speeded up, for which we propose two solutions. One is to perform *network reduction* such that the total number of nodes is reduced. The other is to perform *node-grouping* in order to parallelise the process for checking perturbations, making selections, and updating nodes. For the first solution, we analyse the PBN structure to identify those nodes that can be removed and remove them to reduce the network size; while for the second solution, we analyse the PBN structure to divide nodes into groups and perform the operations for nodes in a group simultaneously. We combine the two solutions together and refer to this simulation technique as *structure-based parallelisation*. We formalise the two solutions in the following three steps: the first solution is described in Step 1 and the second solution is described in Steps 2 and 3.

> Step 1. Remove unnecessary nodes from the PBN.
> Step 2. Parallelise the perturbation process.
> Step 3. Parallelise updating a PBN state with predictor functions.

We describe these three steps in the following subsections.

3.1 Removing Unnecessary Nodes

We first identify those nodes that can be removed and perform network reduction. When simulating a PBN without perturbations, if a node does not affect any other node in the PBN, the states of all other nodes will not be affected after removing this node. If this node is *not of interest* of the analysis, e.g., we are not interested in analysing its steady-state, then this node is dispensable in a PBN without perturbations. We refer to such a dispensable node as a *leaf node* in a PBN and define it as follow:

Algorithm 1. Checking perturbations of leaf nodes in a PBN
1: **procedure** CHECKLEAFNODES(p, ℓ)
2:     $t = pow(1 - p, \ell)$;   *// the probability that no perturbation happens in leaves*
3:     **if** $rand() > t$ **then return** *true*;
4:     **else return** *false*;
5:     **end if**
6: **end procedure**

Definition 1 (Leaf node). *A node in a PBN is a* leaf *node (or* leaf *for short) if and only if either (1) it is not of interest and has no child nodes or (2) it is not of interest and has no other children after iteratively removing all its child nodes which are leaf nodes.*

According to the above definition, leaf nodes can be simply removed without affecting the simulation of the remaining nodes in a PBN without perturbations. In the case of a PBN with perturbations, perturbations in the leaf nodes need to be considered. Updating states with Boolean functions will only be performed when there is no perturbation in both the leaf nodes and the non-leaf nodes. Perturbations of the leaf nodes can be checked in constant time irrespective of the number of leaf nodes as describe in Algorithm 1. The input p is the perturbation probability for each node and ℓ is the number of leaf nodes in the PBN. Then, the probability that no perturbation happens in all the leaf nodes is given by $t = (1 - p)^\ell$. With the consideration of their perturbations, the leaf nodes can be removed without affecting the simulation of the non-leaf nodes also in a PBN with perturbations. Since the leaves are not of interest, results of analyses performed on the simulated trajectories of the reduced network, i.e., containing only non-leaf nodes, will be the same as performed on trajectories of the original network, i.e., containing all the nodes.

3.2 Performing Perturbations in Parallel

The second step of our method speeds up the process of determining perturbations. Normally, perturbations are checked for nodes one by one. In order to speed up the simulation of a PBN, we perform perturbations for k nodes simultaneously instead of one by one. For those k nodes, there are 2^k different perturbation situations. We calculate the probability for each situation and construct an alias table based on the resulting distribution. With the alias table, we make a choice c among 2^k choices and perturb the corresponding nodes based on the choice. The choice c is an integer in $[0, 2^k)$ and for the whole network the perturbation can then be performed k nodes by k nodes using the logical bitwise *exclusive or* operation, denoted $|$. To save memory, the alias table can be reused for all the groups since the perturbation probability p for each node is the same. It might happen that the number of nodes in the last perturbation round will be less than k nodes. Assume there is k' nodes in the last round and $k' < k$. For those k' nodes, we can reuse the same alias table to make the selection in order to save memory. After getting the choice c, we perform $c = c\&m$, where $\&$ is a bitwise *and* operation and m is a mask constructed by setting the first k' bits of m's binary representation to 1 and the remaining bits to 0.

Theorem 1. *The above process for determining perturbations for the last k' nodes guarantees that the probability for each of the k' nodes to be perturbed is still p.*

Proof. Without loss of generality, we assume that in the last k' nodes, t nodes should be perturbed and the positions of the t nodes are fixed. The probability for those t fixed nodes to be perturbed is $p^t(1-p)^{k'-t}$. When we make a selection from the alias table for k nodes, there are $2^{k-k'}$ different choices corresponding to the case that t fixed position nodes in the last k' nodes are perturbed. The sum of the probabilities of the $2^{k-k'}$ different choices is $[p^t(1-p)^{k'-t}] \cdot \sum_{i=0}^{k-k'} p^i(1-p)^{k-k'-i} = p^t(1-p)^{k'-t}$. $\qquad\square$

We present the procedures for constructing groups and performing perturbations based on the groups in Algorithm 2, where n is the given number of nodes,[1] k is the maximum number of nodes that can be perturbed simultaneously and s is the PBN's current state which is represented by an integer. To obtain more balanced groups, k can be decreased in line 2. As perturbing one node equals to flipping one bit of s, perturbing nodes in a group is performed via a logical bitwise *exclusive or* operation, denoted $\oplus$ (see line 13 of Algorithm 2). Perturbing k nodes simultaneously requires 2^k double numbers to store the probabilities of 2^k different choices. The size of k is therefore restricted by the available memory.[2]

3.3   Updating Nodes in Parallel

The last step to speed up PBN simulation is to update a number of nodes simultaneously in accordance with their predictor functions. For this step, we need an initialisation process to divide the n nodes into m groups and construct *combined predictor functions* for each group. After this, we can select a combined predictor function for each group based on a sampled random number and apply this combined function to update the nodes in the group simultaneously.

We first describe how predictor functions of two nodes are combined. The combination of functions for more than two nodes can be performed iteratively. Let x_α and x_β be the two nodes to be considered. Their predictor functions are denoted as $F_\alpha = \{f_1^{(\alpha)}, f_2^{(\alpha)}, \ldots, f_{\ell(\alpha)}^{(\alpha)}\}$ and $F_\beta = \{f_1^{(\beta)}, f_2^{(\beta)}, \ldots, f_{\ell(\beta)}^{(\beta)}\}$. Further, the corresponding selection probability distributions are denoted as $C_\alpha = \{c_1^{(\alpha)}, c_2^{(\alpha)}, \ldots, c_{\ell(\alpha)}^{(\alpha)}\}$ and $C_\beta = \{c_1^{(\beta)}, c_2^{(\beta)}, \ldots, c_{\ell(\beta)}^{(\beta)}\}$. After the grouping, due to

---

[1] In our methods, it is clear that Step 2 and Step 3 are independent of Step 1. Thus, we consistently use n to denote the number of nodes in a PBN.

[2] For the experiments, we set k to 16 and k could be bigger as long as the memory allows. However, a larger k requires larger table to store the 2^k probabilities and the performance of a CPU drops when accessing an element of a much larger table due to the large cache miss rate.

Algorithm 2. The group perturbation algorithm

1: **procedure** PREPAREPERTURBATION(n, k)
2:      $g = \lceil n/k \rceil$; $k = \lceil n/g \rceil$; $k' = n - k * (g - 1)$;
3:      construct the alias table $\mathbb{A}_p$; $mask = 0$; $i = 0$;
4:      **repeat** $mask = mask \mid (1 \ll i)$; $i + +$;
5:      **until** $i = k'$;
6:      **return** $[\mathbb{A}_p, mask]$;
7: **end procedure**
8: **procedure** PERTURBATION$(\mathbb{A}_p, mask, s)$
9:      $perturbed = false$;
10:     **for** $(i = 0; i < g; i + +)$ **do**
11:         $c = Next(\mathbb{A}_p)$;   //$Next(\mathbb{A}_p)$ returns a random integer based on $\mathbb{A}_p$
12:         **if** $c \neq 0$ **then**
13:             $s = s \oplus (c \ll (i * k))$;   //shift c to flip only the bits of current group
14:             $perturbed = true$;
15:         **end if**
16:     **end for**
17:     $c = Next(\mathbb{A}_p)$ & $mask$;
18:     **if** $c \neq 0$ **then**
19:         $s = s \oplus (c \ll (i * k))$; $perturbed = true$;
20:     **end if**
21:     **return** $[s, perturbed]$;
22: **end procedure**

---

the assumed independence, the number of combined predictor functions is $\ell(\alpha) * \ell(\beta)$. We denote the set of combined predictor functions as $\bar{F}_{\alpha\beta} = \{f_1^{(\alpha)} \cdot f_1^{(\beta)}, f_1^{(\alpha)} \cdot f_2^{(\beta)}, \dots, f_{\ell(\alpha)}^{(\alpha)} \cdot f_{\ell(\beta)}^{(\beta)}\}$, where for $i \in [1, \ell(\alpha)]$ and $j \in [1, \ell(\beta)]$, $f_i^{(\alpha)} \cdot f_j^{(\beta)}$ is a combined predictor function that takes the input nodes of functions $f_i^{(\alpha)}$ and $f_j^{(\beta)}$ as its input and combines the Boolean output of functions $f_i^{(\alpha)}$ and $f_j^{(\beta)}$ into integers as output. The combined integers range in $[0, 3]$ and their 2-bit binary representations (from right to left) represent the values of nodes x_α and x_β. The selection probability for function $f_i^{(\alpha)} \cdot f_j^{(\beta)}$ is $c_i^{(\alpha)} * c_j^{(\beta)}$. It holds that $\sum_{i=1}^{\ell(\alpha)} \sum_{j=1}^{\ell(\beta)} c_i^{(\alpha)} * c_j^{(\beta)} = 1$. With the selection probabilities, we can compute the alias table for each group so that the selection of combined predictor function in each group can be performed in constant time.

We now describe how to divide the nodes into groups. Our aim is to have as few groups as possible so that the updating of all the nodes can be finished in as few rounds as possible. However, fewer groups lead to many more nodes in a group, which will result in a huge number of combined predictor functions in the group. Therefore, the number of groups has to been chosen properly so that the number of groups is as small as possible, while the combined predictor functions can be stored within the memory limit of the computer performing the simulation. Besides, nodes with only one predictor function should be considered separately since selections of predictor functions for those nodes are not needed.

In the rest of this section, we first formulate the problem for dividing nodes with more than one predictor function and give our solution afterwards; then we discuss how to treat nodes with only one predictor function.

Problem formulation. Let S be a list of n items $\{\mu_1, \mu_2, \ldots, \mu_n\}$. For $i \in [1, n]$, item μ_i represents a node in a PBN with n nodes. Its weight is assigned by a function $\omega(\mu_i)$, which returns the number of predictor functions of node μ_i. We aim to find a minimum integer m to distribute the nodes into m groups such that the sum of the numbers of combined predictor functions of the m groups will not exceed a memory limit θ. This is equivalent to finding a minimum m and an m-partition $S_1, S_2, \ldots, S_m$ of S, i.e., $S = S_1 \cup S_2 \cup \cdots \cup S_m$ and $S_k \cap S_\ell = \emptyset$ for $k, \ell \in \{1, 2, \ldots, m\}$, such that $\sum_{i=1}^{m} \left(\prod_{\mu_j \in S_i} \omega(\mu_j) \right) \leq \theta$.

Solution. The problem has two outputs: an integer m and an m-partition. We first try to estimate a potential value of m, i.e., the lower bound of m that could lead to an m-partition of S which satisfies $\sum_{i=1}^{m} \left(\prod_{\mu_j \in S_i} \omega(\mu_j) \right) \leq \theta$. With this estimate, we then try to find an m-partition satisfying the above requirements.

Denote the *weight* of a sub-list S_i as w_i, where $w_i = \prod_{\mu_j \in S_i} \omega(\mu_j)$. The inequality in the problem description can be rewritten as $\sum_{i=1}^{m} w_i \leq \theta$. We first compute the minimum value of $\hat{m}$, denoted as $\hat{m}_{min}$, satisfying the following inequality:

$$\hat{m} \cdot \sqrt[\hat{m}]{\prod_{i=1}^{n} \omega(\mu_i)} \leq \theta. \tag{1}$$

Theorem 2. $\hat{m}_{min}$ *is the lower bound on* m *that allows a partition to satisfy* $\sum_{i=1}^{m} w_i \leq \theta$.

Proof. We proceed by showing that for any $k \in \{1, 2, \ldots, \hat{m}_{min} - 1\}$, $\hat{m}_{min} - k$ will make the inequality unsatisfied, i.e., $\sum_{i=1}^{\hat{m}_{min} - k} w_i' > \theta$, where w_i' is the weight of the ith sub-list in an arbitrary partition of S into $\hat{m}_{min} - k$ sub-lists. Since $\hat{m}_{min}$ is the minimum value of $\hat{m}$ that satisfies Inequality (1), we have $(\hat{m}_{min} - k) \cdot {}^{(\hat{m}_{min} - k)}\sqrt{\prod_{i=1}^{n} \omega(\mu_i)} > \theta$. Hence,

$$(\hat{m}_{min} - k) \cdot \sqrt[(\hat{m}_{min} - k)]{\prod_{i=1}^{\hat{m}_{min} - k} w_i'} > \theta. \tag{2}$$

Based on the inequality relating arithmetic and geometric means, we have

$$\sum_{i=1}^{\hat{m}_{min} - k} w_i' \geq (\hat{m}_{min} - k) \cdot \sqrt[(\hat{m}_{min} - k)]{\prod_{i=1}^{\hat{m}_{min} - k} w_i'}. \tag{3}$$

Combining Inequality (2) with Inequality (3) gives $\sum_{i=1}^{\hat{m}_{min} - k} w_i' > \theta$. □

Algorithm 3. The greedy algorithm

1: **procedure** FINDPARTITIONS(S, m)
2:     sort S with descending orders based on the weights of items in S;
3:     initialise A, an array of m lists;   //initially, each $A[i]$ is an empty list
4:     **for** $(j = 0; j < S.size(); j + +)$ **do**//S.size() returns the number of items in S
5:         among the m elements of A,   //the weight of $A[i]$ is $w_i = \prod_{\mu_j \in A[i]} \omega(\mu_j)$
6:             find the one with the smallest weight and add $S[j]$ to it;
7:     **end for**
8:     **return** A;
9: **end procedure**

Starting from the lower bound, we try to find a partition of S into m sub-lists that satisfies $\sum_{i=1}^{m} w_i \leq \theta$. Since the arithmetic and geometric means of non-negative real numbers are equal if and only if every number is the same, we get the heuristic that the weight of the m sub-lists should be as equal as possible so that the sum of the weights is as small as possible. Our problem then becomes similar to the NP-hard multi-way number partition problem: to divide a given set of integers into a collection of subsets, so that the sum of the numbers in each subset are as nearly equal as possible. We adapt the greedy algorithm (see Algorithm 3 for details) for solving the multi-way number partition problem, by modifying the sum to multiplication, to solve our partition problem.[3] If the m-partition we find satisfies the requirement $\sum_{i=1}^{m} w_i \leq \theta$, then we get a solution to our problem. Otherwise, we need to increase m by one and try to find a new m-partition. We repeat this process until the condition $\sum_{i=1}^{m} w_i \leq \theta$ is satisfied. The whole partition process for all the nodes is described in Algorithm 4.

Nodes with only one predictor function are treated in line 8. We divide such nodes into groups based on their parent nodes, i.e., we put nodes sharing the most common parents into the same group. In this way, the combined predictor function size can be as small as possible such that the limited memory can handle more nodes in a group. The number of nodes in a group is also restricted by the combined predictor function size, i.e., the number of parent nodes in this group.[4] The partition is performed with an algorithm similar to Algorithm 3. The difference is that in each iteration we always add a node into a group which shares most common parent nodes with this node.

3.4   The New Simulation Method

We describe our new method for simulating PBNs in Algorithm 5. The procedure PREPARATION describes the whole preparation process of the three steps

---

[3] There exist other algorithms to solve the multi-way number partition problem and we choose the greedy algorithm for its efficiency.

[4] In our experiments, the maximum number of parent nodes in one group is set to 18. Similar to the value of k in Step 2, the number can be larger as long as the memory can handle. However, the penalty from large cache miss rate will diminish the benefits by having fewer groups when the number of parent nodes is too large.

Algorithm 4. Partition n nodes into groups.

1: **procedure** PARTITION(G, θ) *//S' contains nodes with*
2:      compute two lists S and S' based on G;   *// one predictor function*
3:      compute the lower bound $\hat{m}$ according to Inequality (1); $m = \hat{m}$;
4:      **repeat**
5:          $A_1 = $ FINDPARTITIONS(S, m);
6:          $sum = \sum_{i=1}^{m} \left(\prod_{\mu_j \in A_1[i]} \omega(\mu_j) \right)$; $m = m + 1$;   *//compute the sum of*
7:      **until** $sum < \theta$;   *//weights*
8:      divide S' into A_2;   *//using modified Algorithm 3: in each iteration, a node is*
9:      merge A_1 and A_2 into A;   *//put in a list which shares most common parent*
10:     **return** A;   *//nodes with this node*
11: **end procedure**

(network reduction for Step 1, and node-grouping for Step 2 and Step 3). The three inputs of the procedure PREPARATION are the PBN network G, the memory limit θ, and the maximum number k of nodes that can be put in a group for perturbation. The PREPARATION procedure performs network reduction and node grouping. The reduced network and the grouped nodes information are then provided for the PARALLELSIMULATION procedure via seven parameters: $\mathbb{A}_p$ and $mask$ are the alias table and mask used for performing perturbations of non-leaf nodes as explained in Algorithm 2; l is the number of leaf nodes; p is the perturbation rate; $\mathbb{A}$ is an array containing the alias tables for predictor functions in all groups; F is an array containing predictor functions of all groups; and cum is an array storing the cumulative number of nodes in each group, i.e., $cum[0] = 0$ and $cum[i] = \sum_{j=0}^{i-1} \tau_j$ for $i \in [1, m]$, where m is the number of groups and τ_j is the number of nodes in group j. Procedure PARALLELSIMULATION simulates one step of a PBN by first checking perturbation and then updating PBNs with combined predictor functions. Perturbations for leaf nodes and non-leaf nodes have been explained in Algorithms 1 and 2. We now explain how nodes in a group are simultaneously updated with combined predictor function. It is performed via the following three steps: 1) a random combined predictor function is selected from F based on the alias table A; 2) the output of the combined predictor function is obtained according to the current state s; 3) the nodes in this group are updated based on the output of the combined predictor function. To save memory, states are stored as integers and updating a group of nodes is implemented via a logical bitwise *or* operation. To guarantee that the update is performed on the required nodes, a shift operation is needed on the output of the selected function (line 22). The number of bits to be shifted for the current group is in fact the cumulative number of nodes of all its previous groups, which is stored in the array cum.

4   Evaluation

The evaluation of our new simulation method is performed on both randomly generated networks and a real-life biological network. All the experiments are

---

Algorithm 5. Structure-based PBN simulation.

1: **procedure** PREPARATION(G, θ, k)
2:    perform network reduction for G and store the reduced network in G';
3:    get the number of nodes n and perturbation probability p from G;
4:    get the number of nodes n' from G'; $\quad \ell = n - n'$;
5:    $[\mathbb{A}_p, mask]$ =PREPAREPERTURBATION(n', k);
6:    PA =PARTITION(G', θ);
7:    for each group in PA, compute its combined functions
8:       and put them as a list in array F, and compute its alias table in array $\mathbb{A}$;
9:    compute cum as $cum[0] = 0$ and $cum[i] = \sum_{j=0}^{i-1} \tau_j$ for $i \in [1, m]$, where m is
10:    the number of groups in PA and τ_j is the number of nodes in group j;
11:    **return** $[\mathbb{A}_p, mask, \ell, p, \mathbb{A}, F, cum]$;
12: **end procedure**
13: **procedure** PARALLELSIMULATION($\mathbb{A}_p, mask, \mathbb{A}, F, cum, \ell, p, s$)
14:    $[s, perturbed]$ =PERTURBATION($\mathbb{A}_p, mask, s$);   //perturb by group
15:    **if** $perturbed$ || CHECKLEAFNODES(p, ℓ) **then**   //check perturbations of leaves
16:        **return** s;
17:    **else** $s' = 0$; $count = size(\mathbb{A})$; //size($\mathbb{A}$): the number of elements in array $\mathbb{A}$
18:        **for** $(i = 0; i < count - 1; i + +)$ **do**
19:            $index = Next(\mathbb{A}[i])$; //select a random integer based on the alias table
20:            $f = F[i].get(index)$; //obtain the predictor function at the given index
21:            $v = f[s]$;   //$f[s]$ returns the integer output of f based on state s
22:            $s' = s' \mid (v \ll cum[i])$;   // update only nodes in the current group
23:        **end for**
24:    **end if**
25:    **return** s';
26: **end procedure**

---

performed on high performance computing (HPC) machines, each of which contains a CPU of Intel Xeon X5675 @ 3.07 GHz. The program is written in Java and the initial and maximum Java virtual machine heap size is set to 4 GB and 5.89 GB, respectively.

4.1 Randomly Generated Networks

With the evaluation on randomly generated networks, we aim not only to show the efficiency of our method, but also to answer how much speedup our method is likely to provide for a given PBN.

The first step of our new simulation method performs a network reduction technique, which is different from the node-grouping techniques in the later two steps. Therefore, we evaluate the contribution of the first step and the other two steps to the performance of our new simulation method separately. We consider the original simulation method as the reference method and we name it $Method_{ref}$. The simulation method applying the network reduction technique is referred to as $Method_{reduction}$ and the simulation method applying both the network reduction and node-grouping techniques as $Method_{new}$.

$Method_{reduction}$ and $Method_{new}$ require pre-processing of the PBN under study, which leads to a certain computational overhead. However, the proportion of the pre-processing time in the whole computation decreases with the increase of the sample size. In our evaluation, we first focus on comparisons without taking pre-processing into account to evaluate the maximum potential performance of our new simulation method; we then show how different sample sizes will affect the performance when pre-processing is considered.

How does our method perform? Intuitively, the speedup due to the network reduction technique is influenced by how much a network can be reduced and the performance of node-grouping is influenced by both the density and size of a given network. Hence, the evaluation is performed on a large number of randomly generated PBNs covering different types of networks. In total, we use 2307 randomly generated PBNs with different percentages of leaves ranging between 0 % and 90 %; different densities ranging between 1 and 8.1; and different network sizes from the set $\{20, 50, 100, 150, 200, 250, 300, 350, 400, 450, 500, 550,$ $600, 650, 700, 750, 800, 850, 900, 950, 1000\}$. The networks are generated randomly using the tool ASSA-PBN [2], by providing the following information: the number of nodes, the maximum (minimum) number of predictor functions for the nodes, and the maximum (minimum) number of parent nodes for the predictor functions. Thus, the generation of these networks' density and percentage of leaves cannot be fully controlled. In other words, density and percentage of leaves for these 2307 PBNs are not uniformly distributed. We simulate 400 million steps for each of the 2307 PBNs with the three different simulation methods and compare their time costs. For the network reduction technique the speedups are calculated as the ratio between the time of $Method_{reduction}$ and the time of $Method_{ref}$, where the pre-processing time of the former method is excluded. The obtained speedups are between 1.00 and 10.90. For node-grouping, the speedups are calculated as the ratio between the time of $Method_{new}$ and the time of $Method_{reduction}$ without considering the required pre-processing times. We have obtained speedups between 1.56 and 4.99. We plot in Fig. 1 the speedups of the network reduction and node-grouping techniques with respect to their related parameters. For the speedups achieved with network reduction, the related parameters are the percentage of leaves and the density. In fact, there is little influence from density to the speedup resulting from network reduction as the speedups do not change much with the different densities (see Fig. 1a). The determinant factor is the percentage of leaves. The more leaves a PBN has, the more speedup we can obtain for the network. For the speedups obtained from node-grouping, the related parameters are the density and the network size after network reduction, i.e., the number of non-leave nodes. Based on Fig. 1b, the speedup with node-grouping is mainly determined by the network density: a smaller network density could result in a larger speedup contributed from the node-grouping technique. This is mainly due to the fact that sparse network has a relatively small number of predictor functions in each node and therefore, the nodes will be partitioned into fewer groups. Moreover, while the performance of network reduction is largely influenced by the percentage of leaves, the node-grouping technique tends to provide a rather stable speedup. Even for large dense networks, the technique can reduce the time cost almost by half.

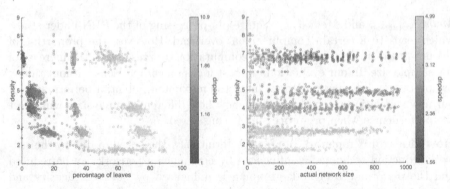

(a) Simulation time of $Method_{reduction}$ over simulation time of $Method_{ref}$.

(b) Simulation time of $Method_{new}$ over simulation time of $Method_{reduction}$.

Fig. 1. Speedups obtained with network reduction and node-grouping techniques. The pre-processing time is excluded from the analysis.

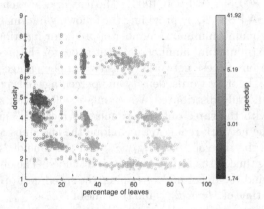

Fig. 2. Speedups of $Method_{new}$ with respect to $Method_{ref}$.

The combination of these two techniques results in speedups (time of $Method_{new}$ over time of $Method_{ref}$) between 1.74 and 41.92. We plot in Fig. 2 the speedups in terms of the percentage of leaves and density. The figure shows a very good performance of our new method on sparse networks with large percentage of leaves.

What is the influence of sample size? We continue to evaluate the influence of sample size on our proposed new PBN simulation method. The pre-processing time for the network reduction step is relatively very small. Therefore, our evaluation focuses on the influence of the total pre-processing time of all the three steps on the speedup of $Method_{new}$ with respect to $Method_{ref}$. We select 9 representative PBNs from the above 2307 PBNs, with respect to their densities, percentages of leaves and the speedups we have obtained. We simulate the 9 PBNs for different sample sizes using both $Method_{ref}$ and $Method_{new}$. We show

Table 1. Influence of sample sizes on the speedups of $Method_{new}$ with respect to $Method_{ref}$.

network #	size	percentage of leaves	density	average pre-processing time (second)	speedup with different sample sizes (million)			
					1	10	100	400
1	900	1.11	6.72	28.12	0.65	1.49	1.71	1.73
2	950	0.84	6.96	32.35	0.59	1.47	1.73	1.75
3	1000	0.30	7.00	33.72	0.58	1.45	1.71	1.73
4	600	67.83	4.25	162.21	0.13	1.08	4.51	6.89
5	800	68.38	3.94	43.17	0.66	3.05	6.75	7.69
6	900	68.00	3.89	36.58	0.69	3.56	6.90	7.70
7	450	89.78	1.60	0.23	21.44	37.59	41.62	41.84
8	550	88.55	1.72	0.24	20.26	35.94	36.47	36.62
9	1000	89.10	1.75	1.08	10.04	31.83	35.09	37.19

the average pre-processing time of $Method_{new}$ and the obtained speedups with $Method_{new}$ (taking into account pre-processing time costs) with different sample sizes in Table 1. As expected, with the increase of the sample size, the influence of pre-processing time becomes smaller and the speedup increases. In fact, in some cases, the pre-processing time is relatively so small that its influence becomes negligible, e.g., for networks 7 and 8, where the sample size is equal or greater than 100 million. Moreover, often with a sample size larger than 10 million, the effort spent in pre-processing can be compensated by the saved sampling time (simulation speedup).

Performance prediction. To predict the speedup of our method for a given network, we apply regression techniques on the results of the 2307 PBNs to fit a prediction model. We use the normalised percentage of leaves and the network density as the predictor variables and the speedup of $Method_{new}$ with respect to $Method_{ref}$ as the response variable in the regression model. We do not consider network size as based on the plotted figures it does not directly affect the speedup. In the end, we obtain a polynomial regression model shown in Equation (4), which can fit 90.9 % of the data:

$$y = b_1 + b_2 * x_1 + b_3 * x_1^2 + b_4 * x_2 + b_5 * x_2^2 \tag{4}$$

where $[b_1, b_2, b_3, b_4, b_5] = [2.89, 2.71, 2.40, -1.65, 0.71]$, y represents the speedup, x_1 represents the percentage of leaves and x_2 represents the network density. The result of a 10-fold cross-validation of this model supports this prediction rate. Hence, we believe this model does not overfit the given data. Based on this model, we can predict how much speedup is likely to be obtained with our proposed method for a given PBN.

4.2 An Apoptosis Network

In this section, we evaluate our method on a real-life biological network, i.e., an apoptosis network of 91 nodes [5]. This network has a density of 1.78 and 37.5 % of the nodes are leaves, which is suitable for applying our method to gain speedups. The network has been analysed in [3]. In one of the analyses, i.e., the long-term influences [8] on complex2 from each of its parent nodes: RIP-deubi, complex1, and FADD, seven steady-state probabilities of the network need to be computed. In this evaluation, we compute the seven steady-state probabilities using our proposed structure-based simulation method ($Method_{new}$) and compare it with the original simulation method ($Method_{ref}$). The precision and confidence level of all the computations, as required by the two-state Markov chain approach [4], are set to 10^{-5} and 0.95, respectively. The results of this computation are shown in Table 2. The computed probabilities using both methods are comparable, i.e., for the same set of states, the differences of the computed probabilities are within the precision requirements. The sample sizes required by both methods for computing the same steady-state probabilities are very close to each other. Note that the speedups are computed based on the accurate data, which are slightly different from the truncated and rounded data shown in Table 2. We have obtained speedups ($Method_{new}$ with respect to $Method_{ref}$) between 7.67 and 10.28 for computing those seven probabilities. In total, the time cost is reduced from 1.5 hours to about 10 min.

Table 2. Performance of $Method_{ref}$ and $Method_{new}$ on an apoptosis network.

#	$Method_{ref}$			$Method_{new}$				speedup
	sample size (million)	time (m)	probability	pre-processing time (s)	sample size (million)	total time (m)	probability	
1	147.50	9.51	0.003243	4.57	147.82	1.05	0.003236	9.09
2	452.35	28.65	0.990049	3.10	452.25	2.79	0.990058	10.28
3	253.85	14.88	0.005583	3.42	253.99	1.74	0.005587	8.54
4	49.52	2.96	0.001087	3.38	50.39	0.36	0.001078	8.31
5	315.06	17.73	0.993293	4.40	305.43	2.05	0.993298	8.39
6	62.22	3.69	0.001088	3.13	50.28	0.39	0.001087	7.67
7	255.88	16.74	0.005621	4.01	256.61	1.70	0.005623	9.88

5 Conclusion

We propose a structure-based method for speeding up simulations of PBNs. Using network reduction and node-grouping techniques, our method can significantly improve the simulation speed of PBNs. We show with experiments that our method is especially efficient in the case of analysing sparse networks with a large number of leaf nodes.

The node-grouping technique gains speedups by using more memory. Theoretically, as long as the memory can handle, the group number can be made as small as possible. However, this causes two issues in practice. First, the pre-processing time increases dramatically with the group number decreasing. Second, the performance of the method drops a lot when operating on large memories due to the increase of cache miss rate. Therefore, in our experiments we do not explore all the available memory to maximise the groups. Reducing the pre-processing time cost and the cache miss rate would be two future works to further improve the performance of our method. We plan to apply our method for the analysis of real-life large biological networks.

References

1. Kauffman, S.A.: Homeostasis and differentiation in random genetic control networks. Nature **224**, 177–178 (1969)
2. Mizera, A., Pang, J., Yuan, Q.: ASSA-PBN: an approximate steady-state analyser of probabilistic Boolean networks. In: Finkbeiner, B., Pu, G., Zhang, L. (eds.) ATVA 2015. LNCS, vol. 9364, pp. 214–220. Springer, Heidelberg (2015). doi:10.1007/978-3-319-24953-7_16
3. Mizera, A., Pang, J., Yuan, Q.: Reviving the two-state Markov chain approach (Technical report) (2015). http://arxiv.org/abs/1501.01779
4. Raftery, A.E., Lewis, S.: How many iterations in the Gibbs sampler? Bayesian Stat. **4**, 763–773 (1992)
5. Schlatter, R., Schmich, K., Vizcarra, I.A., Scheurich, P., Sauter, T., Borner, C., Ederer, M., Merfort, I., Sawodny, O.: ON/OFF and beyond - a Boolean model of apoptosis. PLOS Comput. Biol. **5**(12), e1000595 (2009)
6. Shmulevich, I., Gluhovsky, I., Hashimoto, R., Dougherty, E., Zhang, W.: Steady-state analysis of genetic regulatory networks modelled by probabilistic Boolean networks. Comp. Funct. Genomics **4**(6), 601–608 (2003)
7. Shmulevich, I., Dougherty, E.R.: Probabilistic Boolean Networks: The Modeling and Control of Gene Regulatory Networks. SIAM Press, New York (2010)
8. Shmulevich, I., Dougherty, E.R., Kim, S., Zhang, W.: Probabilistic Boolean networks: a rule-based uncertainty model for gene regulatory networks. Bioinformatics **18**(2), 261–274 (2002)
9. Shmulevich, I., Dougherty, E.R., Zhang, W.: Control of stationary behavior in probabilistic Boolean networks by means of structural intervention. J. Biol. Syst. **10**(04), 431–445 (2002)
10. Shmulevich, I., Dougherty, E.R., Zhang, W.: From Boolean to probabilistic Boolean networks as models of genetic regulatory networks. Proc. IEEE **90**(11), 1778–1792 (2002)
11. Shmulevich, I., Dougherty, E.R., Zhang, W.: Gene perturbation and intervention in probabilistic Boolean networks. Bioinformatics **18**(10), 1319–1331 (2002)
12. Trairatphisan, P., Mizera, A., Pang, J., Tantar, A.A., Sauter, T.: optPBN: an optimisation toolbox for probabilistic Boolean networks. PLOS ONE **9**(7), e98001 (2014)
13. Trairatphisan, P., Mizera, A., Pang, J., Tantar, A.A., Schneider, J., Sauter, T.: Recent development and biomedical applications of probabilistic Boolean networks. Cell Commun. Signal. **11**, 46 (2013)
14. Walker, A.: An efficient method for generating discrete random variables with general distributions. ACM Trans. Math. Soft. **3**(3), 253–256 (1977)

Formal Quantitative Analysis of Reaction Networks Using Chemical Organisation Theory

Chunyan Mu[1], Peter Dittrich[2], David Parker[1(✉)], and Jonathan E. Rowe[1]

[1] School of Computer Science, University of Birmingham, Birmingham, UK
d.a.parker@cs.bham.ac.uk
[2] Institute of Computer Science, Friedrich-Schiller-University Jena, Jena, Germany

Abstract. Chemical organisation theory is a framework developed to simplify the analysis of long-term behaviour of chemical systems. An organisation is a set of objects which are closed and self-maintaining. In this paper, we build on these ideas to develop novel techniques for formal quantitative analysis of chemical reaction networks, using discrete stochastic models represented as continuous-time Markov chains. We propose methods to identify organisations, to study quantitative properties regarding movement between these organisations and to construct an organisation-based coarse graining of the model that can be used to approximate and predict the behaviour of the original reaction network.

1 Introduction

In this paper, we study reaction networks and chemical organisation theory, in particular, investigating the applicability of probabilistic model checking to their analysis. Reaction networks are widely used in modelling chemical phenomena. They describe the dynamical interaction between processes of living systems in a formal way. Reaction networks can be difficult to understand and analyse since they can represent complex interaction behaviour over large state spaces.

Chemical organisation theory [7,9] provides a way to analyse complex dynamical networks and reason about the long-term behaviour of chemical systems. The complex network is decomposed into a set of sub-networks called "organisations". An *organisation* is a set of objects (for example, the species or molecules in a reaction system) which are closed and self-maintaining. Informally, closed means that no new object can be produced by the interactions within the set, and self-maintaining means that no object of the set disappears from the system, i.e., every consumed object of the set can be generated within the set. The concept of organisation allows us to lift the complex reaction network to a hierarchic structure including all stable states and states depicting accumulating molecules regarding to the organisations. The dynamics of the complex state space of the reaction network can then be mapped to movements among the set of organisations. Building a chemical organisation-based model thus helps us to model the structure and behaviour of complex reaction networks, and to simplify the dynamical analysis of the overall system.

In order to study the evolution of reaction networks, we apply probabilistic model checking, a formal verification technique for modelling and analysis of

© Springer International Publishing AG 2016
E. Bartocci et al. (Eds.): CMSB 2016, LNBI 9859, pp. 232–251, 2016.
DOI: 10.1007/978-3-319-45177-0_15

systems with stochastic behaviour. It has been used to study models across a wide range of application domains, including chemical and biological systems. Probabilistic model checking is based on the exhaustive construction and analysis of a state-based probabilistic model, typically a Markov chain or variant. In this work, we model the reaction networks as continuous-time Markov chains. Quantitative properties of interest about the system being analysed are formally specified using temporal logic. Here we use CSL (Continuous Stochastic Logic) [2] with rewards, a quantitative extension of the temporal logic CTL.

Specifically, in this work, we use CSL model checking of continuous-time Markov chains to investigate connections between chemical organisations using model decompositions into strongly connected components (SCCs). We develop an algorithm to automatically find organisations, and then perform a quantitative dynamical analysis in terms of organisations, asking, for example, "what is the probability of moving from one organisation to another?" or "what is the expected time to leave an organisation?" A coarse grained Markov chain model of hierarchic organisations for a given reaction network is then constructed as a result. We implement our techniques as an extension of the probabilistic model checking tool PRISM [15], and illustrate the approach on a set of example reaction networks. Approximating and predicting the system behaviour over time evolution is a direct application of our coarse grained model.

Related work. There are various approaches to modelling the dynamics of reaction networks. Feinberg and Horn [8] proposed methods to identify *positive* stationary states in which *all* molecular species are present in a network. Heinrich and Schuster [12] studied network structure based on *flux modes*, each of which specifies a set of reaction rules that can take place at a steady state and thus implies a set of species participating in those reactions. Species regarding to a flux modes were not required to be *self-maintaining* or *closed* however. We are more interested in the stationary states in which a *subset* of species are present, which is formalised in organisation theory [7]. In that area, the focus was typically on *qualitative* properties, and ODEs [6], approximating the evolution of reaction networks in *continuous* dynamical systems. Kreyssig et al. [14] studied the effects of small particle numbers on long-term behaviours in *discrete* biochemical systems. We build on their notion of discrete organisation but focus on *quantitative* analysis of the transitive dynamics among the organisations, which was not considered in [14]. Other approaches for approximate analysis of discrete models of reaction networks include the use of Linear Noise Approximation [4], the Central Limit Approximation [3] and "sliding window" abstractions [17].

2   Probabilistic Model Checking

Probabilistic model checking is a variant of *model checking* [5], a well-established formal method to automatically verify the correctness of real-life systems. Classical model checking answers the question of whether the behaviour of a given system satisfies a property or not. It thus requires two inputs: a description of the system and a specification of one or more required properties of that system, normally in temporal logic (such as CTL or LTL).

In probabilistic model checking, the models are extended with information about the likelihood that transitions take place. In practice, these models are usually Markov chains or Markov decision processes. In this work, we model the reaction systems as continuous-time Markov chains (CTMCs), which are widely used in fields such as performance analysis or systems biology to model systems with stochastic real-time behaviour. Formally, we define them as follows.

Definition 1 (CTMC). *A CTMC is a tuple $\mathcal{A} = (Q, Q_0, \Delta, L)$, where: Q is a finite set of states; $Q_0 \subseteq Q$ is the set of initial states; $\Delta : Q \times Q \to \mathbb{R}_{\geq 0}$ is the transition rate matrix; $L : Q \to 2^{AP}$ is a labelling function assigning, to each state $q \in Q$, a set of atomic propositions, from a set AP, that are true in q.*

The transition rate matrix Δ assigns a *rate* to each pair of states in the CTMC, which is used as the parameter of an exponential distribution.

In this work, the probabilistic temporal logic CSL (Continuous Stochastic Logic) is used to formally represent properties of reaction networks. It was originally introduced by Aziz et al. [1] and extended by Baier et al. [2]. The extended version allows for the specification of reward (or cost) properties, to reason about rewards (or costs) that have been attached to a CTMC. The extended version of CSL that we use allows us to represent properties such as "the probability of all of species A degrading within t time units is at most 0.1" or "the expected time elapsed before a B molecule first appears is at most 10".

Definition 2 (CSL syntax). *An (extended) CSL formula is an expression Ψ derived from the grammar:*

$$\Psi ::= \texttt{true} \mid p \mid \neg\Psi \mid \Psi \wedge \Psi \mid \mathsf{P}_{\bowtie\lambda}(\Psi \; U^I \; \Psi) \mid \mathsf{R}_{\bowtie r}[\Diamond\Psi]$$

where $p \in AP$ an atomic proposition, $\lambda \in [0,1]$ is a probability threshold, $r \in \mathbb{R}_{\geq 0}$ is a reward threshold, $\bowtie \in \{<, \leq, \geq, >\}$ and I is an interval of $\mathbb{R}_{\geq 0}$.

CSL formulas are described over the states of a Markov chain. A state q satisfies $\mathsf{P}_{\bowtie\lambda}(\psi)$ if the probability of taking a path from q satisfying ψ is in the interval specified by $\bowtie \lambda$. Here, the path formula ψ is an "until" operator: $\Psi \; U^I \; \Psi'$ asserts that Ψ' is satisfied at some future time point within interval I, and that Ψ is true up until that point. We omit the interval I when $I = [0, \infty)$. Common derived operators include: "eventually" $\Diamond^I\Psi := \texttt{true} \; U^I \; \Psi$ and "always" $\square^I\Psi := \neg\Diamond^I\neg\Psi$. For example, $\mathsf{P}_{\leq\lambda}(\square^I\Psi) \equiv \mathsf{P}_{\geq 1-\lambda}(\Diamond^I\neg\Psi)$. The R operator is used for reward properties: $\mathsf{R}_{\bowtie r}[\Diamond\Psi]$ is true from state q if the expected reward cumulated before a state satisfying Ψ is reached meets the bound $\bowtie r$. We also use *numerical queries*, e.g., $\mathsf{R}_{=?}[\Diamond\Psi]$, which return the actual expected reward (or probability), rather than check it against a bound. Rewards and costs are treated identically: here, we will use the R operator to formalise properties about the expected time elapsing before an event's occurrence. We omit a full definition of the semantics of CSL with respect to a Markov chain. Full details can be found in, for example, [2].

3   Modelling Reaction Networks with CTMCs

A reaction network consists of a set of molecules (or, molecular species to be more precise) and a set of reaction rules.

Definition 3. *A reaction network is a pair $(\mathcal{M}, \mathcal{R})$ consisting of a set of possible molecular species $\mathcal{M}$, and a set $\mathcal{R} \subseteq \mathcal{P}_M(\mathcal{M}) \times \mathcal{P}_M(\mathcal{M})$ of possible reactions among those species, where $\mathcal{P}_M(C)$ denotes the set of all multisets of elements over the set C. For a reaction $(R, P) \in \mathcal{R}$, the multisets R and P denote the reactants and products of the reaction, respectively, and we write $R(s)$ and $P(s)$ for the number of molecules of species s consumed by (reactants) and produced by (products) the reaction, respectively.*

For simplicity, we write $s_1 + s_2 + \cdots + s_n \rightarrow s'_1 + s'_2 + \cdots + s'_{n'}$ instead of $(\{s_1, s_2, \ldots, s_n\}, \{s'_1, s'_2, \ldots, s'_{n'}\}) \in \mathcal{R}$ to denote the existence of a reaction.

There are multiple ways in which we can obtain a dynamical model given a reaction network. One way is to consider (real-valued) concentrations of each molecular species and then represent the (deterministic) behaviour of the reactions as a set of ordinary differential equations. Here, we take a *discrete, stochastic* view of the network, modelling the (integer-valued) population count of each species and considering its evolution as a stochastic process, and in particular as a continuous-time Markov chain [11]. The latter is particularly appropriate when the numbers of molecules can be assumed to be relatively small in practice, and is the approach that we take in this work.

Furthermore, we will assume also that the reaction network is executing within a finite volume, which is modelled by limiting the total number $N_{\max} \in \mathbb{N}$ of molecules that can be present at any given time [14]. We also need to define the rates at which reaction events occur in the CTMC. To retain a general approach, we allow an arbitrary function $\mathtt{rate}_r$ from reactant populations to rate values for each reaction r.

Definition 4 (CTMC for reaction network). *Given a reaction network $\langle \mathcal{M}, \mathcal{R} \rangle$, a volume limit $N_{max} \in \mathbb{N}$ and a rate function $\mathtt{rate}_r : \mathbb{N}^{\mathcal{M}} \rightarrow \mathbb{R}_{\geq 0}$ for each $r \in \mathcal{R}$, we define the corresponding CTMC $\mathcal{A} = (Q, Q_0, \Delta, L)$ where:*

- $Q = \{q : \mathcal{M} \rightarrow \mathbb{N} \mid \sum_{s \in \mathcal{M}} q(s) \leq N_{max}\}$

is the set of population counts of $\mathcal{M}$ and Δ is defined as follows. For states $q, q' \in Q$ and reaction $(R, P) \in \mathcal{R}$, we write $q \xrightarrow{(R,P)} q'$ if and only if, for each species $s \in \mathcal{M}$, we have $q(s) \geq R(s)$ and $q'(s) = q(s) - R(s) + P(s)$, and $\sum_{s \in \mathcal{M}} q'(s) \leq N_{max}$. Then, for any $q, q' \in Q$, we have:

- $\Delta(q, q') = \sum\{\mid \mathtt{rate}_r(q) \mid r \in \mathcal{R} \text{ and } q \xrightarrow{r} q'\}$, and we call r the transition label of $q \xrightarrow{r} q'$.

Q_0 can be any subset of Q representing initial configurations of interest, and L can be any labelling function over Q that identifies states with relevant properties.

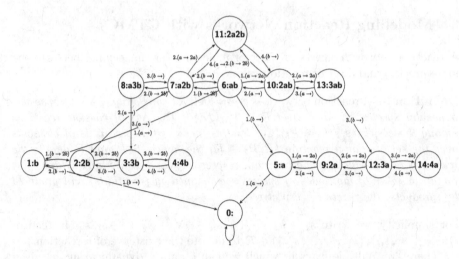

Fig. 1. State transition graph of the CTMC for Example 1. State labels show index and population count, e.g., 11 : 2a2b denotes that there are 2a and 2b in state 11.

For our examples we usually follow the general law of mass-action by setting $\mathrm{rate}_r(q) = \lambda_r \cdot \prod_{s \in R} q(s)$ with λ_r being a kinetic rate constant for reaction r (and assuming the stoichimetric coefficient of each reactant is at most one).

Each state $q \in Q$ of the CTMC gives the number $q(s)$ of molecules of each species $s \in \mathcal{M}$ that are currently present. For a state q, we also write $\phi(q)$ to denote the set of molecular species that are present, i.e., $\phi(q) = \{s \mid q(s) > 0\}$, and define $\phi(Q') = \cup_{q \in Q'} \phi(q)$ for a set of states $Q' \subseteq Q$. We let $Acc(q) \subseteq Q$ denote the set of states that are reachable from q.

Example 1. Consider the reaction network $\mathcal{A}$ with species $\mathcal{M} = \{a, b\}$ and reactions $\mathcal{R} = \{a + b \rightarrow a + 2b, a \rightarrow 2a, b \rightarrow 2b, a \rightarrow \emptyset, b \rightarrow \emptyset\}$. Assume the volume of the system is $N_{\max} = 4$, and the rate of each reaction is of second order, i.e., $\sharp a \cdot \sharp b, \sharp a \cdot \sharp a, \sharp b \cdot \sharp b, \sharp a \cdot \sharp a, \sharp b \cdot \sharp b$, respectively, where $\sharp a$ denotes the number of molecules of species a. We obtain a CTMC with 15 states (Fig. 1). Note that throughout this work, without loss of generality, we use second-order kinetic laws in order to avoid any complicated issues regarding units and scaling.

4   Chemical Organisation Theory and SCC Decomposition

Chemical organisation theory [7] provides a way to cope with the complex "constructive" dynamics of a reaction network by deriving a set of organisations [10], and then mapping the movement through state space to a movement between organisations. Such an abstract view allows us to analyse and predict the dynamical behaviour of a complex reaction network more easily. An organisation is a set of molecules that is algebraically *closed* and *self-maintaining*. A subset $C \subseteq \mathcal{M}$ is called "closed" if no molecules outside C can be produced by applying all reactions possible in C to multisets over C; a subset $S \subseteq \mathcal{M}$ is "self-maintaining" if all reactions that are able to fire in S can occur at certain strictly positive rates without reducing the amount of any species of S. We say that a reaction $(R, P) \in \mathcal{R}$ can *fire* in a set of species S, if S contains each reactant species from R.

Definition 5 (Organisation [7]). *A subset of $O \subseteq M$ is a chemical organisation if it is closed and self-maintaining, that is, if for all $(R, P) \in \mathcal{R}, R \subseteq O$ implies $P \subseteq O$ (closure), and there exists a strictly positive flux vector $v > 0$ such that $N_O \cdot v \geq 0$ with N_O being the stoichiometric matrix of the reactions that can fire in O (self-maintenance). An entry $n_{i,r}$ of the stoichiometric matrix $N_O = (n_{i,r})$ denotes the number of molecules of species $i \in M$ produced when firing reaction r once. The product with a flux vector, $N \cdot v$, results in a vector of the net-production rates for each species for the respective reaction rates v.*

Note that the set of organizations is defined with respect to a reaction network and thus independent from an initial state. However, given an initial state, there is in general only a subset of organizations reachable.

As discussed above, we model the dynamics of a reaction network as a Markov chain. A state is defined as the number of each molecular species and, with a limited total number of molecules, cases of both too few or too many molecules can prevent reaction rules being fired. As a consequence, we need to define *discrete organisations*, and the states contributing to generate them. From now on, $\mathcal{R}_q$ denotes the reactions firing in any state reachable from a state q.

Definition 6 (Discrete organisation and internal generator [14]). *Let $(M, \mathcal{R})$ be a reaction network. A subset of species $D \subseteq M$ is called a* discrete *organisation if there is a state $q \in Q$ such that: (i) $\phi(Acc(q)) = D$ (closure); and (ii) there is a sequence of transition labels $(r_1, \ldots, r_k)$ where $r_i \in \mathcal{R}$ such that $\cup_{i=1}^{k} \{r_i\} = \mathcal{R}_q$ and $q' = (r_k \circ \cdots \circ r_1)(q)$ satisfies $\forall s \in D : q'(s) \geq q(s)$ (self-maintenance), where $\circ$ denotes a composition operator, i.e., $r_j \circ r_i(q_i) = r_j(q_j)$ for $q_j = r_i(q_i)$. Such a state q is called an* internal generator *of the discrete organisation.*

Definition 7 (Generator). *A state $q' \in Q$ is called a* generator *of organisation D iff $\exists q \in Acc(q')$ such that q is an internal generator of D.*

Note that, in general, the organisation D generated by a state q' is not unique. However, if q is an internal generator, there is only one organisation it generates. Unless specifically stated otherwise, we say *organisation* rather than *discrete organisation* in the rest of the paper.

Example 2. The discrete organisations for Example 1 are: $\{a, b\}$, $\{a\}$, $\{b\}$, $\{\}$ and the corresponding generators are, respectively (cf. Fig. 1):
$\{6, 7, 8, 10, 11, 13\}$, $\{5, 6, 7, 8, 9, 10, 11, 12, 13, 14\}$, $\{1, 2, 3, 4, 6, 7, 8, 10, 11, 13\}$, $\{0, 1, \ldots, 14\}$.

In order to analyse the system behaviour and perform an organisation-based quantitative analysis of the reaction network, we study the connections between chemical organisations and the decompositions into strongly connected components (SCCs) of the Markov chain.

Definition 8 (SCC [13]). *A strongly connected component (SCC) of a Markov chain is a maximal set of states T such that, for every pair of states q and q', there is a path from q to q'.*

Intuitively, in the Markov chain for a reaction network, SCCs are important for an organisation-based analysis. However, some but not all SCCs correspond to organisations. In the next section, we will describe an algorithm to find organisations based on a decomposition into SCCs and then identifying those self-maintaining a set of species. We first note that *bottom* strongly connected components do relate to organisations.

Definition 9 (BSCC). *A bottom strongly connected component (BSCC) is an SCC T from which no state outside T is reachable from T.*

Proposition 1. *Each BSCC corresponds to a (unique) organisation, which is generated (uniquely) by any state of that BSCC.*

However, there are organisations whose internal generators are *not* contained in any BSCC. In order to also include such organisations, we call SCCs that correspond to an organisation *good SCCs*.

Definition 10 (Good SCC). *An SCC T is called* good *if it contains a cycle of the firing of every "possible" reaction rule, i.e., those whose reactants R appear in the SCC ($R \subseteq \{\phi(q) \mid q \in T\}$).*

Example 3. All SCCs are good in Example 1.

Clearly, some generators can contribute to multiple organisations. This makes it more difficult to decompose the Markov into its sets of generators. However, internal generators located in good SCCs contribute uniquely to an organisation.

Proposition 2. *A generator g is an internal generator of organisation D iff it is located in a good SCC T such that: $g \in T \ \wedge \ \bigcup_{q \in T} \phi(q) = D$.*

Proposition 3. *Given a good SCC T, let $A = \phi(T)$, if A is closed, then A is a discrete organisation, then $\{q \mid q \in T\}$ is the set of internal generators of A.*

Example 4. In Example 1, the internal generators of organisations $\{a, b\}$, $\{a\}$, $\{b\}$ and $\{\}$ are $\{6, 7, 8, 10, 11, 13\}$, $\{5, 9, 12, 14\}$, $\{1, 2, 3, 4\}$ and $\{0\}$, respectively.

5   Organisation-Based Analysis of Reaction Networks

In this section, we propose techniques for quantitative organisation-based analysis of reaction networks. We first introduce an algorithm to find the set of organisations for a specific reaction network. We then use probabilistic model checking to analyse quantitative properties regarding the dynamics of the network with respect to its organisations. Such organisation-based quantitative analyses can be used to construct the structure of organisation-based coarse-grained model, and provide a framework to approximate the complex dynamical behaviours of the original reaction networks in our next step.

5.1   Finding Organisations

Computing the organisations of a reaction network requires an analysis of the strongly connected components of its Markov chain's underlying transition graph. Since every state in a good SCC is an internal generator of an organisation, we identify good SCCs to find the organisations of the reaction network. Algorithm 1 presents the procedure for finding organisations of a given reaction network modelled as a CTMC. It is based on the following procedures:

– Tarjan($\mathcal{A}$) returns the set of strongly connected components of the Markov chain $\mathcal{A}$, using Tarjan's SCC algorithm [16] on the underlying digraph;
– findGoodSCCs(SCC) returns the "good" part SCC_G of $\mathcal{A}$ in which each possible reaction rule is able to be fired;
– find a set of closed molecules appearing in each $scc \in SCC_G$, and its relevant internal generators i.e., states in scc which generate the organisation.

Algorithm 1. Finding organisations of a reaction network

Data: CTMC $\mathcal{A}$ of reaction network $(\mathcal{M}, \mathcal{R})$
Result: $\mathcal{O}$ as a set of organisations, $\mathcal{G} : \mathcal{O} \to \mathcal{P}(Q)$ as a mapping from
             organisations to sets of internal generators
$\mathcal{O} = \{\}$;
$\mathcal{G} = \{\}$;
SCC $\leftarrow$ Tarjan($\mathcal{A}$);
$SCC_G \leftarrow$ findGoodSCCs(SCC) $\cup$ BSCC;
for $scc \in SCC_G$ **do**
$\quad M_g \leftarrow \{\phi(q) \mid q \in scc\}$;
$\quad$ **if** M_g *is closed* **then**
$\quad\quad$ **if** $M_g \notin \mathcal{O}$ **then**
$\quad\quad\quad \mathcal{O} \leftarrow \mathcal{O} \cup M_g$ /* add new organisation */ ;
$\quad\quad\quad \mathcal{G}(M_g) \leftarrow \{q \mid q \in scc\}$ /* add new internal generators */ ;
$\quad\quad$ **else**
$\quad\quad\quad \mathcal{G}(M_g) \leftarrow \mathcal{G}(M_g) \cup \{q \mid q \in scc\}$ /* update generators */ ;
$\quad\quad$ **end**
$\quad$ **end**
end
return $\mathcal{O}, \mathcal{G}$.

5.2   Organisation-Based Probabilistic Analysis

We now illustrate, via several examples, how we derive quantitative organisation-based properties of reaction networks. We implemented the organisation and generator detection process described above in the PRISM model checker, along with a translator that converts descriptions of reaction networks into the PRISM modelling language to allow construction of the corresponding CTMC. Organisation-based properties of the network, such as probabilities (bounds or average) of the movements among organisations, or the expected time to leave or stay at an organisation, are computed using CSL formulae.

Example 5. Consider the reaction network with molecular species $\mathcal{M} = \{a, b\}$ and reactions rules include: $\{a + b \to a, a \to 2a, b \to 2b, a \to \emptyset, b \to \emptyset\}$ with stochastic rates: $\sharp a \cdot \sharp b$, $(\sharp a)^2$, $(\sharp b)^2$, $(\sharp a)^2$, $(\sharp b)^2$ respectively.

Letting $N_{\max} = 10$, the resulting CTMC has 66 states and 201 transitions, and there are 4 SCCs ($\{a > 0, b > 0\}, \{a > 0, b = 0\}, \{a = 0, b > 0\}, \{a = b = 0\}$) with 1 BSCC ($\{a = b = 0\}$). For reference, we show both the PRISM language model and the CTMC for this example in the Appendix (Figs. 5 and 6).

The first property we consider is the probability of moving between organisations. Specifically, the probability of moving from O_1 to O_2 can be specified in CSL as: $\mathbf{P}_{=?}[o_1\ U\ o_2]$, where o_1 and o_2 are atomic propositions labelling states which represent internal generators of organisations O_1 and O_2. In this example, all SCCs are good and each (good) SCC generates exactly one organisation. To visualise the movement between organisation, we analyse the property above for each pair of organisations and construct the abstract transition graph shown in Fig. 2 (left). Blocks are labelled with organisations and, for each possible transition between organisations, we show the range of probabilities (over all states in the source organisation) and the average value (over the same set of states).

We also consider the expected time to leave (the generators of) each organisation. The CSL property to specify this, for some organisation O_i, is: $\mathbf{R}_{=?}[\Diamond \neg o_i]$, where o_i is an atomic proposition as above, $\neg$ denotes negation and we assign a state reward of 1 to every state of the CTMC, indicating the amount of reward that is accumulated per unit time until $\neg o_i$ is satisfied. This value is also shown for each organisation in Fig. 2 (left), inside the block for the corresponding organisation.

Finally, we consider the effect of making some constructive perturbation to the reaction network, by adding rules to create species with a small rate. Figure 2 (right) shows the results of the same analysis described above for the following constructive perturbation: $\{\emptyset \to a, \emptyset \to b\}$ both with reaction rate $\gamma = 0.01$. The result shows that, generating a and b with a small rate can cause an upward movement and slightly affect the system's behaviour. Note that the upward flow introduced by the constructive perturbation leads to a smoother flow.

Example 6. Consider now the reaction network with $\mathcal{M} = \{a, b, c, d\}$ and $\mathcal{R} = \{a + b \to a + 2b,\ a + d \to a + 2d,\ b + c \to 2c,\ c \to b,\ b + d \to c,\ b \to \emptyset,\ c \to \emptyset,\ d \to \emptyset\}$. We will consider two groups of rates for a purpose of comparison in Sect. 7: R2:: $\sharp a * \sharp b, \sharp b * \sharp c, \sharp c * \sharp c, \sharp b * \sharp d, \sharp b * \sharp b, \sharp c * \sharp c, \sharp d * \sharp e$ and R1:: $\sharp a * \sharp b, \sharp b * \sharp c, \sharp c, \sharp b * \sharp d, \sharp b * \sharp b, \sharp c, \sharp d$. We only use R1 in this section. Figure 7 (in the Appendix) shows the structure of the CTMC for $N_{\max} = 5$. Even for a small volume $N_{\max} = 5$, the structure is quite complex: 126 states, 386 transitions, 28 SCCs and 6 BSCCs.

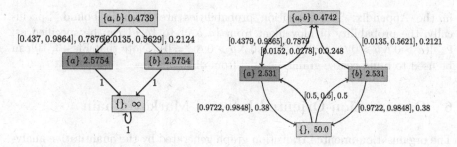

Fig. 2. Organisation-based transition model without/with constructive perturbation

Figure 8 (in the Appendix) illustrates, in the same fashion as above, the transition probabilities between *all* SCCs of the CTMC, and the expected time to leave them. Note that not all SCCs are good SCCs in this example: we highlight good SCCs in colour in Fig. 8. For instance, the SCC labelled as $(99, 105; 0.25)$ is not a good one. There are two states in this SCC: state 99 $(a = 2, b = 0, c = 1, d = 1)$ and state 105 $(a = 2, b = 1, c = 1, d = 1)$. The set of molecules appearing in this node is closed, but reaction rules such as $c \to \emptyset$ and $d \to \emptyset$ cannot be fired within the SCC and it is therefore not good. In addition, the SCC labelled as $(12, 27; 0.25)$ is also not a good one. It contains state 12 $(a = 0, b = 0, c = 2, d = 1)$ and state 27 $(a = 0, b = 1, c = 1, d = 1)$. The set of molecules appeared in this node is closed, but reaction rule $c \to \emptyset$ is unable to be fired locally, i.e., this decay will only introduce transitions to other SCCs. Similar cases can happen for some of the other reaction rules.

Figure 9 (in the Appendix) presents the transition probabilities between good SCCs only, and the expected time to leave them. Note that multiple good SCCs can contribute to the generation of one organisation. For instance, both good SCCs labelled 65... and 98... contribute to organisation $\{a, b, c\}$. Based on this graph, we can build up the transition graph over organisations. Figure 3 presents the transition probabilities between (internal generators of) organisations, and the expected time to leave each of them. It helps us to understand the movement between organisations and can be viewed as an abstract model capturing the behaviour of the reaction network at the level of organisations.

In addition, we also present transition graphs over the lattice of molecules (states

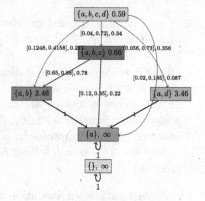

Fig. 3. Transition probabilities (bounds and average) between generators of organisations for Example 6 with $N_{\max} = 5$ and the expected time to leave them.

in which a set of molecules in the lattice with positive numbers) for a quantitative analysis for organisations from a different point of view, see Fig. 10

in the Appendix. The transition probabilities are given in bound. Specifically, the probability of movement from $\{a, b, c\}$ to $\{a, b\}$ can be specified as: $P_{=?}[(a > 0 \wedge b > 0 \wedge c > 0) \ U \ (a > 0 \wedge b > 0 \wedge c = 0)]$. Note that Figs. 3, 10 can be used to build coarse-grained model from different view.

6   Organisation-Oriented Interval Markov Chain

The organisation-oriented transition graph generated by the quantitative analysis can be used to build a coarse-grained model (with either interval based or average based probabilistic transitions). Such a coarse-grained model can mimic the complex reaction behaviours of the reaction network in an abstract way, whose state space and movement structure are much smaller. We can then perform approximation, prediction, and quantitative analysis upon the coarse-grained model instead of the complex concrete model. This section focuses on formalising the interval-based organisation coarse-grained model. Specifically, our quantitative analysis computes an organisation-oriented interval Markov chain, in which each abstract state is specified by a set of internal generators of an organisation, and the abstract transition provides the information about the uncertainty of the abstract behaviours of the system. Probabilities of moving from one abstract state to another are given by the lower and upper bounds, which provides the under and over approximation of the concrete probabilities.

Definition 11 (Organisation-oriented interval Markov chain). *An organ-isation-oriented interval Markov chain is a tuple* $\mathcal{A}_I^\sharp = (Q^\sharp, Q_0^\sharp, \Delta^\sharp, L)$, *where*

- $Q^\sharp$ *is a finite set of abstract states,*
  each of which $q^\sharp \in Q^\sharp$ *is a set of internal generators of an organisation* o: $q^\sharp \subseteq \mathcal{G}_I(o)$;
- $Q_0^\sharp \subseteq Q^\sharp$ *is the set of initial abstract states;*
- $\Delta^\sharp : Q^\sharp \times Q^\sharp \to [lb, ub]$ *is the abstract transition matrix, s.t.* $\Delta^\sharp(q^\sharp, q^{\sharp\prime}) = [lb, ub]$, *where* lb *and* ub *are the lower and upper bound of a set of concrete probabilistic transitions:* $\{\Delta(q, q') \mid q \in q^\sharp, q' \in q^{\sharp\prime}\}$ *specified in the relevant concrete model* $\mathcal{A}$ *respectively;*
- $L : Q^\sharp \to 2^{AP}$ *is a labelling function over* $Q^\sharp$ *that identifies properties of interest.*

An *abstract path* is an execution of the organisation-oriented interval Markov chain.

Definition 12 (Abstract path). *An* abstract path $\omega^\sharp$ *is a non-empty sequence of states* $q_0^\sharp q_1^\sharp \ldots$, *where* $q_i^\sharp \in Q^\sharp$ *and* $\forall i. \Delta^\sharp(q_i^\sharp, q_{i+1}^\sharp) \subseteq (0, k]$ *where* $0 < k \leq 1$. *The set of all finite and infinite paths of* $\mathcal{A}_I^\sharp$ *starting in state* $q^\sharp$ *are denoted as:* $\mathrm{Path}_{\mathrm{fin}}^{\mathcal{A}_I^\sharp}(q^\sharp)$ *and* $\mathrm{Path}^{\mathcal{A}_I^\sharp}(q^\sharp)$ *respectively.*

Definition 13 (Probability bounds of abstract paths). *The lower (*Prob$^-$*) and upper bound (*Prob$^+$*) of the probability of a finite abstract path $\omega^\sharp_{fin}$ starting from state $q^\sharp$ are respectively:*

$$\text{Prob}^-_{q^\sharp}(\omega^\sharp_{fin}) \triangleq \begin{cases} 1 & \text{if } n = 0 \\ \text{Prob}^-_{q^\sharp}(\omega^\sharp_0, \omega^\sharp_1) \times \cdots \times \text{Prob}^-_{q^\sharp}(\omega^\sharp_{n-1}, \omega^\sharp_n) & \text{otherwise} \end{cases}$$

$$\text{Prob}^+_{q^\sharp}(\omega^\sharp_{fin}) \triangleq \begin{cases} 1 & \text{if } n = 0 \\ \text{Prob}^+_{q^\sharp}(\omega^\sharp_0, \omega^\sharp_1) \times \cdots \times \text{Prob}^+_{q^\sharp}(\omega^\sharp_{n-1}, \omega^\sharp_n) & \text{otherwise} \end{cases}$$

where n denotes the length of the abstract path, $\omega^\sharp_i$ denotes the ith element of $\omega^\sharp$.

We focus on the reachability properties, for instance, the probability bounds of reaching or moving to an organisation of interests from another.

Definition 14 (Reachability properties). *Let $\mathcal{A}^\sharp_I$ be an organisation-based interval Markov chain. The lower and upper bound of the probability of reaching an abstract state $q^{\sharp\prime}$ from $q^\sharp$ are computed by:*

$$\text{Reach}^-_{\mathcal{A}^\sharp_I}(q^\sharp, q^{\sharp\prime})$$

$$\triangleq \min \left\{ \sum_{\omega^\sharp \in \text{Path}^{\mathcal{A}^\sharp_I}_{fin}(q^\sharp)} \{\text{Prob}^-_{q^\sharp}(\omega^\sharp) \mid \omega^\sharp_0 = q^\sharp \wedge \exists i \geq 0.\omega^\sharp_i = q^{\sharp\prime}\}, 1 \right\}$$

$$\text{Reach}^+_{\mathcal{A}^\sharp_I}(q^\sharp, q^{\sharp\prime})$$

$$\triangleq \min \left\{ \sum_{\omega^\sharp \in \text{Path}^{\mathcal{A}^\sharp_I}_{fin}(q^\sharp)} \{\text{Prob}^+_{q^\sharp}(\omega^\sharp) \mid \omega^\sharp_0 = q^\sharp \wedge \exists i \geq 0.\omega^\sharp_i = q^{\sharp\prime}\}, 1 \right\}.$$

Our organisation-oriented interval Markov chain should safely approximate the concrete CTMC describing the probabilistic behaviours of the system.

Theorem 1 (Soundness of the abstract semantics). *Let $\mathcal{A}^\sharp_I$ and $\mathcal{A}$ be the coarse-grained model and the relevant concrete model of a reaction network respectively, $\forall q^\sharp = Q, q^{\sharp\prime} = Q' \in Q^\sharp \subseteq Q$, we have:*

$$\text{Reach}^-_{\mathcal{A}^\sharp_I}(q^\sharp_1, q^\sharp_2) \leq \text{Reach}^-_{\mathcal{A}}(Q, Q'), \quad \text{Reach}^+_{\mathcal{A}^\sharp_I}(q^\sharp_1, q^\sharp_2) \geq \text{Reach}^+_{\mathcal{A}}(Q, Q').$$

Proof. Let $\omega^\sharp$ denote an abstract path starting from $q^\sharp$ and reaching $q^{\sharp\prime}$. For any $\omega^\sharp \in \text{Path}_{\mathcal{A}^\sharp_I}(q^\sharp, q^{\sharp\prime})$, such as $\omega^\sharp_0 = q^\sharp$, $\omega^\sharp_{|\omega^\sharp|} = q^{\sharp\prime}$, assume $|\omega^\sharp| = n \in \mathbb{N}$, and let $\omega \in \text{Path}_{\mathcal{A}}(q, q')$ denote a concrete path starting from a state in Q and reaching a state Q', we have:

$$\texttt{Reach}^-(q^\sharp, q^{\sharp\prime}) = \sum_{\omega^\sharp} \texttt{Prob}^-_{q^\sharp}(\omega^\sharp) = \sum_{\omega^\sharp} \left(\texttt{Prob}^-(\omega^\sharp_0, \omega^\sharp_1) \times \cdots \times \texttt{Prob}^-(\omega^\sharp_{n-1}, \omega^\sharp_n) \right)$$

$$= \sum_{\omega^\sharp} \left(\prod_{i=0}^{n-1} \inf\{\texttt{Prob}(q_i, q_{i+1}) | q_i \in \omega^\sharp_i, q_{i+1} \in \omega^\sharp_{i+1}\} \right)$$

$$\leq \sum_{\omega^\sharp} \inf\{\texttt{Prob}(q_0, q_n) | q_0 \in \omega^\sharp_0, q_n \in \omega^\sharp_n\}$$

$$= \sum_{\omega} \inf\{\texttt{Prob}(\omega_0, \omega_n) | \omega_0 \in Q, \omega_n \in Q'\} = \texttt{Reach}^-(Q, Q').$$

Similarly, we have $\texttt{Reach}^+(q^\sharp, q^{\sharp\prime}) \geq \texttt{Reach}^+(Q, Q')$.   □

Example 7. Consider again the reaction network described in Example 6:

- by applying the coarse-grained model shown in Fig. 3, we can calculate the probability of movement from $q^\sharp_{\{a,b,c,d\}}$ to $q^\sharp_{\{a,b\}}$ is: $[0.1506, 1]$; the probability of movement from $q^\sharp_{\{a,b,c,d\}}$ to $q^\sharp_{\{a\}}$ is: $[0.2314, 1]$;
- by applying the concrete model shown in Fig. 7, we obtain the probability of movement from $Q_{\{a,b,c,d\}}$ to $Q_{\{a,b\}}$ is: $[0.1776, 0.8268]$; the probability of movement from $Q_{\{a,b,c,d\}}$ to $Q_{\{a\}}$ is: $[1, 1]$.

Note that our abstract model safely approximates the concrete one.

7   An Application of the Coarse-Grained Model

This section presents an application of our organisation-oriented coarse grained model. We address the following problem: given a reaction network and a fixed number of the maximum population of the system, construct the average-based organisation coarse-grained model $\bar{\mathcal{A}}^\sharp$ (focus on the average transition probabilities between abstract states for simplicity and intuition, this can be replaced by interval-based transitions directly), can we predict the behaviour of the system at any future time using $\bar{\mathcal{A}}^\sharp$?

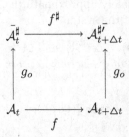

The diagram to the left captures the idea of using the organisation-based coarse grained model to approximate the concrete one. In the concrete world, $\mathcal{A}_t$ denotes the concrete model at time t, f denotes the dynamical transition function over $\mathcal{A}_t$, and $\mathcal{A}_{t+\Delta t}$ denotes the concrete model after Δt time units; g_o denotes the organisation based coarse graining function, which maps the concrete model $\mathcal{A}_t$ (c.f. $\mathcal{A}_{t+\Delta t}$) to the average coarse-grained model $\bar{\mathcal{A}}^\sharp_t$ (c.f. $\bar{\mathcal{A}}^\sharp_{t+\Delta t}$); $f^\sharp$ denotes the coarse-graining dynamical transition function on $\bar{\mathcal{A}}^\sharp_t$.

We apply the traditional "master equation" approach to calculate the stochastic time evolution of the reaction network. We briefly review the main features of the master equation formalism for our purpose of calculating the prediction of an reaction network at any future time. The probability function

$P(X_1, X_2, \ldots, X_n; t)$ defines the probability of number of X_i molecules of species S_i for $i \in \{1, \ldots, n\}$ at time t. This function thus describes the "stochastic state" of the system at time t. The master equation is the time-evolution equation for the function $P(t)$. Function $P(X_1, \ldots, X_n; t+dt)$ can be viewed as the sum of the probabilities of different ways that the system can reach the state $X_1, \ldots, X_n$ at time $t+dt$: $P(X_1, \ldots, X_n; t+dt) = P(X_1, \ldots, X_n; t)(1 - \sum_{i=1}^{m} \alpha_i dt) + \sum_{j=1}^{n} \beta_j dt$, where the quantity $\beta_j dt$ denotes the probability that the system is entering the state $(X_1, \ldots, X_n)$ at time $t + dt$, and the quantity $\alpha_i dt$ denotes the probability that is leaving $(X_1, \ldots, X_n)$ at time t. To avoid confusion, we use $P(t)$ as a short notation of $P(X_1, \ldots, X_n; t)$. Consider a coarse-grained model $\mathcal{A}^\sharp$, and any abstract state $q_i^\sharp$, let α_i denote the average rate of leaving state $q_i^\sharp$, i.e., $\frac{dP_i(t)}{dt} = -\alpha_i P_i(t)$, E_i denote the expected time to leave state $q_i^\sharp$, we have: $E_i = \int_0^\infty P_i(t)dt = \int_0^\infty e^{-\alpha_i t}dt = \frac{1}{\alpha_i}$, i.e., $\alpha_i = \frac{1}{E_i}$ is the rate of leaving $q_i^\sharp$. In addition, for any $j \neq i$ and $\Delta^\sharp(q_j^\sharp, q_i^\sharp) > 0$, similarly, $\beta_j = \frac{1}{E_j}$ is the rate of coming to $q_i^\sharp$ from $q_j^\sharp$. Therefore, for all $i \in \{1, \ldots, n\}$, we have:

$$\frac{dP_i(t)}{dt} = -\frac{1}{E_i}P_i(t) + \sum_{j=0, j \neq i, \Delta^\sharp(q_j^\sharp, q_i^\sharp) > 0}^{n} \frac{1}{E_j}P_j(t).$$

We therefore build a set of equations for all i. By solving the set of equations, we can obtain the distributions of the system at any future time.

Example 8. Consider again Example 6. Due to the coarse-grained model shown in Fig. 3 ($N_{\max} = 5$), we construct the master equations as follows:

$$\begin{cases} \frac{dP_{\{a,b,c,d\}}(t)}{dt} = -\frac{1}{0.59}P_{\{a,b,c,d\}}(t) \\ \frac{dP_{\{a,b,c\}}(t)}{dt} = -\frac{1}{0.66}P_{\{a,b,c\}}(t) + 0.34 * \frac{1}{0.59}P_{\{a,b,c,d\}}(t) \\ \frac{dP_{\{a,b\}}(t)}{dt} = -\frac{1}{3.46}P_{\{a,b\}}(t) + 0.217 * \frac{1}{0.59}P_{\{a,b,c,d\}}(t) + 0.78 * \frac{1}{0.66}P_{\{a,b,c\}}(t) \\ \frac{dP_{\{a,d\}}(t)}{dt} = -\frac{1}{3.46}P_{\{a,d\}}(t) + 0.356 * \frac{1}{0.59}P_{\{a,b,c,d\}}(t) \\ \frac{dP_{\{a\}}(t)}{dt} = \frac{0.087}{0.59}P_{\{a,b,c,d\}}(t) + \frac{0.22}{0.66}P_{\{a,b,c\}}(t) + \frac{1}{3.46}P_{\{a,b\}}(t) + \frac{1}{3.46}P_{\{a,d\}}(t) \end{cases}$$

By solving the above equation systems, Fig. 4 presents a comparison between the time evolution of the reaction network via master equation simulation through the organisation-based average coarse-grained model (left) and the exact evolution of the system through the original concrete model (right).

Figures 11 and 12 (in Appendix) present experimental results for the case of $N_{max} = 10$ with rates R1 and R2 respectively. Note that our prediction produces a similar pattern of the concrete behaviours with time evolution for this case. We focus on the average-based approximation here for the purpose of presenting and comparing the pattern of the system behaviours with time evolution more clearly and intuitively. Our further experiments also demonstrate that the interval-based prediction can safely approximate the concrete model. The precision of the results varies regarding to different models and rates of reaction rules, however the basic pattern of behaviours can be captured. Further note that the

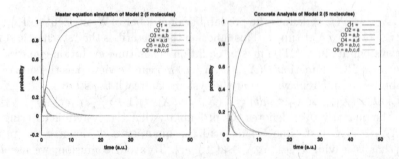

Fig. 4. Organisation dynamics predication via the average coarse-grained model(left) and the concrete model(right) of Example 6, for $N_{max} = 5$. (Color figure online)

coarse-grained model can be used to predict qualitative dynamical properties of the original model, like the absence (or presence) of asymptotically stable attractors inside organizations that have small (or large) time to leave probabilities.

8   Conclusions

This paper investigates the combination of chemical organisation theory and probabilistic model checking for the analysis of reaction networks modelled as continuous-time Markov chains. We use model decompositions into strongly connected components (SCCs), and study the problem of how to analyse the model in terms of organisations. We have presented an algorithm to compute a coarse-grained Markov chain model of hierarchic organisations for a given reaction network. The algorithm computes chemical organisations by identifying a set of good SCCs which can contribute to generating organisations, and building an interval Markov chain based on the organisation-based quantitative analysis.

Experiments with our method on a set of example reaction network models demonstrate that the movements between organisations and the expected time spent in them can approximate the concrete long-term behaviour of the reaction network. The organisation-based coarse grained model helps to summarise and reason about the structure and behaviour of the complex model by focusing on stable states featuring accumulating species.

We also demonstrate how our model can be used to approximate the system behaviour with time evolution. The experiments show that our prediction can mimic the main pattern of concrete behaviour in the long run, but the interval-based organisation coarse graining may suffer from over-estimation. We apply an average-based organisation coarse graining and compute its stochastic time evolution. Our experiments show that the precision of the prediction and approximation varies regarding to different models and rates of their reaction rules. As future work, to improve the precision of the approximation and predictions, we plan to develop algorithms to selectively refine the coarse-grained models.

Acknowledgements. The authors acknowledge support from the European Union through funding under FP7–ICT–2011–8 project HIERATIC (316705). We also thank the anonymous reviewers for their helpful and detailed comments.

APPENDIX: Supplementary Details for Examples 5 and 6

```
ctmc

const int N_MAX = 10;
const double rA = N_MAX; // rA
const double rB = N_MAX; // rB
formula total = a + b;
init total <= N_MAX endinit

module RN
  a : [0..N_MAX];  // range value of species a
  b : [0..N_MAX];  // range value of species b
  c : [0..N_MAX];  // range value of species c

  // r1: a+b -> a
  [r1] (a*b > 0) & (a > 0) & (b > 0) & (total<= N_MAX) -> a*b : (a'=a-1) & (b'=b);

  // r2: a -> 2a
  [r2] (rA*a > 0) & (a > 0) & (total+1<= N_MAX) -> rA*a : (a'=a+1) ;

  // r3: b -> 2b
  [r3] (rB*b > 0) & (b > 0) & (total<= N_MAX) -> rB*b :  (b'=b+1) ;

  // r4: a -> 0
  [r4] (a*a > 0) & (total<= N_MAX) -> a*a : (a'=a-1);

  // r5: b -> 0
  [r5] (a*b > 0) & (total<= N_MAX) -> a*b : (b'=b-1);

endmodule
```

Fig. 5. Example 5 in the PRISM modelling language

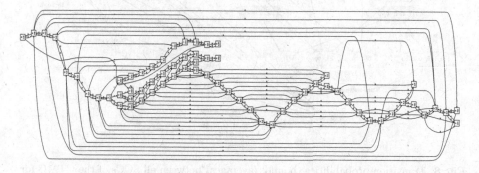

Fig. 6. Example 5: CTMC model with 4 SCCs and 1 BSCC

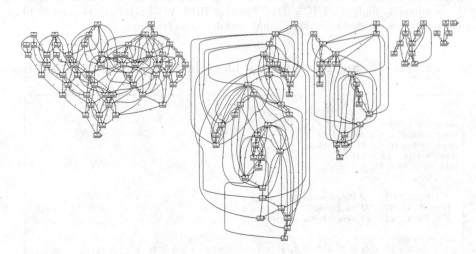

Fig. 7. CTMC for the reaction network from Example 6, with 28 SCCs and 6 BSCCs.

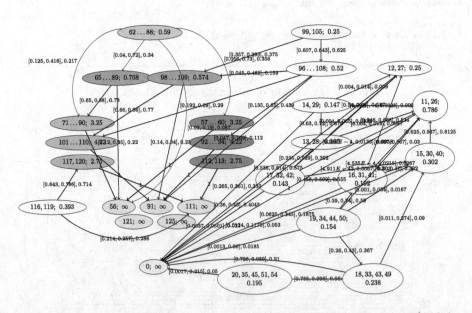

Fig. 8. Transition probabilities (bounds/averages) between all SCCs of the CTMC for Example 6 and expected leaving times.

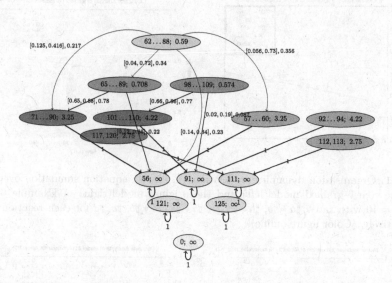

Fig. 9. Transition probabilities (bounds and average) between good SCCs for Example 6 and the expected time to leave them.

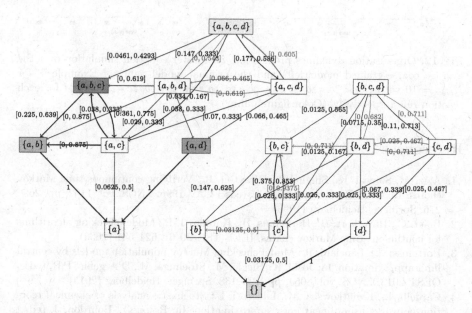

Fig. 10. Example 6: transition probabilities in bounds among the lattice of molecules

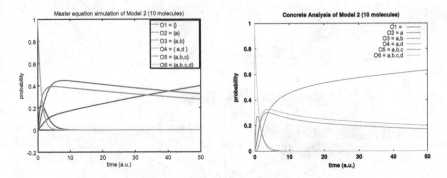

Fig. 11. Organisation dynamics predication via master equation simulation over the average coarse-grained model(left) and the original model(right) of Example 6, for $N_{max} = 10$ with rates: $\sharp a * \sharp b$, $\sharp b * \sharp c$, $\sharp c$, $\sharp b * \sharp d$, $\sharp b * \sharp b$, $\sharp c$, $\sharp d$ for each reaction rule respectively. (Color figure online)

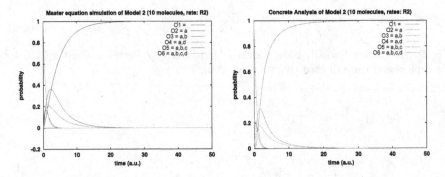

Fig. 12. Organisation dynamics predication via master equation simulation over the average coarse-grained model(left) and the original model(right) of Example 6, for $N_{max} = 10$ with rates: $\sharp a * \sharp b$, $\sharp b * \sharp c$, $\sharp c * \sharp c$, $\sharp b * \sharp d$, $\sharp b * \sharp b$, $\sharp c * \sharp c$, $\sharp d * \sharp d$ for each reaction rule respectively. (Color figure online)

References

1. Aziz, A., Sanwal, K., Singhal, V., Brayton, R.: Verifying continuous time Markov chains. In: Alur, R., Henzinger, T.A. (eds.) CAV 1996. LNCS, vol. 1102, pp. 269–276. Springer, Heidelberg (1996)
2. Baier, C., Haverkort, B., Hermanns, H., Katoen, J.-P.: Model-checking algorithms for continuous-time Markov chains. IEEE TSE **29**(6), 524–541 (2003)
3. Bortolussi, L., Lanciani, R.: Model checking Markov population models by central limit approximation. In: Joshi, K., Siegle, M., Stoelinga, M., D'Argenio, P.R. (eds.) QEST 2013. LNCS, vol. 8054, pp. 123–138. Springer, Heidelberg (2013)
4. Cardelli, L., Kwiatkowska, M., Laurenti, L.: Stochastic analysis of chemical reaction networks using linear noise approximation. In: Roux, O., Bourdon, J. (eds.) CMSB 2015. LNCS, vol. 9308, pp. 64–76. Springer, Heidelberg (2015)
5. Clarke, E., Grumberg, O., Peled, D.: Model Checking. The MIT Press, Cambridge (2000)

6. Dittrich, P., Ziegler, J., Banzhaf, W.: Artificial chemistries-a review. Artif. Life **7**(3), 225–275 (2001)
7. Dittrich, P., di Fenizio, P.: Chemical organisation theory. Bull. Math. Biol. **69**(4), 1199–1231 (2007)
8. Feinberg, M., Horn, F.J.M.: Dynamics of open chemical systems and the algebraic structure of the underlying reaction network. Chem. Eng. Sci. **29**(3), 775–787 (1973)
9. Fontana, W.: Algorithmic chemistry. In: Artificial Life II. Addison Wesley (1992)
10. Fontana, W., Buss, L.W.: "The arrival of the fittest": toward a theory of biological organization. Bull. Math. Biol. **56**(1), 1–64 (1994)
11. Gillespie, D.: Exact stochastic simulation of coupled chemical reactions. J. Phys. Chem. **81**(25), 2340–2361 (1977)
12. Heinrich, R., Schuster, S.: The Regulation of Cellular Systems. Chapman and Hall, New York (1996)
13. Kleinberg, J., Tardos, É.: Algorithm Design. Addison-Wesley, Boston (2006)
14. Kreyssig, P., Wozar, C., Peter, S., Veloz, T., Ibrahim, B., Dittrich, P.: Effects of small particle numbers on long-term behaviour in discrete biochemical systems. Bioinformatics **30**(17), 475–481 (2014)
15. Kwiatkowska, M., Norman, G., Parker, D.: PRISM 4.0: verification of probabilistic real-time systems. In: Gopalakrishnan, G., Qadeer, S. (eds.) CAV 2011. LNCS, vol. 6806, pp. 585–591. Springer, Heidelberg (2011)
16. Tarjan, R.: Depth first search and linear graph algorithms. SIAM J. Comput. **1**(2), 146–160 (1972)
17. Wolf, V., Goel, R., Mateescu, M., Henzinger, T.: Solving the chemical master equation using sliding windows. BMC Syst. Biol. J. **4**, 42 (2010)

Goal-Oriented Reduction of Automata Networks

Loïc Paulevé$^{(\boxtimes)}$

LRI UMR 8623, Univ. Paris-Sud – CNRS, Université Paris-Saclay,
91405 Orsay, France
loic.pauleve@lri.fr

Abstract. We consider networks of finite-state machines having local transitions conditioned by the current state of other automata. In this paper, we introduce a reduction procedure tailored for reachability properties of the form "from global state s, there exists a sequence of transitions leading to a state where an automaton g is in a local state $\top$". By analysing the causality of transitions within the individual automata, the reduction identifies local transitions which can be removed while preserving *all* the minimal traces satisfying the reachability property. The complexity of the procedure is polynomial with the total number of local transitions, and exponential with the maximal number of local states within an automaton. Applied to Boolean and multi-valued networks modelling dynamics of biological systems, the reduction can shrink down significantly the reachable state space, enhancing the tractability of the model-checking of large networks.

1 Introduction

Automata networks model dynamical systems resulting from simple interactions between entities. Each entity is typically represented by an automaton with few internal states which evolve subject to the state of a narrow range of other entities in the network. Richness of emerging dynamics arises from several factors including the topology of the interactions, the presence of feedback loop, and the concurrency of transitions.

Automata networks, which subsume Boolean and multi-valued networks, are notably used to model dynamics of biological systems, including signalling networks or gene regulatory networks (e.g., [1, 10, 15, 21, 31–33, 38]). The resulting models can then be confronted with biological knowledge, for instance by checking if some time series data can be reproduced by the computational model. In the case of models of signalling or gene regulatory networks, such data typically refer to the possible activation of a transcription factor, or a gene, from a particular state of the system, which reflects both the environment and potential perturbations. Automata networks have also been used to infer targets to control the behaviour of the system. For instance, in [1, 32], the author use Boolean networks to find combinations of signals or combinations of mutations that should alter the cellular behaviour.

From a formal point of view, numerous biological properties can be expressed in computation models as reachability properties: from an initial state, or set

© Springer International Publishing AG 2016
E. Bartocci et al. (Eds.): CMSB 2016, LNBI 9859, pp. 252–272, 2016.
DOI: 10.1007/978-3-319-45177-0_16

of states, the existence of a sequence of transitions which leads to a desired state, or set of states. For instance, an initial state can represent a combination of signals/perturbations of a signalling network; and the desired states the set of states where the concerned transcription factor is active. One can then verify the (im)possibility of such an activation, possibly by taking into account mutations, which can be modelled, for instance, as the freezing of some automata to some fixed states, or by the removal of some transitions.

Due to the increasing precision of biological knowledge, models of networks become larger and larger and can gather hundreds to thousands of interacting entities making the formal analysis of their dynamics a challenging task: the reachability problem in automata networks/bounded Petri nets is PSPACE-complete [7], which limits its scalability.

Facing a model too large for a raw exhaustive analysis, a natural approach is to reduce its dynamics while preserving important properties. Multiple approaches, often complementary, have been explored since decades to address such a challenge in dynamical and concurrent systems [22,24,36]. In the scope of rule-based models of biological networks, efficient static analysis methods have been developed to lump numerous global states of the systems based on the fragmentation of interacting components [14]; and to *a posteriori* compress simulated traces to obtain compact witnesses of dynamical properties [12]. Reductions preserving the attractors of dynamics (long-term/steady-state behaviour) have also been proposed for chemical reaction networks [25] and Boolean networks [26]. The latter approach applies to formalisms close to automata networks but does not preserve reachability properties. On Petri nets, different structural reductions have been proposed to reduce the size of the model specification while preserving bisimulation [34], or liveness and LTL properties [4,17]. Procedures such as the cone of influence reduction [5] or relevant subnet computation [37] allow to identify variables/transitions which have no influence on a given dynamical property. Our work has a motivation similar to the two latter approaches.

Contribution. We introduce a reduction of automata networks which identifies transitions that do not contribute to a given reachability property and hence can be ignored. The considered automata networks are finite sets of finite-state machines where transitions between their local states are conditioned by the state of other automata in the network. We use a general concurrent semantics where any number of automata can apply one transition within one step. We call a *trace* a sequential interleaved execution of steps.

Our reduction preserves all the minimal traces satisfying reachability properties of the form "from state s there exist successive steps that lead to a state where a given automaton g is in local state $g_\top$". A trace is *minimal* if no step nor transition can be removed from it and resulting in a sub-trace that satisfies the concerned reachability property. The complexity of the procedure is polynomial in the number of local transitions, and exponential in the maximal size of automata. Therefore, the reduction is scalable for networks of multiple automata, where each have a few local states.

The identification of the transitions that are not part of any minimal trace is performed by a static analysis of the causality of transitions within automata. It extends previous static analysis of reachability properties by abstract interpretation [28,29]. In [29], necessary or sufficient conditions for reachability are derived, but they do not allow to capture all the (minimal) traces towards a reachability goal. In [28], the static analysis extracts local states, referred to as cut-sets, which are necessarily reached prior to a given reachability goal. The results presented here are orthogonal: we identify transitions that are never part of a minimal trace for the given reachability property. It allows us to output a reduced model where all such transitions are removed while preserving all the minimal traces for reachability. Hence, whereas [28] focuses on identifying necessary conditions for reachability, this article focuses on preserving sufficient conditions for reachability.

The effectiveness of our goal-oriented reduction is experimented on actual models of biological networks and show significant shrinkage of the dynamics of the automata networks, enhancing the tractability of a concrete verification. Compared to other model reductions, our goal is similar to the cone of influence reduction [5] or relevant subnet computation [37] mentioned above, which identify variables/transitions that do not impact a given property. Here, our approach offers a much more fine-grained analysis in order to identify the sufficient transitions and values of variables that contribute to the property, which leads to stronger reductions.

Outline. Section 2 sets up the definition and semantics of the automata networks considered in this paper, together with the local causality analysis for reachability properties, based on prior work. Section 3 first depicts a necessary condition using local causality analysis for satisfying a reachability property and then introduce the goal-oriented reduction with the proof of minimal traces preservation. Section 4 shows the efficiency of the reduction on a range of biological networks. Finally, Sect. 5 discusses the results and motivates further work.

Notations. Integer ranges are noted $[m;n] \triangleq \{m, m+1, \cdots, n\}$. Given a finite set A, $|A|$ is the cardinality of A; 2^A is the power set of A. Given $n \in \mathbb{N}$, $x = (x^i)_{i \in [1;n]}$ is a sequence of elements indexed by $i \in [1;n]$; $|x| = n$; $x^{m..n}$ is the subsequence $(x^i)_{i \in [m;n]}$; $x :: e$ is the sequence x with an additional element e at the end; ε is the empty sequence.

2   Automata Networks and Local Causality

2.1   Automata Networks

We declare an Automata Network (AN) with a finite set of finite-state machines having transitions between their local states conditioned by the state of other automata in the network. An AN is defined by a triple (Σ, S, T) (Definition 1) where Σ is the set of automata identifiers; S associates to each automaton a finite set of local states: if $a \in \Sigma$, $S(a)$ refers to the set of local states of a; and T associates to each automaton its local transitions. Each local state is written of the form a_i, where $a \in \Sigma$ is the automaton in which the state belongs to, and i is a unique identifier; therefore given $a_i, a_j \in S(a)$, $a_i = a_j$ if and only if a_i and a_j refer to the same local state of the automaton a. For each automaton $a \in \Sigma$, $T(a)$ refers to the set of transitions of the form $t = a_i \xrightarrow{\ell} a_j$ with $a_i, a_j \in S(a)$, $a_i \neq a_j$, and ℓ the enabling condition of t, formed by a (possibly empty) set of local states of automata different than a and containing at most one local state of each automaton. The *pre-condition* of transition t, noted ${}^\bullet t$, is the set composed of a_i and of the local states in ℓ; the *post-condition*, noted $t^\bullet$ is the set composed of a_j and of the local states in ℓ.

Definition 1 (Automata Network (Σ, S, T)). *An* Automata Network *(AN) is defined by a tuple (Σ, S, T) where*

- Σ *is the finite set of automata identifiers;*
- *For each $a \in \Sigma$, $S(a) = \{a_i, \ldots, a_j\}$ is the finite set of local states of automaton a; $S \overset{\Delta}{=} \prod_{a \in \Sigma} S(a)$ is the finite set of global states;*
  LS $\overset{\Delta}{=} \bigcup_{a \in \Sigma} S(a)$ *denotes the set of all the local states.*
- $T = \{a \mapsto T_a \mid a \in \Sigma\}$, *where $\forall a \in \Sigma, T_a \subseteq S(a) \times 2^{\mathbf{LS} \setminus S(a)} \times S(a)$ with $(a_i, \ell, a_j) \in T_a \Rightarrow a_i \neq a_j$ and $\forall b \in \Sigma, |\ell \cap S(b)| \leq 1$, is the mapping from automata to their finite set of local transitions.*

We note $a_i \xrightarrow{\ell} a_j \in T \overset{\Delta}{\Leftrightarrow} (a_i, \ell, a_j) \in T(a)$ and $a_i \to a_j \in T \overset{\Delta}{\Leftrightarrow} \exists \ell \in 2^{\mathbf{LS} \setminus S(a)}, a_i \xrightarrow{\ell} a_j \in T$. Given $t = a_i \xrightarrow{\ell} a_j \in T$, $\mathrm{orig}(t) \overset{\Delta}{=} a_i$, $\mathrm{dest}(t) \overset{\Delta}{=} a_j$, $\mathrm{enab}(t) \overset{\Delta}{=} \ell$, ${}^\bullet t \overset{\Delta}{=} \{a_i\} \cup \ell$, and $t^\bullet \overset{\Delta}{=} \{a_j\} \cup \ell$.

At any time, each automaton is in one and only one local state, forming the global state of the network. Assuming an arbitrary ordering between automata identifiers, the set of global states of the network is referred to as S as a shortcut for $\prod_{a \in \Sigma} S(a)$. Given a global state $s \in S$, $s(a)$ is the local state of automaton a in s, i.e., the a-th coordinate of s. Moreover we write $a_i \in s \overset{\Delta}{\Leftrightarrow} s(a) = a_i$; and for any $ls \in 2^{\mathbf{LS}}$, $ls \subseteq s \overset{\Delta}{\Leftrightarrow} \forall a_i \in ls, s(a) = a_i$.

In the scope of this paper, we allow, but do not enforce, the parallel application of transitions in different automata. This leads to the definition of a *step* as a set of transitions, with at most one transition per automaton (Definition 2). For notational convenience, we allow empty steps. The pre-condition (resp. post-condition) of a step τ, noted ${}^\bullet \tau$ (resp. $\tau^\bullet$), extends the similar notions on transitions: the pre-condition (resp. post-condition) is the union of the pre-conditions (resp. post-conditions) of composing transitions. A step τ is *playable* in a state $s \in S$ if and only if ${}^\bullet \tau \subseteq s$, i.e., all the local states in the pre-conditions of transitions are in s. If τ is playable in s, $s \cdot \tau$ denotes the state after the applications of all the transitions in τ, i.e., where for each transition $a_i \xrightarrow{\ell} a_j \in \tau$, the local state of automaton a has been replaced with a_j.

Definition 2 (Step). *Given an AN (Σ, S, T), a step τ is a subset of local transitions T such that for each automaton $a \in \Sigma$, there is at most one local transition $T(a)$ in τ ($\forall a \in \Sigma, |(\tau \cap T(a))| \leq 1$).*

We note $^\bullet \tau \triangleq \bigcup_{t \in \tau} {}^\bullet t$ and $\tau^\bullet \triangleq \bigcup_{t \in \tau} t^\bullet \setminus \{\text{orig}(t) \mid t \in \tau\}$.

Given a state $s \in S$ where τ is playable ($^\bullet \tau \subseteq s$), $s \cdot \tau$ denotes the state where $\forall a \in \Sigma$, $(s \cdot \tau)(a) = a_j$ if $\exists a_i \to a_j \in \tau$, and $(s \cdot \tau)(a) = s(a)$ otherwise.

Remark that $\tau^\bullet \subseteq s \cdot \tau$ and that this definition implicitly rules out steps composed of incompatible transitions, i.e., where different local states of a same automaton are in the pre-condition.

A *trace* (Definition 3) is a sequence of successively playable steps from a state $s \in S$. The pre-condition $^\bullet \pi$ of a trace π is the set of local states that are required to be in s for applying π ($^\bullet \pi \subseteq s$); and the post-condition $\pi^\bullet$ is the set of local states that are present in the state after the full application of π ($\pi^\bullet \subseteq s \cdot \pi$).

Definition 3 (Trace). *Given an AN (Σ, S, T) and a state $s \in S$, a trace π is a sequence of steps such that $\forall i \in [1; |\pi|]$, $^\bullet \pi^i \subseteq (s \cdot \pi^1 \cdots \pi^{i-1})$.*

The pre-condition $^\bullet \pi$ and the post-condition $\pi^\bullet$ are defined as follows: for all $n \in [1; |\pi|]$, for all $a_i \in {}^\bullet \pi^n$, $a_i \in {}^\bullet \pi \triangleq \forall m \in [1; n-1], S(a) \cap \pi^m = \emptyset$; similarly, for all $n \in [1; |\pi|]$, for all $a_j \in \pi^{n\bullet}$, $a_j \in \pi^\bullet \triangleq \forall m \in [n+1; m], S(a) \cap \pi^{m\bullet} = \emptyset$. If π is empty, $^\bullet \pi = \pi^\bullet = \emptyset$.

The set of transitions composing a trace π is noted $\text{tr}(\pi) \triangleq \bigcup_{n=1}^{|\pi|} \pi^n$.

Given an automata network (Σ, S, T) and a state $s \in S$, the local state $g_T \in \mathbf{LS}$ is *reachable* from s if and only if either $g_T \in s$ or there exists a trace π with $^\bullet \pi \subseteq s$ and $g_T \in \pi^\bullet$.

We consider a trace π for g_T reachability from s is *minimal* if and only if there exists no different trace reaching g_T having each successive step being a subset of a step in π with the same ordering (Definition 4). Say differently, a trace is minimal for g_T reachability if no step or transition can be removed from it without breaking the trace validity or g_T reachability.

Definition 4 (Minimal trace for local state reachability). *A trace π is minimal w.r.t. g_T reachability from s if and only if there is no trace ϖ from s, $\varpi \neq \pi$, $|\varpi| \leq |\pi|$, $g_T \in \varpi^\bullet$, such that there exists an injection $\phi : [1; |\varpi|] \to [1; |\pi|]$ with $\forall i, j \in [1; |\varpi|]$, $i < j \Leftrightarrow \phi(i) < \phi(j)$ and $\varpi^i \subseteq \pi^{\phi(i)}$.*

Automata networks as presented can be considered as a class of 1-safe Petri Nets [3] (at most one token per place) having groups of mutually exclusive places, acting as the automata, and where each transition has one and only one incoming and out-going arc and any number of read arcs. The semantics considered in this paper where transitions within different automata can be applied simultaneously echoes with Petri net step-semantics and concurrent/maximally concurrent semantics [19,20,30]. In the Boolean network community, such a semantics is referred to as the asynchronous generalized update schedule [2].

2.2   Local Causality

Locally reasoning within one automaton a, the reachability of one of its local state a_j from some global state s with $s(a) = a_i$ can be described by a (local) *objective*, that we note $a_i \rightsquigarrow a_j$ (Definition 5).

Definition 5 (Objective). *Given an automata network (Σ, S, T), an objective is a pair of local states $a_i, a_j \in S(a)$ of a same automaton $a \in \Sigma$ and is denoted $a_i \rightsquigarrow a_j$. The set of all objectives is referred to as $\mathbf{Obj} \overset{\Delta}{=} \{a_i \rightsquigarrow a_j \mid (a_i, a_j) \in S(a) \times S(a), a \in \Sigma\}$.*

Given an objective $a_i \rightsquigarrow a_j \in \mathbf{Obj}$, local-paths$(a_i \rightsquigarrow a_j)$ is the set of local acyclic paths of transitions $T(a)$ within automaton a from a_i to a_j (Definition 6).

Definition 6 (local-paths). *Given $a_i \rightsquigarrow a_j \in \mathbf{Obj}$, if $i = j$, local-paths$(a_i \rightsquigarrow a_i) \overset{\Delta}{=} \{\varepsilon\}$; if $i \neq j$, a sequence η of transitions in $T(a)$ is in local-paths$(a_i \rightsquigarrow a_j)$ if and only if $|\eta| \geq 1$, $\mathrm{orig}(\eta^1) = a_i$, $\mathrm{dest}(\eta^{|\eta|}) = a_j$, $\forall n \in [1; |\eta| - 1]$, $\mathrm{dest}(\eta^n) = \mathrm{orig}(\eta^{n+1})$, and $\forall n, m \in [1; |\eta|], n > m \Rightarrow \mathrm{dest}(\eta^n) \neq \mathrm{orig}(\eta^m)$.*

As stated by Property 1, any trace reaching a_j from a state containing a_i uses all the transitions of at least one local acyclic path in local-paths$(a_i \rightsquigarrow a_j)$.

Property 1. For any trace π, for any $a \in \Sigma$, $a_i, a_j \in S(a)$, $1 \leq n \leq m \leq |\pi|$ where $a_i \in {}^{\bullet}\pi^n$ and $a_j \in \pi^{m\bullet}$, there exists a local acyclic path $\eta \in$ local-paths$(a_i \rightsquigarrow a_j)$ that is a sub-sequence of $\pi^{n..m}$, i.e., there is an injection $\phi : [1; |\eta|] \to [n; m]$ with $\forall u, v \in [1; |\eta|], u < v \Leftrightarrow \phi(u) < \phi(v)$ and $\eta^u \in \pi^{\phi(u)}$.

A local path is not necessarily a trace, as transitions may be conditioned by the state of other automata that may need to be reached beforehand. A local acyclic path being of length at most $|S(a)|$ with unique transitions, the number of local acyclic paths is polynomial in the number of transitions $T(a)$ and exponential in the number of local states in a.

Example 1. Let us consider the automata network (Σ, S, T), graphically represented in Fig. 1, where:

$$\Sigma = \{a, b, c, d\}$$

$$S(a) = \{a_0, a_1\} \quad T(a) = \{a_0 \xrightarrow{\{b_0\}} a_1, a_1 \xrightarrow{\emptyset} a_0\}$$

$$S(b) = \{b_0, b_1\} \quad T(b) = \{b_0 \xrightarrow{\{a_1\}} b_1, b_1 \xrightarrow{\{a_0\}} b_0\}$$

$$S(c) = \{c_0, c_1, c_2\} \quad T(c) = \{c_0 \xrightarrow{\{a_1\}} c_1, c_1 \xrightarrow{\{b_1\}} c_0, c_1 \xrightarrow{\{b_0\}} c_2, c_0 \xrightarrow{\{d_1\}} c_2\}$$

$$S(d) = \{d_0, d_1\} \quad T(d) = \emptyset$$

The local paths for the objective $c_0 \rightsquigarrow c_2$ are local-paths$(c_0 \rightsquigarrow c_2) = \{c_0 \xrightarrow{\{a_1\}} c_1 \xrightarrow{\{b_0\}} c_2, c_0 \xrightarrow{\{d_1\}} c_2\}$. From the state , A_0, b_0, c_0, d_0, instances of traces are $\{a_0 \xrightarrow{\{b_0\}} a_1\} :: \{b_0 \xrightarrow{\{a_1\}} b_1, c_0 \xrightarrow{\{a_1\}} c_1\} :: \{a_1 \xrightarrow{\emptyset} a_0\} :: \{b_1 \xrightarrow{\{a_0\}} b_0\} :: \{c_1 \xrightarrow{\{b_0\}} c_2\};$
$\{a_0 \xrightarrow{\{b_0\}} a_1\} :: \{c_0 \xrightarrow{\{a_1\}} c_1\} :: \{c_1 \xrightarrow{\{b_0\}} c_2\};$
the latter only being a minimal trace for c_2 reachability.

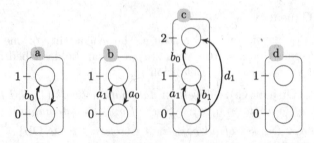

Fig. 1. An example of automata network. Automata are represented by labelled boxes, and local states by circles where ticks are their identifier within the automaton – for instance, the local state a_0 is the circle ticked 0 in the box a. A transition is a directed edge between two local states within the same automaton. It can be labelled with a set of local states of other automata. In this example, all the transitions are conditioned by at most one other local state.

3   Goal-Oriented Reduction

Assuming a global AN (Σ, S, T), an initial state $s \in S$ and a reachability goal $g_\top$ where $g \in \Sigma$ and $g_\top \in S(g)$, the goal-oriented reduction identifies a subset of local transitions T that are sufficient for producing all the minimal traces leading to $g_\top$ from s. The reduction procedure takes advantage of the local causality analysis both to fetch the transitions that matter for the reachability goal and to filter out objectives that can be statically proven impossible.

3.1   Necessary Condition for Local Reachability

Given an objective $a_i \rightsquigarrow a_j$ and a global state $s \in S$ where $s(a) = a_i$, prior work has demonstrated necessary conditions for the existence of a trace leading to a_j from s [28,29]. Those necessary conditions rely on the local causality analysis defined in previous section for extracting necessary steps that have to be performed in order to reach the concerned local state.

Several necessary conditions have been established in [29], taking into account several features captured by the local paths (dependencies, sequentiality, partial order constraints, ...). The complexity of deciding most of these necessary conditions is polynomial in the total number of local transitions and exponential in the maximum number of local states within an automaton.

In this section, we consider a generic reachability over-approximation predicate **valid**$_s$ which is false only when applied to an objective that has no trace concretizing it from s: a_j is reachable from s with $s(a) = a_i$ only if **valid**$_s(a_i \rightsquigarrow a_j)$.

Definition 7 (valid$_s$). *Given any objective* $a_i \rightsquigarrow a_j \in \mathbf{Obj}$, **valid**$_s(a_i \rightsquigarrow a_j)$ *if there exists a trace* π *from* s *such that* $\exists m, n \in [1; |\pi|]$ *with* $m \leq n$, $a_i \in {}^\bullet\pi^m$, *and* $a_j \in \pi^{n\bullet}$.

For the sake of self-consistency, we give in Proposition 1 an instance implementation of such a predicate. It is a simplified version of a necessary condition for reachability demonstrated in [29]. Essentially, the set of valid objectives Ω is built as follows: initially, it contains all the objectives of the form $a_i \leadsto a_i$ (that are always valid); then an objective $a_i \leadsto a_j$ is added to Ω only if there exists a local acyclic path $\eta \in$ local-paths$(a_i \leadsto a_j)$ where all the objectives from the initial state s to the enabling conditions of the transitions are already in Ω: if $b_k \in$ enab(η^n) for some $n \in [1; |\eta|]$, then the objective $b_0 \leadsto b_k$ is already in the set, assuming $s(b) = b_0$.

Proposition 1. *For all objective $P \in \mathbf{Obj}$, valid$_s(P) \overset{\triangle}{\Leftrightarrow} P \in \Omega$ where Ω is the least fixed point of the monotonic function* $\mathrm{F} : 2^{\mathbf{Obj}} \to 2^{\mathbf{Obj}}$ *with*

$$\mathrm{F}(\Omega) \overset{\triangle}{=} \{a_i \leadsto a_j \in \mathbf{Obj} \mid \exists \eta \in \text{local-paths}(a_i \leadsto a_j) :$$
$$\forall n \in [1; |\eta|], \forall b_k \in \text{enab}(\eta^n), s(b) \leadsto b_k \in \Omega\}.$$

Applied to the AN of Fig. 1, if $s = \langle a_0, b_0, c_0, d_0 \rangle$, valid$_s(c_0 \leadsto c_2)$ is true because $c_0 \xrightarrow{a_1} c_1 \xrightarrow{b_0} c_2 \in$ local-paths$(c_0 \leadsto c_2)$ with valid$_s(a_0 \leadsto a_1)$ true and valid$_s(b_0 \leadsto b_0)$ true. On the other hand, valid$_s(d_0 \leadsto d_1)$ is false.

Note that Proposition 1 is an instance of valid$_s$ implementation; any other implementation satisfying Definition 7 can be used to apply the reduction proposed in this article. In [29], more restrictive over-approximations are proposed.

3.2 Reduction Procedure

This section depicts the goal-oriented reduction procedure which aims at identifying transitions that do not take part in any minimal trace from the given initial state to the goal local state $g_\top$. The reduction relies on the local causality analysis to delimit local paths that may be involved in the goal reachability: any local transitions that is not captured by this analysis can be removed from the model without affecting the minimal traces for its occurrence.

The reduction procedure (Definition 8) consists of collecting a set $\mathcal{B}$ of objectives whose local acyclic paths may contribute to a minimal trace for the goal reachability. To ease notations, and without loss of generality, we assume that any automaton a is in state a_0 in s. Given an objective, only the local paths where all the enabling conditions lead to valid objectives are considered (local-paths$_s$). The local transitions corresponding to the objectives in $\mathcal{B}$ are noted tr$(\mathcal{B})$.

Initially starting with the main objective $g_0 \leadsto g_\top$ (Definition 8(1)), the procedure iteratively collects objectives that may be involved for the enabling conditions of local paths of already collected objectives. If a transition $b_j \xrightarrow{\ell} b_k$ is in tr$(\mathcal{B})$, for each $a_i \in \ell$, the objective $a_0 \leadsto a_i$ is added in $\mathcal{B}$ (Definition 8(2)); and for each other objective $b_\star \leadsto b_i \in \mathcal{B}$, the objective $b_k \leadsto b_i$ is added in $\mathcal{B}$ (Definition 8(3)). Whereas the former criteria references the objectives required for concretizing a local path from the initial state, the later criteria accounts for the possible interleaving and successions of local paths within a same automaton: e.g., $g_\top$ reachability may require to reach b_k and b_i in some (undefined) order, we then consider 4 objectives: $b_0 \leadsto b_k$, $b_k \leadsto b_i$, $b_0 \leadsto b_i$, and $b_i \leadsto b_k$.

Definition 8 ($\mathcal{B}$). *Given an AN (Σ, S, T), an initial state s where, without loss of generality, $\forall a \in \Sigma$, $s(a) = a_0$, and a local state $g_\top$ with $g \in \Sigma$ and $g_\top \in S(g)$, $\mathcal{B} \subseteq \mathbf{Obj}$ is the smallest set which satisfies the following conditions:*

1. $g_0 \rightsquigarrow g_\top \in \mathcal{B}$
2. $b_j \xrightarrow{\ell} b_k \in \mathrm{tr}(\mathcal{B}) \Rightarrow \forall a_i \in \ell, a_0 \rightsquigarrow a_i \in \mathcal{B}$
3. $b_j \xrightarrow{\ell} b_k \in \mathrm{tr}(\mathcal{B}) \wedge b_\star \rightsquigarrow b_i \in \mathcal{B} \Rightarrow b_k \rightsquigarrow b_i \in \mathcal{B}$

$$\text{with} \qquad \mathrm{tr}(\mathcal{B}) \triangleq \bigcup_{P \in \mathcal{B}} \mathrm{tr}(\text{local-paths}_s(P)), \quad \text{where}, \forall P \in \mathbf{Obj},$$

$$\text{local-paths}_s(P) \triangleq \{\eta \in \text{local-paths}(P) \mid \forall n \in [1; |\eta|],$$
$$\forall b_k \in \mathrm{enab}(\eta^n), \mathbf{valid}_s(b_0 \rightsquigarrow b_k)\},$$

$\mathrm{enab}(t)$ *being the enabling condition of local transition t (Definition 1).*

Theorem 1 states that any trace which is minimal for the reachability of $g_\top$ from initial state s is composed only of transitions in $\mathrm{tr}(\mathcal{B})$. The proof is given in Appendix A. It results that the AN $(\Sigma, S, \mathrm{tr}(\mathcal{B}))$ contains less transitions but preserves all the minimal traces for the reachability of the goal.

Theorem 1. *For each minimal trace π reaching $g_\top$ from s, $\mathrm{tr}(\pi) \subseteq \mathrm{tr}(\mathcal{B})$.*

Figure 2 shows the results of the reduction on the example AN of Fig. 1 for the reachability of c_2 from the state where all automata start at 0. Basically, the local path from c_0 to c_2 using d_1 being impossible to concretize (because $\mathbf{valid}_s(d_0 \rightsquigarrow d_1)$ is false), it has been removed, and consequently, so are the transitions involving b_1 as b_1 is not required for c_2 reachability. In this example, the subnet computation for reachability properties proposed in [37] would have removed only the transition $c_0 \xrightarrow{d_1} c_2$ from Fig. 1.

Because the number of objectives is polynomial ($|\mathbf{Obj}| = \sum_{a \in \Sigma} |S(a)|^2$), the computation of $\mathcal{B}$ and $\mathrm{tr}(\mathcal{B})$ is very efficient, both from a time and space complexity point of view. The sets $\mathcal{B} \subseteq \mathbf{Obj}$ and $\mathrm{tr}(\mathcal{B}) \subseteq T$ can be built iteratively, from the empty sets: when a new objective $b_\star \rightsquigarrow b_i$ is inserted in $\mathcal{B}$, each transition in $\mathrm{tr}(\text{local-paths}_s(b_\star \rightsquigarrow b_i))$ is added in $\mathrm{tr}(\mathcal{B})$, if not already in; and for each

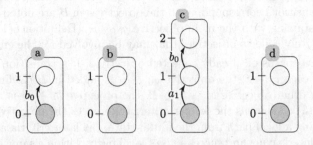

Fig. 2. Reduced automata network from Fig. 1 for the reachability of c_2 from initial state indicated in grey.

transition $b_j \to b_k$ currently in $\mathrm{tr}(\mathcal{B})$, the objective $b_k \leadsto b_i$ is added in $\mathcal{B}$, if not already in. When a new transition $b_j \xrightarrow{\ell} b_k$ is added in $\mathrm{tr}(\mathcal{B})$, for each $a_i \in \ell$, the objective $a_0 \leadsto a_i$ is added in $\mathcal{B}$, if not already in; and for each objective $b_\star \leadsto b_i$ currently in $\mathcal{B}$, the objective $b_k \leadsto b_i$ is added in $\mathcal{B}$, if not already in.

Putting aside the $\mathrm{tr}(\text{local-paths}_s)$ computation, the above steps require a polynomial time and a linear space with respect to the number of transitions and objectives. The computation of $\mathrm{tr}(\text{local-paths}_s(a_i \leadsto a_j))$ requires a time exponential with the number of local states in automaton a ($|S(a)|$), due to the number of acyclic local paths (Sect. 2.2), but a quadratic space: indeed, each individual local acyclic path does not need to be stored, only its set of local transitions, without conditions. Then, $\mathbf{valid}_s$ is called at most once per objective. We assume that the complexity of $\mathbf{valid}_s$ is polynomial with the number of automata and transitions and exponential with the maximum number of local states within an automaton (it is the case of the one presented in Sect. 3.1)

Overall, the reduction procedure has a polynomial space complexity ($|\mathbf{Obj}| + |T|$) and time complexity polynomial with the total number of automata and local transitions, and exponential with the maximum number k of local states within an automaton ($k = \max_{a \in \Sigma} |S(a)|$). Therefore, assuming $k \ll |\Sigma|$, the goal-oriented reduction offers a very low complexity, especially with regard to a full exploration of the $k^{|\Sigma|}$ states.

4   Experiments

We experimented the goal-oriented reduction on several biological networks and quantify the shrinkage of the reachable state space. Then, we illustrate potential applications with the verification of simple reachability, and of cut sets. In both cases, the reduction drastically increases the tractability of those applications.

4.1   Results on Model Reduction

We conducted experiments on Automata Networks (ANs) that model dynamics of biological networks. For different initial states, and for different reachability goals, we compared the number of local transitions in the AN specifications ($|T|$), the number of reachable states, and the size of the so-called complete finite prefix of the unfolding of the net [13]. This latter structure is a finite partial order representation of all the possible traces, which is well studied in concurrency theory. It aims at offering a compact representations of the reachable state spaces by exploiting the concurrency between transitions: if t_1 and t_2 are playable in a given state and are not in conflict (notably when ${}^\bullet t_1 \cap {}^\bullet t_2 = \emptyset$), a standard approach would consider 4 global transitions (t_1 then t_2, and t_2 then t_1), whereas a partial order structure would simply declare t_1 and t_2 as concurrent, imposing no ordering between them. Hence, unfoldings drop part of the combinatorial explosion of the state space due to the interleaving of concurrent transitions.

The selected networks are models of signalling pathways and gene regulatory networks: two Boolean models of Epidermal Growth Factor receptors (EGF-r) [32,33], one Boolean model of tumor cell invasion (Wnt) [10], two Boolean models of T-Cell receptor (TCell-r) [21,31], one Boolean model of Mitogen-Activated Protein Kinase network (MAPK) [15], one multi-valued model of fate determination in the Vulval Precursor Cells (VPC) in C. elegans [38], one Boolean model of T-Cell differentiation (TCell-d) [1], and one Boolean models of cell cycle regulation (RBE2F) [11]. The ANs result from automatic translation from the logical network specifications in the above references; for most models using the `logicalmodel` tool [16]. Note that the obtained ANs are bisimilar to the logical networks [6]. For each of these models, we selected initial states and nodes for which the activation will be the reachability goal[1]. Typically, the initial states correspond to various input signal combinations in the case of signalling cascades, or to pluripotent states for gene networks; and goals correspond to transcription factors or genes of importance for the model (output nodes for signalling cascades, key regulators for gene networks).

Table 1 sums up the results before and after the goal-oriented reduction. The number of reachable states is computed with `its-reach` [23] using a symbolic representation, and the size of the complete finite prefix (number of instances of transitions) is computed with `Mole` [35]. The goal-oriented reduction is performed using `Pint` [27]. In each case, the reduction step took less than 0.1s, thanks to its very low complexity when applied to logical networks.

There is a substantial shrinkage of the dynamics for the reduced models, which can turn out to be drastic for large models. In some cases, the model is too large to compute the state space without reduction. For some large models, the unfolding is too large to be computed, whereas it can provide a very compact representation compared to the state space for large networks exhibiting a high degree of concurrency (e.g., TCell-d, RBE2F). In the case of first profile of TCell-d and EGF-r (104) the reduction removed all the transitions, resulting in an empty model. Such a behaviour can occur when the local causality analysis statically detect that the reachability goal is impossible, i.e., the necessary condition of Sect. 3.1 is not satisfied. On the other hand, a non-empty reduced model does not guarantee the goal reachability. Appendix B show additional results with the reduction made without the filtering $valid_s$ (Sect. 3.1).

4.2   Example of Application: Goal Reachability

In order to illustrate practical applications of the goal-oriented model reduction, we first systematically applied model-checking for the goal reachability on the initial and reduced model (Table 1).

We compared two different softwares: NuSMV [8] which combines Binary Decision Diagrams and SAT approaches for synchronous systems, and `its-reach` [23] which implements efficient decision diagram data structures [18]. In both cases, the transition systems specified as input of these tools is an exact encoding of the

---

[1] Scripts and models available at http://loicpauleve.name/gored-suppl.zip.

Table 1. Comparisons before (normal font) and **after** (bold font) the goal-oriented AN reduction. Each model is identified by the system, the number of automata (within parentheses), and a profile specifying the initial state and the reachability goal. $|T|$ is the number of local transitions in the AN specification; "#states" is the number of reachable global states from the initial state; "—unf—" is the size of the complete finite prefix of the unfolding. "KO" indicates an execution running out of time (30 min) or memory. When applied to goal reachability, we show the total execution time and memory used by the tools NuSMV and its-reach. Computation times where obtained on an Intel® Core™ i7 3.4 GHz CPU with 16 GB RAM. *For each case, the reduction procedure took less than 0.1 s*

Model	$\|T\|$	# states	|unf|	NuSMV		its-reach	
				Verification of goal reachability			
EGF-r (20)	68	4,200	1,749	0.2s	10Mb	0.17	7Mb
	43	**722**	**336**	**0.1**	**8Mb**	**0.1s**	**5Mb**
Wnt (32)	197	7,260,160	KO	30s	48Mb	0.3s	18Mb
	117	**241,060**	**217,850**	**0.9s**	**32Mb**	**0.5s**	**17Mb**
TCell-r (40)	90	≈ $1.2 \cdot 10^{11}$	KO	KO		1.1s	52Mb
	46	**25,092**	**14,071**	**3.8s**	**36Mb**	**0.6s**	**15Mb**
MAPK (53) profile 1	173	≈ $3.8 \cdot 10^{12}$	KO	KO		0.9s	60Mb
	113	**≈ $4.5 \cdot 10^{10}$**	**KO**	**KO**		**2s**	**48Mb**
MAPK (53) profile 2	173	8,126,465	KO	63s	83Mb	0.2s	15Mb
	69	**269,825**	**155,327**	**1.5s**	**36Mb**	**0.4s**	**18Mb**
VPC (88)	332	KO	KO	KO		1s	50Mb
	219	**$1.8 \cdot 10^9$**	**43,302**	**236s**	**156Mb**	**0.8s**	**21Mb**
TCell-r (94)	217	KO	KO	KO		KO	
	42	**54.921**	**1,017**	**0.4**	**23Mb**	**0.26s**	**14Mb**
TCell-d (101) profile 1	384	≈ $2.7 \cdot 10^8$	257	3s	40Mb	0.5s	24Mb
	0	**1**	**1**				
TCell-d (101) profile 2	384	KO	KO	KO		0.5s	23Mb
	161	**75,947,684**	**KO**	**474s**	**260Mb**	**0.3s**	**19Mb**
EGF-r (104) profile 1	378	9,437,184	47,425	7s	35Mb	0.6s	23Mb
	0	**1**	**1**				
EGF-r (104) profile 2	378	≈ $2.7 \cdot 10^{16}$	KO	KO		1.36s	60Mb
	69	**62,914,560**	**KO**	**11s**	**33Mb**	**0.3s**	**17Mb**
RBE2F (370)	742	KO	KO	KO		KO	
	56	**2,350,494**	**28,856**	**5s**	**377Mb**	**5s**	**170Mb**

asynchronous semantics of the automata networks, where steps (Definition 2) are always composed of only one transition. For NuSMV, the reachability property is specified with CTL [9] ("EF g_T", g_T being the goal local state, and EF the *exists eventually* CTL operator). It is worth noting that NuSMV implements the *cone of influence* reduction [5] which removes variables not involved in the property. its-reach is optimized for checking if a state belongs to the reachable state space, and cannot perform CTL checking.

Experiments show a remarkable gain in tractability for the model-checking of reduced networks. For large cases, we observe that the dynamics can be tractable only after model reduction (e.g., TCell-r (94), RBE2F (370)). `its-reach` is significantly more efficient than NuSMV because it is tailored for simple reachability checking, whereas NuSMV handles much more general properties.

Because the goal-reduction preserves all the minimal traces for the goal reachability, it preserves the goal reachability: the results of the model-checking is equivalent in the initial and reduced model.

4.3   Example of Application: Cut Set Verification

The above application to simple reachability does not requires the preservation of *all* the minimal traces. Here, we apply the goal-oriented reduction to the cut sets for reachability, where the *completeness of minimal traces is crucial*.

Given a goal, a *cut set* is a set of local states such that any trace leading to the goal involves, in some of its transitions, one of these local states. Therefore, disabling all the local states of a cut set should make the reachability of the goal impossible. This disabling could be implemented by the knock-out/in of the corresponding species in the biological system: cut sets predict mutations which should prevent a concerned reachability to occur (e.g., active transcription factor). Such cut sets have been studied in [28,32] and are close to intervention sets [21] (which are not defined on traces but on pseudo-steady states).

We focus here on verifying if a (predicted) set of local states is, indeed, a cut set for the goal reachability. In the scope of this experiment, we consider cut sets that are disjoint with the initial state. The cut set property can be expressed with CTL: $\{a_1, b_1\}$ is a cut set for g_T reachability if the model satisfies the CTL property `not E [ (not` a_1 `and not` b_1`) U` g_T `]` (U being the *until* operator). The property states that there exists no trace where none of the local state of the cut set is reached prior to the goal. It is therefore required that *all* the minimal traces to the goal reachability are present in the model: if one is missing, a set of local states could be validated as cut set whereas it may not be involved in the missed trace.

Table 2 compares the model-checking of cut sets properties using NuSMV and `its-ctl` [23] on a range of the biological networks used in the previous sections. Because the dynamical property is much more complex, `its-reach` cannot be used. The cut sets have been computed beforehand with Pint. Because the

Table 2. Comparisons before (normal font) and **after** (bold font) the goal-oriented AN reduction for CTL model-checking of cut sets.

	Wnt (32)	TCell-r (40)	EGF-r (104)	TCell-d (101)	RBE2F (370)
NuSMV	44 s 55 Mb	KO	KO	KO	KO
	9.1 s 27 Mb	**2.4 s 34 Mb**	**13 s 33 Mb**	**600 s 360 Mb**	**6 s 29 Mb**
its-ctl	105 s 2.1 Gb	492 s 10 Gb	KO	KO	KO
	16 s 720 Mb	**11 s 319 Mb**	**21 s 875 Mb**	**KO**	**179 s 1.8 Gb**

goal-oriented reduction preserves all the minimal traces to the goal, the results are equivalent in the reduced models. Similarly to the simple reachability, the goal-oriented reduction drastically improves the tractability of large models.

5   Discussion

This paper introduces a new reduction for automata networks parametrized by a reachability property of the form: from a state s there exists a trace which leads to a state where a given automaton g is in state $g_\top$.

The goal-oriented reduction preserves *all* the minimal traces satisfying the reachability property under a general concurrent semantics which allows at each step simultaneous transitions of an arbitrary number of automata. Those results straightforwardly apply to the asynchronous semantics where only one transition occurs at a time: any minimal trace of the asynchronous semantics is a minimal trace in the general concurrent semantics.

Its time complexity is polynomial in the total number of transitions and exponential with the maximal number of local states within an automaton. Therefore, the procedure is extremely scalable when applied on networks between numerous automata, but where each automaton has a few local states.

Applied to logical models of biological networks, the goal-oriented reduction can lead to a drastic shrinkage of the reachable state space with a negligible computational cost. We illustrated its application for the model-checking of simple reachability properties, but also for the validation of cut sets, which requires the completeness of minimal traces in the reduced model. It results that the goal-oriented reduction can increase considerably the scalability of the formal analysis of dynamics of automata networks.

The goal is expressed as a single local state reachability, which also allows to to support sequential reachability properties between (sub)states using an extra automaton. For instance, the property "reach a_1 and b_1, then reach c_1" can be encoded using one extra automaton g, where $g_0 \xrightarrow{\{a_1,b_1\}} g_1$ and $g_1 \xrightarrow{\{c_1\}} g_\top$.

Further work consider performing the reduction on the fly, during the state space exploration, expecting a stronger pruning. Although the complexity of the reduction is low, such approaches would benefit from heuristics to indicate when a new reduction step may be worth to apply.

A   Proof of Minimal Traces Preservation

We assume a global AN (Σ, S, T) where $g \in \Sigma$, $g_\top \in S(g)$, and $s \in S$ with $s(g) \neq g_\top$.

From Property 1 and Definition 7, any trace reaching first a_i and then a_j uses all the transitions of at least one local path in local-paths$_s(a_i \rightsquigarrow a_j)$.

We first prove with Lemma 2 that the last transition of a minimal trace π for $g_\top$ reachability, of the form $\pi^{|\pi|} = \{g_i \rightarrow g_\top\}$, is necessarily in $\mathrm{tr}(\mathcal{B})$. Indeed, by definition of $\mathcal{B}$, $g_0 \rightsquigarrow g_\top \in \mathcal{B}$; and by Lemma 1, $g_i \rightarrow g_\top \notin$ local-paths$_s(g_0 \rightsquigarrow g_\top)$ implies that reaching g_i requires to reach $g_\top$ beforehand.

266   L. Paulevé

Lemma 1. *Given $a_j \to a_i \in T$, if $a_j \to a_i \notin \mathrm{tr}(\text{local-paths}_s(a_0 \rightsquigarrow a_i))$, then for any trace π from s with $a_j \in \pi^{v\bullet}$ and $a_i \in \pi^{w\bullet}$ for some $v, w \in [1; |\pi|]$, there exists $u < v$ with $a_i \in \pi^{u\bullet}$.*

Proof. Let $\eta \in \text{local-paths}_s(a_0 \rightsquigarrow a_j)$ be an acyclic local path such that $\forall n \in [1; |\eta|]$, $a_i \neq \mathrm{dest}(\eta^n)$. The sequence $\eta :: a_j \to a_i$ is then acyclic and, by definition, belongs to $\text{local-paths}_s(a_0 \rightsquigarrow a_i)$, which is a contradiction. $\square$

Lemma 2. *If π is a minimal trace for g_T reachability from state s, then, necessarily, $\pi^{|\pi|} \subseteq \mathrm{tr}(\mathcal{B})$.*

Proof. As π is minimal for g_T reachability, without loss of generality, we can assume that $\pi^{|\pi|} = \{g_i \to g_T\}$. By definition, $\mathrm{tr}(\text{local-paths}_s(g_0 \rightsquigarrow g_T)) \subseteq \mathrm{tr}(\mathcal{B})$. By Lemma 1, if $g_i \to g_T \notin \mathrm{tr}(\text{local-paths}_s(g_0 \rightsquigarrow g_T))$, then there exists $u < |\pi|$ such that $g_T \in \pi^{u\bullet}$; hence, π would be non minimal. $\square$

The rest of the proof of Theorem 1 is derived by contradiction: if a transition of π is not in $\mathrm{tr}(\mathcal{B})$, we can build a sub-trace of π which preserves g_T reachability, therefore π is not minimal.

Given a transition $a_i \to a_j$ in the q-th step of π that is not in $\mathrm{tr}(\mathcal{B})$, removing $a_i \to a_j$ from π^q would imply to remove any further transition that depend causally on it. Two cases arise from this fact: either all further transitions that depend on a_j must be removed; or $a_i \to a_j$ is part of loop within automaton a, and it is sufficient to remove the loop from π.

Lemma 3 ensures that if $a_z \rightsquigarrow a_k$ is in $\mathcal{B}$ and if a_z occurs before the q-th step and a_k after the q-th step of π, then $a_i \to a_j \notin \mathrm{tr}(\text{local-paths}_s(a_z \rightsquigarrow a_k))$ only if $a_i \to a_j$ is part of a loop, i.e., there are two steps surrounding q where the automaton a is in the same state before their application.

Lemma 3. *Given $a \in \Sigma$ and $u, q, v \in [1; |\pi|]$, $u \leq q < v$, with $a_z \in {}^\bullet\pi^u$, $a_k \in {}^\bullet\pi^v \cup \pi^{v\bullet}$, and $a_i \to a_j \in \pi^q \setminus \mathrm{tr}(\mathcal{B})$, if $a_z \rightsquigarrow a_k \in \mathcal{B}$ then $\exists m, n \in [u; v]$, $m \leq q \leq n$ such that $(\pi^{1..m-1})^\bullet \cap S(a) = (\pi^{1..n})^\bullet \cap S(a)$; and $a_k \in {}^\bullet\pi^v \Rightarrow n < v$.*

Proof. If $a_i \to a_j \notin \mathrm{tr}(\mathcal{B})$ and $a_z \rightsquigarrow a_k \in \mathrm{tr}(\mathcal{B})$, necessarily $a_i \to a_j \notin \mathrm{tr}(\text{local-paths}_s(a_z \rightsquigarrow a_k))$. Therefore $a_i \to a_j$ belongs to a loop of a local path from a_z (at index u in π) to a_k (at index v in π). Hence, $\exists m, n \in [u; v]$ with $m \leq q \leq n$ and $a_h, a_x, a_y \in S(a)$ such that $a_h \to a_x \in \pi^m$ and $a_y \to a_h \in \pi^n$; therefore $(\pi^{1..m-1})^\bullet \cap S(a) = (\pi^{1..n})^\bullet \cap S(a) = a_h$. In the case where $a_k \in {}^\bullet\pi^v$, $a_k \neq a_h$, hence $n < v$. $\square$

Intuitively, Lemma 3 imposes that π has the following form:

$$\pi = \cdots :: \pi^u :: \cdots :: a_h \to a_x :: \cdots :: \mathbf{a_i} \to \mathbf{a_j} :: \cdots :: a_y \to a_h :: \cdots :: \pi^v :: \cdots$$

given that $a_z \rightsquigarrow a_k \in \mathcal{B}$.

The idea is then to remove the transitions forming the loop within automaton a. However, transitions in other automata may depend causally on the transitions that compose the local loop in automaton a within steps m and n, following the notations in Lemma 3.

Lemma 4 establishes that we can always find m and n such that none of the transitions within these steps with an enabling condition depending on automaton a are in $\text{tr}(\mathcal{B})$. Indeed, if a transition in $\text{tr}(\mathcal{B})$ depends on a local state of a, let us call it a_p, the objectives $a_0 \rightsquigarrow a_p$ and $a_p \rightsquigarrow a_k$ are in $\mathcal{B}$, due to the second and third condition in Definition 8. Lemma 3 can then be applied on the subpart of π that contains the transition $a_i \rightarrow a_j$ not in $\text{tr}(\mathcal{B})$ and that concretizes either $a_0 \rightsquigarrow a_p$ or $a_p \rightsquigarrow a_k$ to identify a smaller loop containing $a_i \rightarrow a_j$.

Lemma 4. *Let us assume $a \in \Sigma$ and $q \in [1; |\pi|]$ with $a_i \rightarrow a_j \in \pi^q \setminus \text{tr}(\mathcal{B})$. There exists $m, n \in [1; |\pi|]$ with $m \leq q \leq n$ such that $\forall t \in \text{tr}(\pi^{m+1..n})$, $\text{enab}(t) \cap S(a) \neq \emptyset \Rightarrow t \notin \text{tr}(\mathcal{B})$, and, if $a = g$ or $\exists t \in \text{tr}(\pi^{n+1..|\pi|}) \cap \text{tr}(\mathcal{B})$ with $\text{enab}(t) \cap S(a) \neq \emptyset$, then $(\pi^{1..m-1})^\bullet \cap S(a) = (\pi^{1..n})^\bullet \cap S(a)$.*

Proof. First, let us assume that $a \neq g$ and for any $t \in \pi^{q+1..|\pi|}$, $\text{enab}(t) \cap S(a) \neq \emptyset \Rightarrow t \notin \text{tr}(\mathcal{B})$: the lemma is verified with $m = q$ and $n = |\pi|$.

Then, let us assume there exists $v \in [q + 1; |\pi|]$ such that $\exists t \in \text{tr}(\pi^v) \cap \text{tr}(\mathcal{B})$ with $a_k \in \text{enab}(t)$. By Definition 8, this implies $a_0 \rightsquigarrow a_k \in \mathcal{B}$. By Lemma 3, there exists $m, n \in [1; v - 1]$ with $m \leq q \leq n$ such that $(\pi^{1..m-1})^\bullet \cap S(a) = (\pi^{1..n})^\bullet \cap S(a)$.

Otherwise, $a = g$, and by Lemma 3 with $a_k = g_\top$, there exists $m, n \in [1; |\pi|]$ with $m \leq q \leq n$ and $m \neq n$ such that $(\pi^{1..m-1})^\bullet \cap S(a) = (\pi^{1..n})^\bullet \cap S(a)$. Remark that it is necessary that $n < |\pi|$: if $n = |\pi|$, $g_\top \in (\pi^{1..m-1})^\bullet$, so π would be not minimal.

In both cases, if there exists $r \in [m + 1; n]$ such that $\exists a_p \in S(a)$ and $\exists t \in \pi^r$ with $a_p \in \text{enab}(t)$, then $t \in \text{tr}(\mathcal{B})$ implies that $a_0 \rightsquigarrow a_p \in \mathcal{B}$ and $a_p \rightsquigarrow a_k \in \mathcal{B}$ (Definition 8). If $r > q$, by Lemma 3 with $a_k = a_p$ and $v = r$, there exists $m', n' \in [m+1; n]$ such that $m' \leq q \leq n' < r \leq n$ with $(\pi^{1..m'-1})^\bullet \cap S(a) = (\pi^{1..n'})^\bullet \cap S(a)$. If $r \leq q$, by Lemma 3 with $a_0 = a_p$ and $u = r$, there exists $m', n' \in [m+1; n]$ such that $r \leq m' \leq q \leq n'$ with $(\pi^{1..m'-1})^\bullet \cap S(a) = (\pi^{1..n'})^\bullet \cap S(a)$. Therefore, by induction with Lemma 3, there exists $m, n \in [1; |\pi|]$ such that $\forall t \in \text{tr}(\pi^{m+1..n})$, $\text{enab}(t) \cap S(a) \neq \emptyset \Rightarrow t \notin \text{tr}(\mathcal{B})$. □

Using Lemma 4, we show how we can identify a subset of transitions in π that can be removed to obtain a sub-trace for $g_\top$ reachability. In the following, we refer to the couple (m, n) of Lemma 4 with $\text{cb}(\pi, a, q)$ (Definition 9).

Definition 9 ($\text{cb}(\pi, a, q)$). *Given $a \in \Sigma$, $q \in [1; |\pi|]$ with $t \in \pi^q \setminus \text{tr}(\mathcal{B})$ and $\Sigma(t) = a$, we define $\text{cb}(\pi, a, q) = (m, n)$ where $m, n \in [1; |\pi|]$ such that:*

- $\forall t \in \text{tr}(\pi^{m+1..n})$, $\text{enab}(t) \cap S(a) \neq \emptyset \Rightarrow t \notin \text{tr}(\mathcal{B})$;
- $a = g \lor \exists t \in \text{tr}(\pi^{n+1..|\pi|}) \cap \text{tr}(\mathcal{B})$ with $\text{enab}(t) \cap S(a) \neq \emptyset \Longrightarrow (\pi^{1..m-1})^\bullet \cap S(a) = (\pi^{1..n})^\bullet \cap S(a)$. *Moreover, if $a = g$, then $n < |\pi|$.*

We use Lemma 4 to collect the portions of π to redact according to each automaton. We start from the last transition in π that is not in $\text{tr}(\mathcal{B})$: if $\text{tr}(\pi) \not\subseteq$

$\mathrm{tr}(\mathcal{B})$, there exists $l \in [1; |\pi|]$ such that $\pi^l \not\subseteq \mathrm{tr}(\mathcal{B})$ and $\forall n > l, \pi^n \subseteq \mathrm{tr}(\mathcal{B})$. By Lemma 2, we know that $l < |\pi|$. Let us denote by $b_i \rightarrow b_j$ one of the transitions in π^l which is not in $\mathrm{tr}(\mathcal{B})$.

We define $\Psi \subseteq \Sigma \times [1; |\pi|] \times [1; |\pi|]$ the smallest set which satisfies:

- $(b, m, n) \in \Psi$ if $\mathrm{cb}(\pi, l, b) = (m, n)$
- $\forall (a, m, n) \in \Psi, \forall q \in [m+1; n], \forall t \in \pi^q$, $\mathrm{enab}(t) \cap S(a) \neq \emptyset \Longrightarrow (\Sigma(t), m', n') \in \Psi$ where $\mathrm{cb}(\pi, q, \Sigma(t)) = (m', n')$.

Finally, let us define the sequence of steps ϖ as the sequence of steps π where the transitions delimited by Ψ are removed: for each $(a, m, n) \in \Psi$, all the transitions of automaton a occurring between π^m and π^n are removed. Formally, $|\varpi| = |\pi|$ and for all $q \in [1; |\pi|]$, $\varpi^q \overset{\Delta}{=} \{t \in \pi^q \mid \nexists (a, m, n) \in \Psi : a = \Sigma(t) \wedge m \leq q \leq n\}$.

From Lemma 4 and Ψ definition, ϖ is a valid trace. Moreover, by Lemma 4, there is no $q \in [1; |\pi|]$ such that $(g, q, |\pi|) \in \Psi$, hence $g_T \in \varpi^\bullet$. Therefore, π is not minimal, which contradicts our hypothesis. □

Example 2 Let us consider the reachability of c_2 in the AN of Fig. 1 from state $\langle A_0, b_0, c_0, d_0 \rangle$. The transitions $\mathrm{tr}(\mathcal{B})$ preserved by the reduction for that goal are listed in Fig. 2.

Let π be the following trace in the AN of Fig. 1:

$$\pi = \{a_0 \xrightarrow{\{b_0\}} a_1\} :: \{b_0 \xrightarrow{\{a_1\}} b_1, c_0 \xrightarrow{\{a_1\}} c_1\} :: \{a_1 \xrightarrow{\emptyset} a_0\} :: \{b_1 \xrightarrow{\{a_0\}} b_0\}$$

$$:: \{c_1 \xrightarrow{\{b_0\}} c_2\}.$$

The latest transition not in $\mathrm{tr}(\mathcal{B})$ is $b_1 \xrightarrow{\{a_0\}} b_0$ at step 4. One can compute $\mathrm{cb}(\pi, 4, b) = (2, 4)$, and as there is no transition involving b between steps 3 and 4, $\Psi = \{(b, 2, 4)\}$; therefore, the sequence

$$\varpi = \{a_0 \xrightarrow{\{b_0\}} a_1\} :: \{c_0 \xrightarrow{\{a_1\}} c_1\} :: \{a_1 \xrightarrow{\emptyset} a_0\} :: \{\} :: \{c_1 \xrightarrow{\{b_0\}} c_2\}$$

is a valid sub-trace of π reaching c_2, proving π non-minimality.

In conclusion, if π is a minimal trace for g_T reachability from state s, then, $\mathrm{tr}(\pi) \subseteq \mathrm{tr}(\mathcal{B})$.

B   Experiments with Partial Reduction

The goal-oriented reduction relies on two intertwined analyses of the local causality in ANs: (1) the computation of potentially involved objectives (Sect. 3.2) and (2) the filtering of objective that can be proven impossible (Sect. 3.1). The second part can be considered optional: one could simply define the predicate **valid$_s$** to be always true. In order to appreciate the effect of this second part, we show here the intermediary results of model reduction without the filtering of impossible objectives. It is shown in table below, in the lines in *italic*. As we can see, for some models it has no effect on the reduction, for some others the filtering parts is necessary to obtained important reduction of the state space (e.g., MAPK, TCell-r (94), TCell-d).

Model	# tr	# states	$\|unf\|$
EGF-r (20)	68	4,200	1,749
	43	*722*	*336*
	43	**722**	**336**
Wnt (32)	197	7,260,160	KO
	134	*241,060*	*217,850*
	117	**241,060**	**217,850**
TCell-r (40)	90	$\approx 1.2 \cdot 10^{11}$	KO
	46	*25,092*	*14,071*
	46	**25,092**	**14,071**
MAPK (53) profile 1	173	$\approx 3.8 \cdot 10^{12}$	KO
	147	$\approx 9 \cdot 10^{10}$	*KO*
	113	$\approx \mathbf{4.5 \cdot 10^{10}}$	**KO**
MAPK (53) profile 2	173	8,126,465	KO
	148	*1,523,713*	*KO*
	69	**269,825**	**155,327**
VPC (88)	332	KO	KO
	278	$\approx 2.9 \cdot 10^{12}$	*185,006*
	219	$\mathbf{1.8 \cdot 10^9}$	**43,302**
TCell-r (94)	217	KO	KO
	112	*KO*	*KO*
	42	**54.921**	**1,017**
TCell-d (101) profile 1	384	$\approx 2.7 \cdot 10^8$	257
	275	$\approx 1.1 \cdot 10^8$	*159*
	0	**1**	**1**
TCell-d (101) profile 2	384	KO	KO
	253	$\approx 2.4 \cdot 10^{12}$	*KO*
	161	**75,947,684**	**KO**
EGF-r (104) profile 1	378	9,437,184	47,425
	120	*12,288*	*1,711*
	0	**1**	**1**
EGF-r (104) profile 2	378	$\approx 2.7 \cdot 10^{16}$	KO
	124	$\approx 2 \cdot 10^9$	*KO*
	69	**62,914,560**	**KO**
RBE2F (370)	742	KO	KO
	56	*2,350,494*	*28,856*
	56	**2,350,494**	**28,856**

References

1. Abou-Jaoudé, W., Monteiro, P.T., Naldi, A., Grandclaudon, M., Soumelis, V., Chaouiya, C., Thieffry, D.: Model checking to assess T-helper cell plasticity. Front. Bioeng. Biotechnol. **2**, 86 (2015)
2. Aracena, J., Goles, E., Moreira, A., Salinas, L.: On the robustness of update schedules in Boolean networks. Biosystems **97**(1), 1–8 (2009)
3. Bernardinello, L., De Cindio, F.: A survey of basic net models and modular net classes. In: Rozenberg, G. (ed.) Advances in Petri Nets 1992. LNCS, vol. 609, pp. 304–351. Springer, Heidelberg (1992)
4. Berthelot, G.: Checking properties of nets using transformations. In: Rozenberg, G. (ed.) Advances in Petri Nets 1985. LNCS, vol. 222, pp. 19–40. Springer, Heidelberg (1986)
5. Biere, A., Clarke, E., Raimi, R., Zhu, Y.: Verifying safety properties of a PowerPC^{TM} microprocessor using symbolic model checking without BDDs. In: Halbwachs, N., Peled, D.A. (eds.) CAV 1999. LNCS, vol. 1633, pp. 60–71. Springer, Heidelberg (1999)
6. Chatain, T., Haar, S., Jezequel, L., Paulevé, L., Schwoon, S.: Characterization of reachable attractors using petri net unfoldings. In: Mendes, P., Dada, J.O., Smallbone, K. (eds.) CMSB 2014. LNCS, vol. 8859, pp. 129–142. Springer, Heidelberg (2014)
7. Cheng, A., Esparza, J., Palsberg, J.: Complexity results for 1-safe nets. Theoret. Comput. Sci. **147**(1&2), 117–136 (1995)
8. Cimatti, A., Clarke, E., Giunchiglia, E., Giunchiglia, F., Pistore, M., Roveri, M., Sebastiani, R., Tacchella, A.: NuSMV 2: an opensource tool for symbolic model checking. In: Brinksma, E., Larsen, K.G. (eds.) CAV 2002. LNCS, vol. 2404, pp. 359–364. Springer, Heidelberg (2002)
9. Clarke, E.M., Emerson, E.A.: Design and synthesis of synchronization skeletons using branching-time temporal logic. In: Kozen, D. (ed.) Logic of Programs. LNCS, vol. 131, pp. 52–71. Springer, Heidelberg (1981)
10. Cohen, D.P.A., Martignetti, L., Robine, S., Barillot, E., Zinovyev, A., Calzone, L.: Mathematical modelling of molecular pathways enabling tumour cell invasion and migration. PLoS Comput. Biol. **11**(11), e1004571 (2015)
11. I. Curie/Sysbio. RB/E2F pathway. http://bioinfo-out.curie.fr/projects/rbpathway/
12. Danos, V., Feret, J., Fontana, W., Harmer, R., Hayman, J., Krivine, J., Thompson-Walsh, C.D., Winskel, G.: Graphs, rewriting and pathway reconstruction for rule-based models. In: D'Souza, D., Kavitha, T., Radhakrishnan, J. (eds.) IARCS Annual Conference on Foundations of Software Technology and Theoretical Computer Science, FSTTCS 2012. LIPIcs, vol. 18, pp. 276–288. Schloss Dagstuhl - Leibniz-Zentrum für Informatik (2012)
13. Esparza, J., Heljanko, K.: Unfoldings: A Partial-Order Approach to Model Checking. Monographs in Theoretical Computer Science. An EATCS Series, 1st edn. Springer Publishing Company, Incorporated, New York (2008)
14. Feret, J., Koeppl, H., Petrov, T.: Stochastic fragments: a framework for the exact reduction of the stochastic semantics of rule-based models. Int. J. Softw. Inf. **7**(4), 527–604 (2013)
15. Grieco, L., Calzone, L., Bernard-Pierrot, I., Radvanyi, F., Kahn-Perlès, B., Thieffry, D.: Integrative modelling of the influence of MAPK network on cancer cell fate decision. PLoS Comput. Biol. **9**(10), e1003286 (2013)
16. C. group: Logicalmodel. https://github.com/colomoto/logicalmodel

17. Haddad, S., Pradat-Peyre, J.-F.: New efficient Petri nets reductions for parallel programs verification. Parallel Process. Lett. **16**(1), 101–116 (2006)
18. Hamez, A., Thierry-Mieg, Y., Kordon, F.: Building efficient model checkers using hierarchical set decision diagrams and automatic saturation. Fundam. Inf. **94**(3–4), 413–437 (2009)
19. Janicki, R., Kleijn, J., Koutny, M., Mikulski, Ł.: Step traces. Acta Informatica **53**, 35–65 (2015)
20. Janicki, R., Lauer, P.E., Koutny, M., Devillers, R.: Concurrent and maximally concurrent evolution of nonsequential systems. Theoret. Comput. Sci. **43**, 213–238 (1986)
21. Klamt, S., Saez-Rodriguez, J., Lindquist, J., Simeoni, L., Gilles, E.: A methodology for the structural and functional analysis of signaling and regulatory networks. BMC Bioinform. **7**(1), 56 (2006)
22. Kurshan, R.P.: Computer-Aided Verification of Coordinating Processes: The Automata-Theoretic Approach. Princeton University Press, Princeton (1994)
23. LIP6/Move: Its tools. http://ddd.lip.6.fr/itstools.php
24. Loiseaux, C., Graf, S., Sifakis, J., Bouajjani, A., Bensalem, S., Probst, D.: Property preserving abstractions for the verification of concurrent systems. Formal Methods Syst. Des. **6**(1), 11–44 (1995)
25. Madelaine, G., Lhoussaine, C., Niehren, J.: Attractor equivalence: an observational semantics for reaction networks. In: Fages, F., Piazza, C. (eds.) FMMB 2014. LNCS, vol. 8738, pp. 82–101. Springer, Heidelberg (2014)
26. Naldi, A., Remy, E., Thieffry, D., Chaouiya, C.: Dynamically consistent reduction of logical regulatory graphs. Theoret. Comput. Sci. **412**(21), 2207–2218 (2011)
27. Paulevé, L.: PINT - Static analyzer for dynamics of automata networks. http://loicpauleve.name/pint
28. Paulevé, L., Andrieux, G., Koeppl, H.: Under-approximating cut sets for reachability in large scale automata networks. In: Sharygina, N., Veith, H. (eds.) CAV 2013. LNCS, vol. 8044, pp. 69–84. Springer, Heidelberg (2013)
29. Paulevé, L., Magnin, M., Roux, O.: Static analysis of biological regulatory networks dynamics using abstract interpretation. Math. Struct. Comput. Sci. **22**(04), 651–685 (2012)
30. Priese, L., Wimmel, H.: A uniform approach to true-concurrency and interleaving semantics for Petri nets. Theoret. Comput. Sci. **206**(1–2), 219–256 (1998)
31. Saez-Rodriguez, J., Simeoni, L., Lindquist, J.A., Hemenway, R., Bommhardt, U., Arndt, B., Haus, U.-U., Weismantel, R., Gilles, E.D., Klamt, S., Schraven, B.: A logical model provides insights into T cell receptor signaling. PLoS Comput. Biol. **3**(8), e163 (2007)
32. Sahin, O., Frohlich, H., Lobke, C., Korf, U., Burmester, S., Majety, M., Mattern, J., Schupp, I., Chaouiya, C., Thieffry, D., Poustka, A., Wiemann, S., Beissbarth, T., Arlt, D.: Modeling ERBB receptor-regulated G1/S transition to find novel targets for de novo trastuzumab resistance. BMC Syst. Biol. **3**(1), 1–20 (2009)
33. Samaga, R., Saez-Rodriguez, J., Alexopoulos, L.G., Sorger, P.K., Klamt, S.: The logic of EGFR/ERBB signaling: theoretical properties and analysis of high-throughput data. PLoS Comput. Biol. **5**(8), e1000438 (2009)
34. Schnoebelen, P., Sidorova, N.: Bisimulation and the reduction of Petri nets. In: Nielsen, M., Simpson, D. (eds.) ICATPN 2000. LNCS, vol. 1825, pp. 409–423. Springer, Heidelberg (2000)
35. Schwoon, S.: Mole. http://www.lsv.ens-cachan.fr/~schwoon/tools/mole/

36. Sifakis, J.: Property preserving homomorphisms of transition systems. In: Clarke, E., Kozen, D. (eds.) Logics of Programs. LNCS, vol. 164, pp. 458–473. Springer, Heidelberg (1984)

37. Talcott, C., Dill, D.L.: Multiple representations of biological processes. In: Priami, C., Plotkin, G. (eds.) Transactions on Computational Systems Biology VI. LNCS (LNBI), vol. 4220, pp. 221–245. Springer, Heidelberg (2006)

38. Weinstein, N., Mendoza, L.: A network model for the specification of vulval precursor cells and cell fusion control in Caenorhabditis elegans. Front. Genet. **4**(112) (2013)

Hybrid Reductions of Computational Models of Ion Channels Coupled to Cellular Biochemistry

Jasha Sommer-Simpson[2], John Reinitz[2,3], Leonid Fridlyand[4], Louis Philipson[4], and Ovidiu Radulescu[1]([envelope])

[1] DIMNP UMR CNRS 5235, University of Montpellier, Montpellier, France
ovidiu.radulescu@univ-montp2.fr
[2] Department of Statistics, University of Chicago, Chicago, USA
[3] Department of Ecology and Evolution, Department of Molecular Genetics
and Cell Biology, University of Chicago, Chicago, USA
[4] Department of Medicine and Pediatrics, University of Chicago, Chicago, USA

Abstract. Computational models of cellular physiology are often too complex to be analyzed with currently available tools. By model reduction we produce simpler models with less variables and parameters, that can be more easily simulated and analyzed. We propose a reduction method that applies to ordinary differential equations models of voltage and ligand gated ion channels coupled to signaling and metabolism. These models are used for studying various biological functions such as neuronal and cardiac activity, or insulin production by pancreatic beta-cells. Models of ion channels coupled to cell biochemistry share a common structure. For such models we identify fast and slow sub-processes, driving and slaved variables, as well as a set of reduced models. Various reduced models are valid locally and can change on a trajectory. The resulting reduction is hybrid, implying transitions from one reduced model (mode) to another one.

1 Introduction

Ion channels are essential in biological processes that involve fast modifications of cell physiology. They control the flows and the gradients of ions across the plasma membrane, as well as the membrane potential. Ion channels can open and close as a function of the membrane potential and/or of the concentrations of ligands such as ATP. The multiple control of ion channels implies positive and negative feed-back loops that are responsible for rapid bursts and oscillations of the electrical activity of the cells. Excitability of parts or of entire plasma membrane allows the generation and propagation of action potentials needed for communication between neurons, for muscle contraction, or for endocrine secretion by specialized cells such as pancreatic beta cells [6,7,10]. The dynamics of ion channels can be very intricate. Firstly, there are many types of interacting ion channels, each channel having several subunits that react to voltage and ligands. Secondly, ion channels are integrated in the cell's physiology and interact

© Springer International Publishing AG 2016
E. Bartocci et al. (Eds.): CMSB 2016, LNBI 9859, pp. 273–288, 2016.
DOI: 10.1007/978-3-319-45177-0_17

strongly with metabolism and signaling. For these reasons, models of virtual cells and organs should necessarily include ion channels especially for functions such as nervous impulse transmission, muscle and cardiac activity [13], or insulin secretion by pancreatic islets [8]. Biological function is often not a property of a single cell but an emerging property of interacting cells. Therefore, realistic models of physiology are necessarily multicellular and should contain hundreds or thousands of cells that are coupled together electrically and biochemically. Such models can be computationally expensive if the dynamics of single cells contain many variables and parameters. Model reduction is useful for building virtual physiology models that are both realistic and computationally tractable. By model reduction, one can coarse grain fast variables dynamics whose computation is expensive, while keeping accurate descriptions of the dynamics of slower, driving variables. Such a strategy has already been used to model spiral wave dynamics in cardiac tissue [2].

Several model reduction methods where proposed to reduce ion channel models [1,9,18,26]. All these methods contain at least one ad hoc stage in the choice of fast parameters or of small parameters needed for singular perturbation approximations. Some attempts to develop automatic model reduction techniques based on sensitivity analysis were proposed in [3,4]. In previous work, we have developed model reduction methods for biochemical reactions networks, based on tropical geometry [19–25] allowing the automatic determination of time scales and of small parameters. These methods work for polynomial or rational ordinary differential equations and need to be extended in order to cope with ion channel models that contain transcendental functions.

In this paper we propose an extension to ion channels models of reduction methods based on equilibration and time scales. The concept of equilibration of polynomial dominant terms used in model reduction by tropical approaches is generalized to situations when these terms are rational or even transcendental functions. The possibility of such reductions follows from the property of ion channels dynamics to have multiple time scales ranging from milliseconds to minutes (or even to hours in models involving changes of gene expression) [15]. As is the case for polynomial systems [22], the time scales of variables are state dependent and can change on a trajectory. Therefore, a hybrid reduction is appropriate: the coarse grained dynamics consists in piecewise smooth reduced modes and discrete transitions between modes. The definition of the modes and the transitions between them can be justified in the framework of matched asymptotic expansions from singular perturbations.

To obtain such approximations we will employ a mathematical technique called matched-asymptotic expansion. This singular perturbation technique provides several approximations, neither of which is uniformly valid, but which have overlapping domains of validity [12,16]. The lack of uniformity could pose problems when the solutions of the full and reduced models are compared. For instance, the reduction can slightly change the period of periodic solutions, which leads to large differences at large times. In order to compare the full and reduced model solutions it is therefore appropriate to minimize such discrepancies via parameter optimization of the reduced model. The same method can be used for learning the parameters of the reduced model from a given data set.

2   Ion Channels Coupled with Cell Biochemistry Models

These models contain several types of variables.

Voltage-gated channel variables. These variables are needed for ion channel dynamics. They include gating variables as well as the voltage across the plasma membrane, which controls opening and closing of channels. The simplest case is a channel that has only two states: an open state O letting current pass, and a closed state C which allows for no current. For voltage gated channels, the probabilities per unit time α, β that a channel opens (its state changes from C to O) or closes (its state changes from O to C), respectively, are functions of the membrane potential V. As a consequence, the probability p that the channel is open (p also gives the proportion of open channels) obeys the differential equation

$$\frac{dp}{dt} = \alpha(V)(1-p) - \beta(V)p. \tag{1}$$

Ion channels can have several identical subunits, each of which can be closed or open. The number of different states for a channel with m identical subunits is $m + 1$ (with each state being represented by a number k between 0 and m, where k is the number of subunits that are open). The closing and opening of subunits is modeled as a Markov process with a finite number of states. The Markovian dynamics are described by a system of ordinary differential equations, known as the master equation. For a two-state channel, the master equation is (1). When the channels are identical, the master equation has permutation symmetries. These symmetries can be exploited to obtain exact reductions of the channels dynamics [14]. As a result of the exact reduction, the probabilities p_i that i channels are open (p_i also represents the proportion of channels with i subunits open) are polynomial functions of a smaller number of variables that satisfy linear differential equations [14,15]. For instance, a channel with two identical subunits has three states corresponding to 0,1,or 2 subunits open. The corresponding probabilities are derived from the binomial distribution, namely $p_0 = (1-n)^2$, $p_1 = 2n(1-n)$ and $p_2 = n^2$, where n is a probability satisfying the ordinary differential equation

$$\frac{dn}{dt} = \alpha(V)(1-n) - \beta(V)n, \tag{2}$$

where $\alpha(V), \beta(V)$ are, respectively, the probabilities per unit time that a subunit opens or closes.

Ion currents induce changes of the membrane voltage according to the classical equation

$$C\frac{dV}{dt} = I(t) - \sum_i I_i(t) \tag{3}$$

where I is an input and output current, the variables I_i are currents through channels of type i, and C is the membrane capacitance.

The current-voltage characteristics of a channel of a given type i is affine, the current I_i being proportional to the difference between voltage V and the rest potential V_i (defined as the equilibrium voltage corresponding to the zero current; these depend only on the type of ion and are the same for different channels of the same ion):

$$I_i = g_i p_i (V - V_i), \tag{4}$$

where g_i, p_i are, respectively, the open channel conductance and the proportion of channels of type i that are open.

Metabolic and signaling variables. Metabolic and signaling pathways can be modeled as networks of biochemical reactions. The coupling between metabolic pathways and ion channel dynamics can be performed using the metabolite concentrations. For instance, an increase of the ATP/ADP ratio leads to closure of ATP sensing potassium channels, which in turn triggers plasma membrane depolarization and opening of calcium voltage gated channels. This leads to an increase in cytoplasmic calcium, which triggers important physiological responses such as insulin release in pancreatic beta cells. A few signaling or metabolic and signaling variables (e.g. calcium) have contributions to their dynamics coming from the fluxes fed into and out of the cell by ion channels.

In summary, a general model of ion channels coupled to cell biochemistry is described by the following equations:

$$\frac{dV}{dt} = (V - V_\infty)/\tau_V(h)$$

$$\frac{dh_i}{dt} = (h_i - h_{i,\infty}(V, c))/\tau_{h_i}(V, c)$$

$$\frac{dc_k}{dt} = \sum_j S_{kj} R_j(c) + \sum_{i \in I(k)} g_i p_i(h_i)(V - V_i) \tag{5}$$

where $V_\infty(h, c) = \sum_i V_i g_i p_i(h_i, c)$, $\tau_V(h, c) = \sum_i g_i p_i(h_i, c)$, $h_{i,\infty}(V) = \alpha_i(V)/(\alpha_i(V) + \beta_i(V))$, and $\tau_{h_i}(V) = 1/(\alpha_i(V) + \beta_i(V))$. The R_j are multivariate polynomial or rational functions of the metabolite concentrations c, the p_i are polynomials of the gating variables h_i, the functions $\alpha_i(V), \beta_i(V)$ are combinations of exponential functions of V, the S_{kj} are the entries of the stoichiometric matrix, and the set valued function $I(k)$ denotes the channels feeding the variable c_k.

The voltage V can take negative or positive values but its variations are bounded $-V_1 < V < V_2$. The gating variables h_i are also bounded $0 \leq h_i \leq 1$. The metabolic variables are unbounded, positive concentrations $0 < c_k$; these variables can have various orders of magnitude, from very small to very large.

3   Matched Asymptotic Method

The natural framework for approximations of systems with multiple time scales is the theory of singular perturbations [12]. The classical presentation of this theory starts with the identification of a small parameter η in the problem. We then ask: what are the asymptotic behaviors of the solutions when $\eta \rightarrow 0$? In certain cases, this small parameter η can be obtained by nondimensionalization [27]. Variables and parameters are rescaled via division by other variables and parameters such that the resulting quotient is unitless. A smallest parameter can be chosen among unitless parameters. However, the nondimensionalization can be done in several different ways, and the smallest unitless parameter is not guaranteed to be the correct small singular perturbations parameter (extra conditions are needed to guarantee validity of the approximation). For the time being we suppose that the small parameter has been identified. We will provide automatic methods to do this in the next sections.

The rescaled equations of a system with fast variables x and slow variables y read:

$$\frac{\mathrm{d}x}{\mathrm{d}t} = \frac{1}{\eta} f(x,y), \quad \frac{\mathrm{d}y}{\mathrm{d}t} = g(x,y), \tag{6}$$

where η is a small positive parameter representing the ratio of fast and slow time scales.

The trajectory of a slow/fast system is typically composed of an alternating sequence of slow parts and fast parts. Let us suppose that the slow segments are defined by the time intervals (t_0, t_1), (t_1, t_2), ... The fast segments are interlaced between successive slow segments and can be considered as inner layer solutions placed at $t_1, t_2, \ldots$. In order to describe the fast inner layer solutions, let us define the variable $\tau = (t - t_i)/\eta$, $i = 1, 2, \ldots$. Note that τ changes by one when t changes by η, hence τ corresponds to short time scales inside the inner layers at $t = t_i$. Using this new time variable, the differential equations (6) become

$$\frac{\mathrm{d}x}{\mathrm{d}\tau} = f(x,y), \quad \frac{\mathrm{d}y}{\mathrm{d}\tau} = \eta g(x,y). \tag{7}$$

The *inner layer approximation* is a solution

$$X(\tau) = X_0(\tau) + \eta X_1(\tau) + \ldots,$$
$$Y(\tau) = Y_0(\tau) + \eta Y_1(\tau) + \ldots$$

of the Eq. (7). We find that, at the lowest order in η, the inner layer solution satisfies

$$\frac{\mathrm{d}X_0}{\mathrm{d}\tau} = f(X_0, Y_0), \quad \frac{\mathrm{d}Y_0}{\mathrm{d}\tau} = 0, \tag{8}$$

which is to say that Y_0 is constant.

The *outer layer approximation* is a solution

$$x(t) = x_0(t) + \eta x_1(t) + \ldots,$$
$$y(t) = y_0(t) + \eta y_1(t) + \ldots$$

of the Eq. (6). At the lowest order in η we find

$$0 = f(x_0, y_0), \quad \frac{\mathrm{d}y_0}{\mathrm{d}t} = g(x_0, y_0). \tag{9}$$

In other words, in outer layers the fast variables are "slaved" by the slow variables. At lower order in η, outer layers are thus described by the quasi-stationary approximation. Indeed, the outer layer solution lies on a surface which can be approximated at lowest order by the equation $f(x_0, y_0) = 0$. Provided that the stability condition $Re(Spec(\frac{\partial f}{\partial x}(x_0, y_0))) < 0$ is fulfilled ($\frac{\partial f}{\partial x}$ is the Jacobian matrix of f with respect to rapid variables x) the validity of the quasi-stationarity approximation is guaranteed by the Tikhonov theorem [28]. The surface defined by $f(x_0, y_0) = 0$ also represents the lowest order approximation of an invariant manifold (low dimensional surface that contains the reduced dynamics). The existence of the invariant manifold is guaranteed by Fenichel's results [5]. An invariant manifold can be stable (attractive) or unstable; furthermore, the same invariant manifold can have stable parts that become unstable at bifurcations (for instance of the saddle-node type), when one or several eigenvalues of the Jacobian $\frac{\partial f}{\partial x}$ vanish or touch the pure imaginary axis of the complex plane. The (slow) outer solutions correspond to dynamics on the same or on different stable parts of the invariant manifold. We should emphasize that more complex behaviour can occur as a result of the so-called *canard* phenomena when part of the trajectory can lie for some time on the unstable manifold. Such mechanisms, identified in the Morris-Lecar and FitzHugh-Nagumo models of excitable systems, [29] will not be discussed here.

As well known in singular perturbation theory [16], neither the inner nor the outer solution has uniform validity. However, their domains of validity overlap. As a matter of fact, the two solutions must agree for intermediate time scales $(t - t_i) = -\eta^\alpha$, $0 < \alpha < 1$. Hence, $Y^{(i)}(-\eta^{\alpha-1}) = y^{(i)}(t_i - \eta^\alpha)$, $X^{(i)}(-\eta^{\alpha-1}) = x^{(i)}(t_i - \eta^\alpha)$. By taking the limit $\eta \to 0$, we obtain the matching conditions

$$\lim_{\tau \to -\infty} Y_0^{(i)}(\tau) = y_0^{(i)}(t_i), \quad \lim_{\tau \to -\infty} X_0^{(i)}(\tau) = x_0^{(i)}(t_i), \tag{10}$$

where $(Y_0^{(i)}, X_0^{(i)})$, and $(y_0^{(i)}, x_0^{(i)})$ are the lowest orders of the inner and outer layer solutions at t_i and on (t_{i-1}, t_i), respectively. For a more formal treatment of this result, the reader may consult [16].

Similarly,

$$\lim_{\tau \to \infty} Y_0^{(i)}(\tau) = y_0^{(i+1)}(t_i), \quad \lim_{\tau \to \infty} X_0^{(i)}(\tau) = x_0^{(i+1)}(t_i), \tag{11}$$

where $(y_0^{(i+1)}, x_0^{(i+1)})$ are the lowest orders of the outer layer solutions on (t_i, t_{i+1}).

In other words the boundary conditions of the outer layer solutions are the asymptotic states of the inner layer solutions (corresponding to given fixed values of the slow variables).

A composite lowest order approximation combines inner and outer layers and has to subtract common terms:

$$y(t) \sim \sum [(y_0^{(i)}(t) - y_0^{(i)}(t_i)) \mathbb{1}_{(t_{i-1}, t_i]} + Y_0^i((t - t_i)/\eta)],$$
$$x(t) \sim \sum [(x_0^{(i)}(t) - x_0^{(i)}(t_i)) \mathbb{1}_{(t_{i-1}, t_i]} + X_0^i((t - t_i)/\eta)], \tag{12}$$

where $\mathbb{1}_{(t_{i-1}, t_i]}$ is the indicator function for the interval $(t_{i-1}, t_i]$.

In this paper we will use only the lowest order of the matched asymptotic expansion. However, higher order expansions can be also obtained with this method [12,16].

The solution (12) can be seen as a hybrid approximation. The local dynamics are reduced with respect to the full model and, at lowest order, are given either by the inner layer (solution of (8)) or by the outer layer (solution of (9)) approximation.

4   Algorithm for the Hybrid Reduction

The matched asymptotic expansion method justifies the possibility of hybrid approximations but is not a reduction algorithm per se. Several steps, namely the detection of slow/fast variables and the determination of the inner layer positions need further development.

Numerical determination of time scales and outer layers. In the full (un-reduced) model, each variable satisfies an ordinary differential equation (ODE) $\frac{dx_i}{dt} = f_i(x_i, x^{(i)})$, where $x^{(i)}$ denotes all of the variables other than x_i. Let us denote by $x_i^*(t)$ the solution of $f_i(x_i, x^{(i)}(t)) = 0$ that is closest to $x_i(t)$, where $(x_i(t), x^{(i)}(t))$ is a solution of the given ODE system. We then define two positive indices

$$\tau_i(t) = (|f_i(x_i(t), x^{(i)}(t))/(x_i(t) - x_i^*(t))|)^{-1}$$
$$\text{and } s_i(t) = |x_i(t) - x_i^*(t)|/x_{i,s}, \tag{13}$$

where $x_{i,s}$ is a positive normalizing value (a typical choice is $x_{i,s} = \max |x_i(t) - x_i^*(t)|$). The index τ_i is an estimate of the time scale characteristic of the variable x_i. Note that the voltage and gating variables of the generic ion channel model described by Eq. (5) are $\tau_i = \tau_V$ and $\tau_i = \tau_{h_i}$, respectively. The index $s(t)$ is a measure of the distance between the value of x_i on a trajectory and the imposed value x_i^* that x_i would have as fast variable in an outer layer solution. A low value of s_i indicates that x_i is a fast slaved variable in an outer layer. Variables can be fast, but not slaved, in the inner layers.

Detection of fast species via sorting of timescales. The value $s_i(t)$ is used to determine the intervals of time where fast variables are slaved; these intervals are the outer layer modes. Suppose that an outer layer starts at t_i and ends at t_{i+1}. The values t_i, t_{i+1} depend on the trajectory and will change if initial conditions are changed. We should therefore look for another way to define the limits of the outer layers. A convenient way to do this is to define exit from an outer layer as a condition on the values of one or several variables of the model, using ordinary differential equations together with events that trigger transitions between the modes. This is possible because outer layer solutions belong to invariant manifolds, and the end of a given outer layer is characterized by loss of stability of the invariant manifold (which can be written as a condition on the model's variables).

Let us consider that the species time scales are sorted, so that $\tau_1(t) \leq \tau_2(t) \leq \ldots \leq \tau_n(t)$. The subscripts indexing the values τ_i may change order, depending on time. Suppose that, for a given time t, there is an index k such that the value $\eta = \tau_k(t)/\tau_{k+1}(t)$ is much smaller than one. Then, we can use k to separate fast and slow timescales, and take η as the singular perturbation parameter. Let us note that the multiple timescales situation $\eta = \tau_k(t)/\tau_{k+1}(t) \to 0$ is covered by the second theorem of Tikhonov [28] and also leads to the outer layer approximation (9).

The reduction algorithm is summarized by the following steps:

1. Detect fast species. These have small values of $\tau_i(t)$, separated from the rest of the variables by a gap. To be precise, choose a number g larger than one. This number g shall be called the *gap width*. Define a threshold function $\tau_{th}(t)$ such that
   (a) $\tau_{th}(t) = (\tau_k(t)\tau_{k+1}(t))^{1/2}$,
   (b) $\tau_{k+1}(t)/\tau_k(t) = \eta^{-1} > g$, and
   (c) for each t, k is the smallest index satisfying the condition (b).
   Then all species such that $\tau_k(t) < \tau_{th}(t)$ are declared fast.
2. Detect outer layers. These are defined by small values of s_i, smaller than a fixed threshold.
3. Slow, outer layer modes are defined by Eq. (9). Fast, inner layer modes are described by Eq. (8). A slow mode is followed and/or preceded by a fast mode.
4. Define conditions for exit from outer layers. These conditions depend on the values of variables, and can be implemented as ODE system events triggering the transition between modes.
5. The last step of the algorithm consists in parameter optimization of the reduced model by simulated annealing or by other optimization method. This step is needed for comparison of the full and reduced model solutions, or for learning parameters of the reduced model from data.

5   A Hybrid Approximation of the Hodgkin-Huxley Model

We have applied our algorithm to several ion channel models. To keep the presentation short, we illustrate our results on the well-known Hodgkin-Huxley (HH) model. This model is a four-variable system (14) of ordinary differential equations, fitting into the general framework of Eq. (5) above.

$$
\frac{dV}{dt} = \frac{V - V^*(h_m, h_w, h_n)}{\tau_V(h_m, h_w, h_n)} \qquad \frac{dh_m}{dt} = \frac{h_m - h_m^*(V)}{\tau_m(V)}
$$
$$
\frac{dh_w}{dt} = \frac{h_w - h_w^*(V)}{\tau_h(V)} \qquad \frac{dh_n}{dt} = \frac{h_n - h_n^*(V)}{\tau_n(V)}, \tag{14}
$$

where h_m is the sodium channel activation, h_w is the sodium channel inactivation, and h_n is the potassium channel activation. For each gating variable x in the set $\{m, w, n\}$, the timescale τ_x and the imposed value h_x^* are defined by

$$
\tau_x(V) = \alpha_x(V) + \beta_x(V) \qquad \text{and} \qquad h_x^*(V) = \alpha_x(V)/(\alpha_x(V) + \beta_x(V)),
$$

respectively, where α_x and β_x are as given below.

$$\alpha_m(V) = \frac{0.32(V + 54)}{1 - \exp(-(V + 54)/4)} \qquad \beta_m(V) = \frac{0.28(V + 27)}{\exp((V + 27)/5) - 1}$$

$$\alpha_w(V) = \frac{0.128}{\exp((V + 50)/18)} \qquad \beta_w(V) = \frac{4}{\exp(-(V + 27)/5) + 1}$$

$$\alpha_n(V) = \frac{0.032(V + 52)}{1 - \exp(-(V + 52)/5)} \qquad \beta_n(V) = \frac{0.5}{\exp((V + 57)/40)}$$

The timescale τ_V and the imposed value V^* for the voltage variable are defined below.

$$\tau_V(h_m, h_n) = \frac{C}{g_m h_w h_m^3 + g_n h_n^4 + g_L},$$

$$V^*(h_m, h_n) = \frac{g_m h_w h_m^3 V_m + g_n h_n^4 V_n + g_L V_L + g_I h_I}{g_m h_w h_m^3 + g_n h_n^4 + g_L}.$$

The conductances g_x are defined by $g_m(t) = 100 \cdot h_w(t)$, $g_n = 80$, $g_L = 0.1$, and $g_I = 0.32$. The capacitance was taken to be $C = 1$. Finally, the constants V_x and h_I are set as follows: $V_m = 50$, $V_n = -100$, $V_L = -67$, and $h_I = 1$ The units of conductance are mS/cm^2, those of voltage are mV, those of current are $\mu A/cm^2$, and those of capacitance are $\mu F/cm^2$. This model contains only a voltage variable V and three gating variables h_x, with x in $\{m, w, n\}$. In essence, the equations above are the same as those used in the seminal paper of Hodgkin and Huxley [11] describing action potentials in the squid giant axon; the organization of the equations and parameter values used are adopted from [3].

The analysis and reduction of this model was performed using MATLAB [17].

5.1   Trajectory of the HH Model

The system of ODEs described above will quickly converge to a limit cycle (see Fig. 2(a)). We will see that the voltage V is slow except for during the spikes. Figure 2(a) plots all four variables over one period of this limiting cycle (Fig. 1).

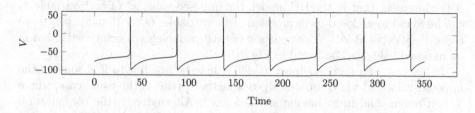

Fig. 1. Above, left: A plot of the voltage variable V from the (non-reduced) HH model, obtained by forward numerical simulation of the Eq. (14). With parameters as above, this limiting cycle has an approximate period of $\Delta t \approx 49.5$.

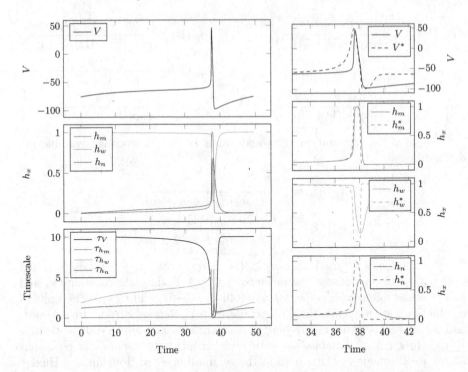

Fig. 2. Above, left: One period of the limiting cycle for the HH model. The voltage variable is plotted on the upper axes, the middle axis displays the trajectories of the gating variables, and the timescales τ_x are plotted on the lower axis. A larger timescale means that the given variable x is slower to follow its imposed value x^*. Above, right: A comparison of each variable x_i from the (non-reduced) HH model with its imposed value x_i^*, in a short window of time surrounding the spike. The imposed value of x is given by the steady state of the equation $\frac{dx_i}{dt} = f_i(x_i, x^{(i)})$ with $x^{(i)}$ held fixed.

5.2   Determination of the Imposed Values

It is convenient that in the HH model, the imposed value x_i^* of each variable x_i can be found as a closed-form function of the variables other than x_i, which are collectively denoted $x^{(i)}$. A comparison of each variable's trajectory with that of its imposed value can be found in Fig. 2(b).

In the case of a more complex model, it may be possible to find some of the imposed values x_i^* via computer algebra methods (this is the usual case, where f_i is a polynomial in x_i having small degree). Alternatively, the computation of the imposed values can be performed via numerical solution of the equation $0 = f_i(x_i^*, x^{(i)})$. In the case of multiple solutions to this equation, we choose the solution which provides the smallest index s_i and which satisfies the physical variables constraints, i.e. whose chemical concentrations are positive and whose gating variables fall in the interval $[0, 1]$.

5.3 Detection of Fast Species

According to Eq. (13), the timescale τ_{x_i} is defined as the absolute value of the ratio $\frac{x_i^* - x_i}{dx_i/dt}$. Intuitively, τ_{x_i} is an amount of time required for the variable x_i to move "significantly" in the direction of its imposed value x_i^*. Indeed, if we hold constant the distance $x_i^* - x_i$ between x_i and its imposed value, then the timescale τ_{x_i} varies inversely with the derivative $\frac{dx_i}{dt}$. Thus the "fast" variables are those x's such that τ_x is smaller than some given threshold, whereas the "slow" variables are the variables x whose timescales τ_x are large.

The timescale threshold $\tau_{th}(t)$, which distinguishes between the slow and fast variables, will be allowed to change with time. Following the algorithm in Sect. 4, we assign a number between 1 and 4 to each of the timescales $\tau_V(t)$, $\tau_m(t)$, $\tau_w(t)$ and $\tau_n(t)$, so that the inequality

$$\tau_1(t) \leq \tau_2(t) \leq \tau_3(t) \leq \tau_4(t)$$

is satisfied for each point in time. At a given point t, we write k for the smallest integer such that $1 \leq k \leq 4$ and such that the ratio $\frac{\tau_{k+1}(t)}{\tau_k(t)}$ is larger than a chosen value g; we have chosen $g = 3.5$. Thus, τ_k is the slowest of the fast variables, and τ_{k+1} is the fastest among the slow variables. The threshold τ_{th} is defined as the geometric mean $(\tau_k(t)\tau_{k+1}(t))^{1/2}$, so that a variable x is fast if and only if τ_x is smaller than τ_{th}.

See the top axis of Fig. 3 for a plot of the timescales τ_x and the timescale threshold τ_{th}. Note that the variable h_m is always fast, the variables h_w and h_n are always slow, and the voltage variable V is fast only during the spike (from $t = 37.4915$ to $t = 38.3835$).

5.4 Detection of Outer Layers

The slowness index s_x of each variable x is obtained by normalizing the difference between x_i and x_i^*, as per Eq. (13). The bottom axis of Fig. 3 displays a plot of the slowness indices s_V and s_m corresponding to fast variables V and h_m, respectively. We have chosen $s_{th} = 0.2$ as the threshold for distinguishing between slaved and unslaved fast variables: say that x_i is slaved at time t if and only if the inequality $s_i(t) \leq s_{th}$ is satisfied.

The position of the outer layer solution is characterized by small values s_i for each of the fast variables x_i. Referring to the solution of the full HH model in Fig. 3, the outer layer occurs from 37.627 to 38.0055 and from 38.2630 to 38.3835 (during which times both V and h_m are fast), as well as before 37.4915 and after 38.3835 (during which times only h_m is fast). See Table 1 for a summary of information concerning the different modes (inner and outer) as well as the transitions between them.

At this point we can compare our results with similar approximations of the HH model. Like [26], we identify the variable h_m as fast everywhere. However, contrary to our approach, [26] considers that h_m is slaved everywhere, which is not true at least for the parameters values that we use (similar to theirs).

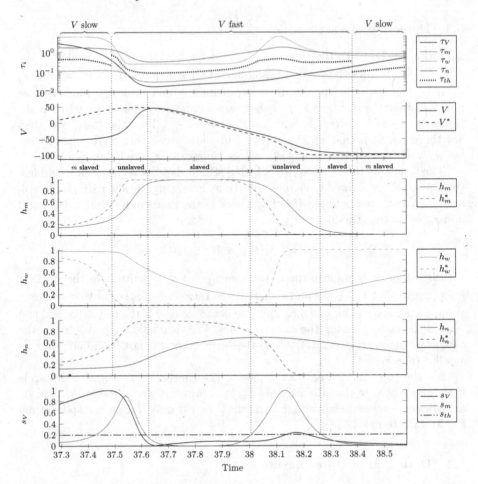

Fig. 3. Delineation of modes of the (non-reduced) HH model. The top axes show timescales τ_{x_i} of each of the variables in a short window of time surrounding the voltage spike. The dotted line (marked as τ_{th} in the legend) is the moving threshold used for distinguishing slow variables from fast ones. This timescale threshold τ_{th} is calculated as the geometric mean $(\tau_k(t)\tau_{k+1}(t))^{1/2}$ of the smallest two adjacent timescales whose ratio is larger than a chosen constant value g. The vertical dotted lines mark timepoints where variables change between slow and fast or between slaved and unslaved.

Thus, with respect to more conventional singular perturbations methods, our approach has two advantages: it detects automatically which type of approximation should be applied and it considers the possibility of inner layers where fast variables are not slaved.

The approach used in [3,4] identifies slow and fast regimes which are equivalent to our outer and inner layers, respectively. However, the method in [3,4] is based on a sensitivity study of the inputs of each variable rather than on direct

Table 1. Summary of the regions and ODEs defining the modes of the hybrid simplification of the HH model. The type I stands for inner layer and O for outer layer. Slaved variables are set to their quasi-stationary values. Unslaved variables follow ODE dynamics. In inner layers slow variables are constant at lowest order approximation and fast variables are unslaved. In outer layers fast variables are slaved and slow variables are unslaved.

Region	Type	V	h_m	h_w	h_n	Exit event
1	O	slow	fast/slaved	slow	slow	$V > V_1$
2	I	fast/unslaved	fast/unslaved	slow	slow	$m > m_1$
3	O	fast/slaved	fast/slaved	slow	slow	$h < h_1$
4	I	fast/unslaved	fast/unslaved	slow	slow	$m < m_2$
5	O	fast/slaved	fast/slaved	slow	slow	$h > h_2$

Region 1	$h_m = h_m^*$	$\dot{x} = (x - x^*)/\tau_x$	for $x \in \{V, h_w, h_n\}$
Regions 2 and 4	h_w and h_n constant	$\dot{x} = (x - x^*)/\tau_x$	for $x \in \{V, h_m\}$
Regions 3 and 5	$h_m = h_m^*$ and $V = V^*$	$\dot{x} = (x - x^*)/\tau_x$	for $x \in \{h_w, h_n\}$

testing of quasi-stationarity as in our approach. The former method requires a complex heuristic to consolidate the results, whereas in our case the mode decomposition is simply controlled by the two thresholds g and s_{th}. We therefore expect the method presented in this paper to be better terms of robustness and precision of the approximation. Indeed, for the HH model, [4] finds six regimes, and the agreement between trajectories simulated with the full and hybrid model is only qualitative.

5.5 Differential Equations of Modes, Exit Events and Parameter Optimization

As discussed in Sect. 4, triggering transitions between modes of the reduced model is performed via thresholds, listed under "Exit event" in Table 1. For example, the event "$V > V_1$" is used to trigger exit from region 1, meaning that there is a threshold V_1 such that exit from region 1 is signaled by the value of V surpassing that of V_1. For each region, the thresholded variable was chosen from the *active* ones, i.e. slow variables trigger exit from outer layers and fast variables trigger exit from inner layers. If several variables are active at a given time, we chose the variable whose logarithmic time derivative has the largest absolute value. The thresholds were refined in order to minimize the L^2 distance between the trajectories generated by the full and hybrid models. Furthermore, the period of the reduced model was adjusted to match that of the full model; this was achieved by adding C to the list of optimized parameters. Local optimization using the function *lsqnonlin* of MATLAB (trust-region-reflective least square algorithm) was enough to obtain a good fit. After optimization, we arrived at the following parameter values: $V_1 = -46.99973$, $m_1 = 0.99496$, $h_1 = 0.03597$,

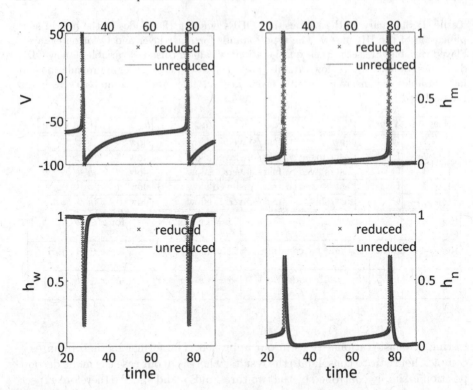

Fig. 4. The trajectory of each of the four variables of the full and reduced HH models.

$m_2 = 0.16298$, $h_2 = 0.72711$, $C = 1.00117$. For completeness, the equations governing the model's evolution in each of the five regions are detailed in Table 1. A juxtaposition of the trajectories of the reduced and full models can be found in Fig. 4.

6   Conclusion and Future Work

We have shown how to obtain hybrid reduced models for differential equations models of ion channels dynamics. These hybrid reductions can be used as simplified units of multiscale models of tissues or organs. In certain cases, hybrid simplifications can relate biochemical parameters to physiological properties analytically. For instance, a matched asymptotic simplification of the HH model with one dimensional description of the slowest outer layer (coarser than the one presented here) can be used to find an approximate analytic expression relating the period of bursting to the model parameters. The details of this application will be presented elsewhere.

In this paper we have presented a trajectory based method for reduction. This method has the advantage of generality and simplicity of implementation, but could, in certain situations, provide a reduction that is valid only locally in the phase and parameter spaces. Tropical geometry approaches, currently applied to polynomial and rational differential equations, do not use trajectory simulations and their robustness is guaranteed by replacing positive real numbers by orders of magnitude, i.e. valuations. In this work we borrowed equilibration ideas from tropical methods but we have not used orders yet. The main difficulty in computing orders is the transcendental nature of some voltage dependent terms. This will be overcome in future work by using an elimination method in which valuations are computed as a function of voltage (considered as a parameter).

Acknowledgments. This work was supported by the University of Chicago and by the FACCTS (France and Chicago Collaborating in The Sciences) program. The authors express their gratitude to the reviewers for their many helpful comments.

References

1. Biktasheva, I., Simitev, R., Suckley, R., Biktashev, V.: Asymptotic properties of mathematical models of excitability. Philos. Trans. R. Soc. Lond. A Math. Phys. Eng. Sci. **364**(1842), 1283–1298 (2006)
2. Bueno-Orovio, A., Cherry, E.M., Fenton, F.H.: Minimal model for human ventricular action potentials in tissue. J. Theoret. Biol. **253**(3), 544–560 (2008)
3. Clewley, R.: Dominant-scale analysis for the automatic reduction of high-dimensional ODE systems. In: ICCS 2004 Proceedings, Complex Systems Institute, New England (2004)
4. Clewley, R., Rotstein, H.G., Kopell, N.: A computational tool for the reduction of nonlinear ODE systems possessing multiple scales. Multiscale Model. Simul. **4**(3), 732–759 (2005)
5. Fenichel, N.: Geometric singular perturbation theory for ordinary differential equations. J. Differ. Equ. **31**(1), 53–98 (1979)
6. Fridlyand, L.E., Jacobson, D., Kuznetsov, A., Philipson, L.H.: A model of action potentials and fast Ca 2+ dynamics in pancreatic β-cells. Biophys. J. **96**(8), 3126–3139 (2009)
7. Fridlyand, L.E., Jacobson, D.A., Philipson, L.: Ion channels and regulation of insulin secretion in human β-cells: a computational systems analysis. Islets **5**(1), 1–15 (2013)
8. Fridlyand, L.E., Philipson, L.H.: Pancreatic beta cell G-protein coupled receptors and second messenger interactions: a systems biology computational analysis. PloS one **11**(5), e0152869 (2016)
9. Grosu, R., Batt, G., Fenton, F.H., Glimm, J., Le Guernic, C., Smolka, S.A., Bartocci, E.: From cardiac cells to genetic regulatory networks. In: Gopalakrishnan, G., Qadeer, S. (eds.) CAV 2011. LNCS, vol. 6806, pp. 396–411. Springer, Heidelberg (2011)
10. Hille, B.: Ion Channels of Excitable Membranes. Sinauer, Sunderland (2001)
11. Hodgkin, A., Huxley, A.: Propagation of electrical signals along giant nerve fibres. Proc. R. Soc. Lond. Ser. B Biol. Sci. **140**, 177–183 (1952)
12. Holmes, M.H.: Introduction to Perturbation Methods, vol. 20. Springer Science & Business Media, New York (2012)

13. Iyer, V., Mazhari, R., Winslow, R.L.: A computational model of the human left-ventricular epicardial myocyte. Biophys. J. **87**(3), 1507–1525 (2004)
14. Keener, J.P.: Invariant manifold reductions for Markovian ion channel dynamics. J. Math. Biol. **58**(3), 447–457 (2009)
15. Keener, J.P., Sneyd, J.: Mathematical Physiology, vol. 1. Springer, New York (1998)
16. Lagerstrom, P., Casten, R.: basic concepts underlying singular perturbation techniques. SIAM Rev. **14**(1), 63–120 (1972)
17. MATLAB: version 1.7.0.11 (R2013b). The MathWorks Inc., Natick, Massachusetts (2013)
18. Murthy, A., Islam, M.A., Bartocci, E., Cherry, E.M., Fenton, F.H., Glimm, J., Smolka, S.A., Grosu, R.: Approximate bisimulations for sodium channel dynamics. In: Gilbert, D., Heiner, M. (eds.) CMSB 2012. LNCS, vol. 7605, pp. 267–287. Springer, Heidelberg (2012)
19. Noel, V., Grigoriev, D., Vakulenko, S., Radulescu, O.: Tropical geometries and dynamics of biochemical networks application to hybrid cell cycle models. Electron. Notes Theoret. Comput. Sci. **284**, 75–91 (2012). In: Feret, J., Levchenko, A. (eds.) Proceedings of the 2nd International Workshop on Static Analysis and Systems Biology (SASB 2011). Elsevier
20. Noel, V., Grigoriev, D., Vakulenko, S., Radulescu, O.: Tropicalization and tropical equilibration of chemical reactions. In: Litvinov, G., Sergeev, S. (eds.) Tropical and Idempotent Mathematics and Applications, Contemporary Mathematics, vol. 616, pp. 261–277. American Mathematical Society (2014)
21. Radulescu, O., Gorban, A.N., Zinovyev, A., Noel, V.: Reduction of dynamical biochemical reactions networks in computational biology. Front. Genet. **3**(131) (2012)
22. Radulescu, O., Swarup Samal, S., Naldi, A., Grigoriev, D., Weber, A.: Symbolic dynamics of biochemical pathways as finite states machines. In: Roux, O., Bourdon, J. (eds.) CMSB 2015. LNCS, vol. 9308, pp. 104–120. Springer, Heidelberg (2015)
23. Radulescu, O., Vakulenko, S., Grigoriev, D.: Model reduction of biochemical reactions networks by tropical analysis methods. Math. Model Nat. Phenom. **10**(3), 124–138 (2015)
24. Samal, S.S., Grigoriev, D., Fröhlich, H., Weber, A., Radulescu, O.: A geometric method for model reduction of biochemical networks with polynomial rate functions. Bull. Math. Biol. **77**(12), 2180–2211 (2015)
25. Soliman, S., Fages, F., Radulescu, O.: A constraint solving approach to model reduction by tropical equilibration. Algorithms Mol. Biol. **9**(1), 1 (2014)
26. Suckley, R., Biktashev, V.N.: The asymptotic structure of the Hodgkin-Huxley equations. Int. J. Bifurcat. Chaos **13**(12), 3805–3825 (2003)
27. Tang, Y., Stephenson, J.L., Othmer, H.G.: Simplification and analysis of models of calcium dynamics based on IP3-sensitive calcium channel kinetics. Biophys. J. **70**(1), 246 (1996)
28. Tikhonov, A.N.: Systems of differential equations containing small parameters in the derivatives. Matematicheskii Sbornik **73**(3), 575–586 (1952)
29. Wechselberger, M., Mitry, J., Rinzel, J.: Canard theory and excitability. In: Kloeden, P.E., Pötzsche, C. (eds.) Nonautonomous Dynamical Systems in the Life Sciences. LIM, vol. 2102, pp. 89–132. Springer, Switzerland (2013)

Formal Modeling and Analysis of Pancreatic Cancer Microenvironment

Qinsi Wang[1(✉)], Natasa Miskov-Zivanov[2], Bing Liu[3], James R. Faeder[3], Michael Lotze[4], and Edmund M. Clarke[1]

[1] Computer Science Department, Carnegie Mellon University, Pittsburgh, USA
qinsiw@cs.cmu.edu
[2] Electrical and Computer Engineering Department, Carnegie Mellon University, Pittsburgh, USA
[3] Department of Computational and Systems Biology, University of Pittsburgh, Pittsburgh, USA
[4] Surgery and Bioengineering, UPMC, Pittsburgh, USA

Abstract. The focus of pancreatic cancer research has been shifted from pancreatic cancer cells towards their microenvironment, involving pancreatic stellate cells that interact with cancer cells and influence tumor progression. To quantitatively understand the pancreatic cancer microenvironment, we construct a computational model for intracellular signaling networks of cancer cells and stellate cells as well as their intercellular communication. We extend the rule-based BioNetGen language to depict intra- and inter-cellular dynamics using discrete and continuous variables respectively. Our framework also enables a statistical model checking procedure for analyzing the system behavior in response to various perturbations. The results demonstrate the predictive power of our model by identifying important system properties that are consistent with existing experimental observations. We also obtain interesting insights into the development of novel therapeutic strategies for pancreatic cancer.

1 Introduction

Pancreatic cancer (PC), as an extremely aggressive disease, is the seventh leading cause of cancer death globally [3]. For decades, extensive efforts were made on developing therapeutic strategies targeting at pancreatic cancer cells (PCCs). However, the poor prognosis for PC remains largely unchanged. Recent studies have revealed that the failure of systemic therapies for PC is partially due to the tumor microenvironment, which turns out to be essential to PC development [13,15,16,25]. As a characteristic feature of PC, the microenvironment includes pancreatic stellate cells (PSCs), immune cells, endothelial cells, nerve cells, lymphocytes, dendritic cells, the extracellular matrix, and other molecules surrounding PCCs, among which, PSCs play key roles during the PC development [25].

This work was partially supported by ONR Award (N00014-13-1-0090), NSF CPS Breakthrough (CNS-1330014), NSF CPS Frontier (CNS-1446725), and NIH award U54HG008540.

© Springer International Publishing AG 2016
E. Bartocci et al. (Eds.): CMSB 2016, LNBI 9859, pp. 289–305, 2016.
DOI: 10.1007/978-3-319-45177-0_18

In this paper, to obtain a system-level understanding of the PC microenvironment, we construct a multicellular model including intracellular signaling networks of PCCs and PSCs respectively, and intercellular interactions among them.

Boolean Networks (BNs) [36] has been widely used to model biological networks [4]. A Boolean network is an executable model that characterizes the status of each biomolecule by a binary variable that related to the abundance or activity of the molecule. It can capture the overall behavior of a biological network and provide important insights and predictions. Recently, it has been found useful to study the signaling networks in PCCs [18,19]. Rule-based modeling language is another successfully used formalism for dynamical biological systems, which allows molecular/kinetic details of signaling cascades to be specified [10,14]. It provides a rich yet concise description of signaling proteins and their interactions by representing interacting molecules as structured objects and by using pattern-based rules to encode their interactions. The dynamics of the underlying system can be tracked by performing stochastic simulations. In this paper, to formally describe our multicellular and multiscale model, we extend the rule-based language BioNetGen [14] to enable the formal specification of not only the signaling network within a single cell, but also interactions among multiple cells. Specifically, we represent the intercellular level dynamics using rules with continuous variables and use BNs to capture the dynamics of intracellular signaling networks, considering the fact that a large number of reaction rate constants are not available in the literature and difficult to be experimentally determined. Our extension saves the virtues of both BNs and rule-based kinetic modeling, while advancing the specification power to multicellular and multiscale models. We employ stochastic simulation NFsim [35] and statistical model checking (StatMC) [24] to analyze the systems properties. The formal analysis results show that our model reproduces existing experimental findings with regard to the mutual promotion between pancreatic cancer and stellate cells. The model also provides insights into how treatments latching onto different targets could lead to distinct outcomes. Using the validated model, we predict novel (poly)pharmacological strategies for improving PC treatment.

Related work. Various mathematical formalisms have been used for the cancer microenvironment modeling (see a recent review [6]). In particular, Gong [17] built a qualitative model to analyze the intracellular signaling reactions in PCCs and PSCs. This model is discrete and focuses on cell proliferation, apoptosis, and angiogenesis pathways. While, our model is able to make quantitative predictions and also considers pathways regulating the autophagy of PCCs and the activation and migration of PSCs, as well as the interplay between PCCs and PSCs. In terms of the modeling language, the ML-Rules [30] is a multi-level rule-based language, which can consider multiple biological levels of organization by allowing objects to be able to contain collections of other objects. This embedding relationship can affect the behavior of both container and contents. ML-Rules uses continuous rate equations to capture the dynamics of intracellular reactions, and thus requires all the rate constants to be known. Instead, our language models intracellular dynamics using BNs, which reduces the difficulty of estimating the values of hundreds of unknown parameters often involved in large models.

The paper is organized as follows. In Sect. 2, we present the multicellular model for the PC microenvironment. We then introduce our rule-based modeling formalism extended from the BioNetGen language in Sect. 3. In Sect. 4, we briefly introduce StatMC that is used to carry out formal analysis of the model. The analysis results are given and discussed in Sect. 5. Section 6 concludes the paper.

2 Signalling Networks Within Pancreatic Cancer Microenvironment

We construct a multicellular model for pancreatic cancer microenvironment based on a comprehensive literature search. The reaction network of the model is summarized in Fig. 1. It consists of three parts that are colored with green, blue, and purple respectively: (i) the intracellular signaling network of PCCs, (ii) the intracellular signaling network of PSCs, and (iii) the signaling molecules (such as growth factors and cytokines) in the extracellular space of the microenvironment, which are ligands of the receptors expressed in PCCs and PSCs. Note that → denotes activation/promotion/up-regulation, and –• represents inhibition/suppression/down-regulation.

2.1 The Intracellular Signaling Network of PCCs

Pathways regulating proliferation

KRas mutation enhances proliferation [8]. Mutations of the KRas oncogene occur in the precancerous stages with a mutational frequency over 90 %. It can lead to the continuous activation of the RAS protein, which then constantly triggers the RAF→MEK cascade, and promotes PCCs' proliferation through the activation of ERK and JNK.

EGF activates and enhances proliferation [32]. Epidermal growth factor (EGF) and its corresponding receptor (EGFR) are expressed in ∼95 % of PCs. EGF promotes proliferation through the RAS→RAF→MEK→JNK cascade. It can also trigger the RAS→RAF→MEK→ERK→cJUN cascade to secrete EGF molecules, which can then quickly bind to overexpressed EGFR again to promote the proliferation of PCCs, which is believed to confer the devastating nature on PCs.

HER2/neu mutation also intensifies proliferation [8]. HER2/neu is another oncogene frequently mutated in the initial PC formation. Mutant HER2 can bind to EGFR to form a heterodimer, which can activate the downstream signaling pathways of EGFR.

bFGF promotes proliferation [9]. As a mitogenic polypeptide, bFGF can promote proliferation through both RAF→MEK→ERK and RAF→MEK→JNK cascades. In addition, bFGF molecules are released through RAF→MEK→ERK pathway to trigger another autocrine signaling pathway in the PC development.

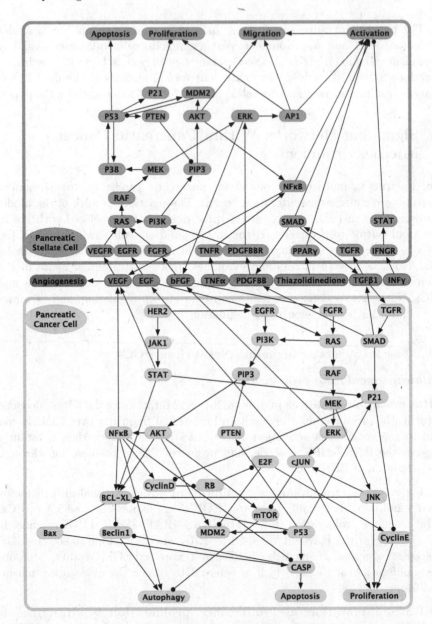

Fig. 1. The pancreatic cancer microenvironment model

Pathways regulating apoptosis

Apoptosis is the most common mode of programmed cell death. It is executed by caspase proteases that are activated by death receptors or mitochondrial pathways.

TGF β1 initiates apoptosis [34]. In PCCs, transforming growth factor β 1 (TGFβ1) binds to and activates its receptor (TGFR), which in turn activates receptor-regulated SAMDs that hetero-oligomerize with the common SAMD3 and SAMD4. After translocating to the nucleus, the complex initiates apoptosis in the early stage of the PC development.

Mutated oncogenes inhibit apoptosis. Mutated KRas and HER2/neu can inhibit apoptosis by downregulating caspases (CASP) through PI3K→AKT → NFκB cascade and by inhibiting Bax (and indirectly CASP) via the PI3K→ PIP3→AKT→···→BCL-XL pathway.

Pathways regulating autophagy. Autophagy is a catabolic process involving the degradation of a cell's own components through the lysosomal machinery. This pro-survival process enables a starving cell to reallocate nutrients from unnecessary processes to essential processes. Recent studies indicate that autophagy is important in the regulation of cancer development and progression and also affects the response of cancer cells to anticancer therapy [21,26].

mTOR regulates autophagy [31]. The mammalian target of rapamycin (mTOR) is a critical regulator of autophagy. In PCCs, the upstream pathway PI3K→PIP3→AKT activates mTOR and inhibits autophagy. The MEK→ERK cascade downregulates mTOR via cJUN and enhances autophagy.

Overexpression of anti-apoptotic factors promotes autophagy [28]. Apoptosis and autophagy can mutually inhibit each other due to their crosstalks. In the initial stage of PC, the upregulation of apoptosis leads to the inhibition of autophagy. Along with the progression of cancer, when apoptosis is suppressed by the highly expressed anti-apoptotic factors (e.g. NFκB and Beclin1), autophagy gradually takes the dominant role and promotes PCC survival.

2.2   Intracellular Signaling Network of PSCs

Pathways regulating activation. PCCs can activate the surrounding inactive PSCs by cancer-cell-induced release of mitogenic and fibrogenic factors, such as PDGFBB and TGFβ1. As a major growth factor regulating cell functions of PSCs, **PDGFBB activates PSCs** [20] through the downstream ERK→AP1 signaling pathway. The **activation of PSCs is also mediated by TGFβ1** [20] via TGFR→SAMD pathway. The autocrine signaling of TGFβ1 maintains the sustained activation of PSCs. Furthermore, the cytokine **TNFα**, which is a major secretion of tumor-associated macrophages (TAMs) in the microenvironment, **is also involved in activating PSCs** [29] through binding to TNFR, which indirectly activates NFκB.

Pathways regulating migration. Migration is another characteristic cell function of PSCs. Activated PSCs move towards PCCs, and form a cocoon around tumor cells, which could protect the tumor from therapies' attacks [7,16].

Growth factors promote migration. Growth factors existing in the microenvironment, including EGF, bFGF, and VEGF, can bind to their receptors on PSCs and activate the migration through the MAPK pathway.

PDGFBB contributes to the migration [33]. PDGFBB regulates the migration of PSCs mainly through two downstream pathways: (i) the PI3K→PIP3 → AKT pathway, which mediates PDGF-induced PSCs' migration, but not proliferation, and (ii) the ERK→AP1 pathway that regulates activation, migration, and proliferation of PSCs.

Pathways regulating proliferation

Growth factors activate proliferation. In PSCs, as key downstream components for several signaling pathways initiated by distinct growth factors, such as EGF and bFGF, the ERK→AP1 cascade activates the proliferation of PSCs. Compared to inactive PSCs, active ones proliferate more rapidly.

Tumor suppressers repress proliferation. Similar to PCCs, P53, P21, and PTEN act as suppressers for PSCs' proliferation.

Pathways regulating apoptosis

P53 upregulates modulator of apoptosis [23]. The apoptosis of PSCs can be initiated by P53, whose expression is regulated by the MAPK pathway.

2.3   Interactions Between PCCs and PSCs

The mechanism underlying the interplay between PCCs and PSCs is complex. In a healthy pancreas, PSCs exist quiescently in the periacinar, perivascular, and periductal space. However, in the diseased state, PSCs will be activated by growth factors, cytokines, and oxidant stress secreted or induced by PCCs, including EGF, bFGF, VEGF, TGFβ1, PDGF, sonic hedgehog, galectin 3, endothelin 1 and serine protease inhibitor nexin 2 [11]. Activated PSCs will then transform from the quiescent state to the myofibroblast phenotype. This results in their losinlipid droplets, actively proliferating, migrating, producing large amounts of extracellular matrix, and expressing cytokines, chemokines, and cell adhesion molecules. In return, the activated PSCs promote the growth of PCCs by secreting various factors, including stromal-derived factor 1, FGF, secreted protein acidic and rich in cysteine, matrix metalloproteinases, small leucine-rich proteoglycans, periostin and collagen type I that mediate effects on tumor growth, invasion, metastasis and resistance to chemotherapy [11]. Among them, EGF, bFGF, VEGF, TGFβ1, and PDGFBB are essential mediators of the interplay between PCCs and PSCs that have been considered in our model.

Autocrine and paracrine involving EGF/bFGF [27]. EGF and bFGF can be secreted by both PCCs and PSCs. In turn, they will bind to EGFR and FGFR respectively on both PCCs and PSCs to activate their proliferation and further secretion of EGF and FGF.

Interplay through VEGF [39]. As a proangiogenic factor, VEGF is found to be of great importance in the activation of PSCs and angiogenesis during the progression of PCs. VEGF, secreted by PCCs, can bind with VEGFR on PSCs to activate the PI3K pathway. It further promotes the migration of PSCs through PIP3→AKT, and suppresses the transcription activity of P53 via MDM2.

Autocrine and paracrine involving TGFβ1 [27]. PSCs by themselves are capable of synthesizing TGFβ1, suggesting the existence of an autocrine loop that may contribute to the perpetuation of PSC activation after an initial exogenous signal, thereby promoting the development of pancreatic fibrosis.

Interplay through PDGFBB [11]. PDGFBB exists in the secretion of PCCs, whose production is regulated by TGFβ1 signaling pathway. PDGFBB can activate PSCs and initiate migration and proliferation as well.

3   The Modeling Language

Rule-based modeling languages are often used to specify protein-to-protein reactions within cells and to capture the evolution of protein concentrations. BioNet-Gen language is a representative rule-based modeling formalism [14], which consists of three components: basic building blocks, patterns, and rules. In our setting, in order to simultaneously simulate the dynamics of multiple cells, interactions among cells, and intracellular reactions, we advance the specifying power of BioNetGen by redefining basic building blocks and introducing new types of rules for cellular behaviors as follows.

Basic building blocks. In BioNetGen, basic building blocks are molecules that may be assembled into complexes through bonds linking components of different molecules. To handle multiscale dynamics (i.e. cellular and molecular levels), we allow the fundamental blocks to be also cells or extracellular molecules. Specifically, a cell is treated as a fundamental block with subunits corresponding to the components of its intracellular signaling network. Furthermore, extracellular molecules (e.g. EGF) are treated as fundamental blocks without subunits.

As we use BNs to model intracellular signaling networks, each subunit of a cell takes binary values (it is straightforward to extend BNs to discrete models). The Boolean values - "True (T)" and "False (F)" - can have different biological meanings for distinct types of components within the cell. For example, for a subunit representing cellular process (e.g. apoptosis), "T" means the cellular process is triggered, and "F" means it is not triggered. For a receptor, "T" means the receptor is bound, and "F" means it is free. For a protein, "T" indicates this protein has a high concentration, and "F" indicates that its concentration level is below the value to regulate downstream targets.

Patterns. As defined in BioNetGen, patterns are used to identify a set of species that share features. For instance, the pattern $C(c_1)$ matches both $C(c_1, c_2 \sim T)$ and $C(c_1, c_2 \sim F)$. Using patterns offers a rich yet concise description in specifying components.

Rules. In BioNetGen, three types of rules are used to specified: binding/unbinding, phosphorylation, and dephosphorylation. Here we introduce nine rules in order to describe the cellular processes in our model and the potential therapeutic interventions. For each type of rules, we present its formal syntax followed by examples that demonstrate how it is used in our model.

Rule 1: Ligand-receptor binding

$$< Lig > + < Cell > (< Rec > \sim F) \rightarrow < Cell > (< Rec > \sim T) \ < binding\_rate >$$

Remark: On the left-hand side, the "F" value of a receptor $< Rec >$ indicates that the receptor is free. When a ligand $< Lig >$ binds to it, the reduction of number of extracellular ligand is represented by its elimination. In the meanwhile, "$< Rec > \sim T$", on the right-hand side, indicates that the receptor is not free any more. Note that, the multiple receptors on the surface of a cell can be modeled by setting a relatively high rate on the following downstream regulating rules, which indicates the rapid "releasing" of bound receptors. An example in our microenvironment model is the binding between EGF and EGFR for PCCs: "$EGF + PCC(EGFR \sim F) \rightarrow PCC(EGFR \sim T) \ 1$".

Rule 2: Mutated receptors form a heterodimer

$$< Cell > (< Rec_1 > \sim F, < Rec_2 > \sim F) \rightarrow$$
$$< Cell > (< Rec_1 > \sim T, < Rec_2 > \sim T) \ < mutated\_binding\_rate >$$

Remark: Unbound receptors can bind together and form a heterodimer. For example, in our model, the mutated HER2 can activate downstream pathways of EGFR by binding with it and forming a heterodimer: "$PCC(EGFR \sim F, HER2 \sim F) \rightarrow PCC(EGFR \sim T, HER2 \sim T) \ 10$".

Rule 3: Downstream signaling transduction
Rule 3.1 (Single parent) upregulation (activation, phosphorylation, etc.)

$$< Cell > (< Act > \sim T, < Tar > \sim F) \rightarrow$$
$$< Cell > (< Act > \sim T, < Tar > \sim T) \ < trate >$$

Rule 3.2 (Single parent) downregulation (inhibition, dephosphorylation, etc.)

$$< Cell > (< Inh > \sim T, < Tar > \sim T) \rightarrow$$
$$< Cell > (< Inh > \sim T, < Tar > \sim F) \ < trate >$$

Rule 3.3 (Multiple parents) Downstream regulation

$$< Cell > (< Inh > \sim F, < Act > \sim T, < Tar > \sim F) \rightarrow$$
$$< Cell > (< Inh > \sim F, < Act > \sim T, < Tar > \sim T) \ < trate >$$

$$< Cell > (< Inh > \sim T, < Tar > \sim T) \rightarrow$$
$$< Cell > (< Inh > \sim T, < Tar > \sim F) \ < trate >$$

Remark: Instead of using kinetic rules (such as in ML-Rules), our language use logical rules of BNs to describe intracellular signal cascades. Downsteam signal transduction rules are used to describe the logical updating functions for all intracellular molecules constructing the signaling cascades. For instance, Rule 3.3 presents the updating function $< Tar >^{(t+1)} = \neg < Inh >^{(t)} \times (< Act >^{(t)} + < Tar >^{(t)})$, where "$< Inh >$" is the inhibitor, and "$< Act >$" is the activator. In this manner, concise rules can be devised to handle complex cases, where there exists multiple regulatory parents. Note that our model follows the biological assumption that inhibitors hold higher priorities than activators with respect to their impacts on the target. "+" and "×" in logical functions represent logical "OR" and "AND" respectively. An example in our model is that, in PCCs, $STAT$ can be activated by $JAK1$: "$PCC(JAK1 \sim T, STAT \sim T) \rightarrow PCC(JAK1 \sim F, STAT \sim T)$ 0.012" and "$PCC(JAK1 \sim T, STAT \sim F) \rightarrow PCC(JAK1 \sim F, STAT \sim T)$ 0.012".

Rule 4: Cellular processes
Rule 4.1 Proliferation

$$< Cell > (Pro \sim T) \rightarrow$$
$$< Cell > (Pro \sim F) + < Cell > (Pro \sim F, \cdots) \quad < pro\_rate >$$

Remark: When a cell proliferates, we keep the current values of subunits for the cell that initiates the proliferation, and assume the new cell to have the default values of subunits. The "$\cdots$" in the rule denotes the remaining subunits with their default values.
Rule 4.2 Apoptosis

$$< Cell > (Apo \sim T) \rightarrow Null() \quad < apop\_rate >$$

Remark: A type "Null()" is declared to represent dead cells or degraded molecules. In our model, the apoptosis of PSCs is described as "$PSC(Apo \sim T) \rightarrow Null()$ $5e - 4$".
Rule 4.3 Autophagy

$$< Cell > (Aut \sim T) \rightarrow < Mol > + \cdots \quad < auto\_rate >$$

Remark: The molecules on the right-hand side of this type of rules will be released into the microenvironment due to autophagy. They are the existing molecules expressed inside this cell when autophagy is triggered.

Rule 5: Secretion

$$< Cell > (< secMol > \sim T) \rightarrow$$
$$< Cell > (< secMol > \sim F) + < Mol > \quad < sec\_rate >$$

Remark: When the secretion of "$< Mol >$" has been triggered, its amount in the microenvironment will be added by 1. Note that, we can differentiate the endogenous and exogenous molecules by labeling the secreted "$< Mol >$" with the cell name. In our model, we have "$PCC(secEGF \sim T) \rightarrow PCC(secEGF \sim F) + EGF$ $2.7e - 4$".

Rule 6: Mutation

$$< Cell > (< Mol > \sim < unmutated >) \rightarrow$$
$$< Cell > (< Mol > \sim < mutated >) \quad < mrate >$$

Remark: For mutant proteins that are constitutively active, we set a very high value to the mutation rate "mrate". In this way, we can almost keep the value of the mutated molecule as what it should be. For example, in our model, the mutation of oncoprotein *Ras* in PCCs is captured by "$PCC(RAS \sim F) \rightarrow PCC(RAS \sim T)$ 10000".

Rule 7: Constantly over-expressed extracellular molecules

$$CancerEvn \rightarrow CancerEvn+ < Mol > \quad < sec\_rate >$$

Remark: We use this type of rules to mimic the situation that the concentration of an over-expressed extracellular molecule stays in a high level constantly.

Rule 8: Degradation of extracellular molecules

$$< Mol > \rightarrow Null() \quad < deg\_rate >$$

Remark: Here, "Null()" is used to represent dead cells or degraded molecules. For instance, *bFGF* in the microenvironment will be degraded via "$bFGF \rightarrow Null()$ 0.05".

Rule 9: Therapeutic intervention

$$< Cell > (< Mol > \sim < untreated >) \rightarrow$$
$$< Cell > (< Mol > \sim < treated >) \quad < treat\_rate >$$

Remark: Given a validated model, intervention rules allow us to evaluate the effectiveness of a therapy targeting at certain molecule(s). Also, the well-tuned value of the intervention rate could, more or less, give indications when deciding the dose of medicine used in this therapy, based on the Law of Mass Action.

Our extension allows the BioNetGen language to be able to model not only the signaling network within a single cell, but also interactions among multiple cells. It also allows one to simulate the dynamics of cell populations, which is crucial to cancer study. Moreover, describing the intracellular dynamics using the style of BNs improves the scalability of our method by overcoming the difficulty of obtaining values of a large amount of model parameters from wet laboratory, which is a common bottleneck of conventional rule-based languages and ML-Rules. Note that, similar to other rule-based languages, our extended one allows different methods for model analysis, since more than one semantics can be defined for the same syntax.

4   Statistical Model Checking

Simulation can recapitulate a number of experimental observations and provide new insights into the system. However, it is not easy to manually analyze a significant amount of simulation trajectories, especially when there is a large set of system properties to be tested. Thus, for our model, we employ statistical model checking (StatMC), which is a fully automated formal analysis technique. In this section, we provide an intuitive and brief description of StatMC. The interested reader can find more details in [24].

Given a system property expressed as a Bounded Linear Temporal Logic (BLTL) [24] formula and the set of simulation trajectories generated by applying the NFsim stochastic simulation to our rule-based model, StatMC estimates the probability of the model satisfying the property (see the supplementary document [2] for a brief introduction of BLTL). In detail, since the underlying semantic model of the stochastic simulation method NFsim that we used for our model is essentially a discrete-time Markov chain, we need to verify stochastic models. StatMC treats the verification problem for stochastic models as a statistical inference problem, using randomized sampling to generate traces (or simulation trajectories) from the system model, and then performing model checking and statistical analysis on those traces. For a (closed) stochastic model and a BLTL property ψ, the probability p that the model satisfies ψ is well defined (but unknown in general). For a fixed $0 < \theta < 1$, we ask whether $p \leq \theta$, or what the value of p is. In StatMC, the first question is solved via hypothesis testing methods, while the second via estimation techniques. Intuitively, hypothesis tests are probabilistic decision procedures, i.e., algorithms with a yes/no reply, and which may give wrong answers. Estimation techniques instead compute (probabilistic) approximations of the unknown probability p. The main assumption of StatMC is that, given a BLTL property ψ, the behavior of a (closed) stochastic model can be described by a Bernoulli random variable of parameter p, where p is the probability that the system satisfies ψ. It is known that discrete-time Markov chains satisfy this requirement [37]. Therefore StatMC can be applied to our setting. More specifically, given σ is a system execution and ψ a BLTL formula, we have that $Prob\{\sigma | \sigma \models \psi\} = p$, and the Bernoulli random variable mentioned above is the following function M defined as follows: $M(\sigma) = 1$ if $\sigma \models \psi$, or $M(\sigma) = 0$ otherwise. Therefore, M will be 1 with probability p and 0 with probability $1 - p$. In general, StatMC works by first obtaining samples of M, and then by applying statistical techniques to such samples to solve the verification problem. The whole checking process is illustrated in the supplementary document [2].

5   Results and Discussion

In this section, we present and discuss formal analysis results for our pancreatic cancer microenvironment model. The model file is available at http://www.cs.cmu.edu/~qinsiw/mpc_model.bngl. All the experiments reported below were conducted on a machine with a 1.7 GHz Intel Core i7 processor and 8 GB RAM, running on Ubuntu 14.04.1 LTS. In our experiments, we use Bayesian sequential estimation with 0.01 as the estimation error bound, coverage probability 0.99, and a uniform prior ($\alpha = \beta = 1$). The time bounds and thresholds given in following properties are determined by considering the model's simulation results. The parameters in our model include initial state (e.g. abundance of extracellular molecules) and reaction rate constants. The initial state was provided by biologists based on wet-lab measurements. The rate constants were estimated based on the general ones in the textbook [5]. The results in scenario I & II demonstrate that using these parameters the model is able to reproduce key observations reported in the literature. We also performed a sensitivity analysis and the results show that the system behavior is robust to most of the parameters (the two sensitive parameters have been labeled in our model file).

Scenario I: mutated PCCs with no treatments

In scenario I, we validate our model by studying the role of PSCs in the PC development.

Property 1: This property aims to estimate the probability that the population of PCCs will eventually reach and maintain in a high level.

$$Prob_{=?} \{(PCCtot = 10) \land F^{1200} G^{100} (PCCtot > 200)\}$$

First, we take a look at the impact from the presence of PSCs on the dynamics of PCC population. As shown in Table 1, with PSCs, the probability of the number of PCCs reaching and keeping in a high level ($Pr = 0.9961$) is much higher than the one when PSCs are absent ($Pr = 0.405$). This indicates that PSCs promote PCCs proliferation during the progression of PC. This is consistent with experimental findings [7,11,39].

Property 2: This property aims to estimate the probability that the number of migrated PSCs will eventually reach and maintain in a high amount.

$$Prob_{=?} \{(MigPSC = 0) \land F^{1200} G^{100} (MigPSC > 40)\}$$

We then study the impacts from PCCs on PSCs. As shown in Table 1, without PCCs, it is quite unlikely ($Pr = 0.1191$) for quiescent PSCs to be activated. While, when PCCs exist, the chance of PSCs becoming active ($Pr = 0.9961$) approaches to 1. This confirms the observation [20] that, during the development of PC, PSCs will be activated by growth factors, cytokines, and oxidant stress secreted or induced by PCCs.

Property 3: This property aims to estimate the probability that the number of PCCs entering the apoptosis phase will be larger than the number of PCCs starting the autophagy process and this situation will be reversed eventually.

$$Prob_{=?} \{F^{400} (G^{300} (ApoPCC > 50 \land AutoPCC < 50)$$
$$\land F^{700} G^{300} (ApoPCC < 50 \land AutoPCC > 50))\}$$

We are also interested in the mutually exclusive relationship between apoptosis and autophagy for PCCs reported in [21,28]. In detail, as PC progresses, apoptosis firstly overwhelms autophagy, and then autophagy takes the leading place after a certain amount of time. This situation is described as property 3 and its estimated probability is close to 1 (see Table 1).

Property 4: This property aims to estimate the probability that, it is always the case that, once the population of activated PSCs reaches a high level, the number of migrated PSCs will also increase.

$$Prob_{=?} \{G^{1600} (ActPSC > 10 \rightarrow F^{100} (MigPSC > 10))\}$$

Table 1. Statistical model checking results for properties under different scenarios

Property	Estimated prob	# Succ	# Sample	Time (s)	Note
Scenario I: mutated PCCs with no treatments					
1	0.4053	10585	26112	208.91	w.o. PSCs
	0.9961	256	256	1.83	w. PSCs
2	0.1191	830	6976	49.69	w.o. PCCs
	0.9961	256	256	1.75	w. PCCs
3	0.9961	256	256	5.21	-
4	0.9961	256	256	4.38	-
Scenario II: mutated PCCs with different exsiting treatments					
5	0.0004	0	2304	17.13	cetuximab and erlotinib
	0.0012	10	9152	68.67	gemcitabine
	0.7810	8873	11360	114.25	nab-paclitaxel
	0.8004	7753	9686	73.83	ruxolitinib
Scenario III: mutated PCCs with blocking out on possible target(s)					
6	0.0792	38363	484128	3727.99	w.o. inhibiting ERK in PSCs
	0.9822	2201	2240	17.37	w. inhibiting ERK in PSCs
7	0.1979	3409	17232	136.39	w.o. inhibiting ERK in PSCs
	0.9961	256	256	2.01	w. inhibiting ERK in PSCs
8	0.2029	2181	10752	92.57	w.o. inhibiting MDM2 in PSCs
	0.9961	256	256	2.18	w. inhibiting MDM2 in PSCs
9	0.0004	0	2304	15.77	w.o. inhibiting RAS in PCCs and ERK in PSCs
	0.9961	256	256	3.15	w. inhibiting RAS in PCCs and ERK in PSCs
10	0.9797	1349	1376	11.98	w.o. inhibiting STAT in PCCs and NFκB in PSCs
	0.1631	1476	9056	81.61	w. inhibiting STAT in PCCs and NFκB in PSCs

One reason why PC is hard to be cured is that activated PSCs will move towards mutated PCCs, and form a cocoon for the tumor cells, which can protect tumor from attacks caused by therapies [7,16]. We investigate this by checking property 4, and obtain an estimated probability approaching to 1 (see Table 1).

Scenario II: mutated PCCs with different existing treatments

Property 5: This property aims to estimate the probability that the population of PCCs will eventually drop to and maintain in a low amount.

$$Prob_{=?} \{(PCCtot = 10) \land F^{1200} G^{400} (PCCtot < 100)\}$$

Property 5 means that, after some time, the population of PCCs can be maintained in a comparatively low amount, implying that PC is under control.

We now consider 5 different drugs that are widely used in PC treatments - cetuximab, erlotinib, gemcitabine, nab-paclitaxel, and ruxolitinib, and estimate the probabilities for them to satisfy property 5. As shown in Table 1, monoclonal antibody targeting EGFR (cetuximab), as well as direct inhibition of EGFR (erlotinib) broadly do not provide a survival benefit in PCs. Inhibition of MAPK pathway (gemcitabine) has also not been promising. These results are consistent with clinical feedbacks from patients [1]. While, strategies aiming at depleting the PSCs in PCs (i.e. nab-paclitaxel) can be successful (with an estimated probability 0.7810). Also, inhibition of Jak/Stat can be very promising (with an estimated probability 0.8004). These results are supported by [38] and [22], respectively.

Scenario III: mutated PCCs with blocking out on possible target(s)

Scenario I and II have demonstrated the descriptive and predictive power of our model. In scenario III, we use the validated model to identify new therapeutic strategies targeting molecules in PSCs. Here we report 4 potential target(s) of interest from our screening.

Property 6: This property aims to estimate the probability that the number of PSCs will eventually drop to and maintain in a low level.

$$Prob_{=?} \{(PSCtot = 5) \wedge F^{1200} \, G^{400} \, (PSCtot < 30)\}$$

Property 7: This property aims to estimate the probability that the population of migrated PSCs will eventually stay in a low amount.

$$Prob_{=?} \{(MigPSC = 0) \wedge F^{1200} \, G^{100} \, (MigPSC < 30)\}$$

The verification results of these two properties (Table 1) suggest that inhibiting ERK in PSCs not only lowers the population of PSCs, but also inhibits PSC migration. The former function can reduce the assistance from PSCs in the progression of PCs indirectly. The later one can prevent PSCs from moving towards PCCs and forming a cocoon to protect PCCs against cancer treatments.

Property 8: This property aims to estimate the probability that the number of PSCs entering the proliferation phase will eventually be less than the number of PSCs starting the apoptosis programme and this situation will maintain.

$$Prob_{=?} \{F^{1200} \, G^{400} \, ((PSCPro - PSCApop) < 0)\}$$

The increased probability (from 0.2029 to 0.9961 as shown in Table 1) indicates that inhibiting MDM2 in PSCs may reduce the number of PSCs by inhibiting PSCs' proliferation and/or promoting their apoptosis. Similar to the former role of inhibiting ERK in PSCs, it can help to treat PCs by alleviating the burden caused by PSCs.

Property 9: This property aims to estimate the probability that the number of bFGF will eventually stay in such a low level.

$$Prob_{=?} \{F^{1200} \, G^{400} \, (bFGF < 100)\}$$

As mentioned in property 5, 6, and 7, inhibiting RAS in PCCs can lower the number of PCCs, and downregulating ERK in PSCs can inhibit their proliferation and migration. Besides these, we find that, when inhibiting RAS in PCCs and ERK in PSCs simultaneously, the concentration of bFGF in the microenvironment drops (see Table 1). As bFGF is a key molecule that induces proliferation of both cell types, targeting RAS in PCCs and ERK in PSCs at the same time could synergistically improve PC treatment.

Property 10: This property aims to estimate the probability that the concentration of VEGF will eventually reach and keep in a high level.

$$Prob_{=?} \{F^{400} \, G^{100} \, (VEGF > 200)\}$$

Furthermore, inhibiting STAT in PCCs and NFκB in PSCs simultaneously postpones and lowers the secretion of VEGF (see Table 1). VEGF plays an important role in the angiogenesis and metastasis of pancreatic tumors. Thus, the combinatory inhibition of STAT in PCCs and NFκB in PSCs may be another potential strategy for PC therapies.

6 Conclusion

We present a multicellular and multiscale model of the PC microenvironment. The model is formally described using the extended BioNetGen language, which can capture the dynamics of multiscale biological systems using a combination of continuous and discrete rules. We carry out stochastic simulation and StatMC to analyze system behaviors under distinct conditions. Our verification results confirm the experimental findings with regard to the mutual promotion between PCCs and PSCs. We also gain insights on how existing treatments latching onto different targets can lead to distinct outcomes. These results demonstrate that our model might be used as a prognostic platform to identify new drug targets. We then identify four potentially (poly)pharmacological strategies for depleting PSCs and inhibiting the PC development. We plan to test our predictions empirically. Another interesting direction is to extend the model by considering spatial information [12] and TAMs in the PC microenvironment.

References

1. Personal communication with Jeffrey M Clarke, MD (Duke University School of Medicine)
2. Supplementary document. http://www.cs.cmu.edu/~qinsiw/cmsb2016/supplementary_doc.pdf
3. World Cancer Report 2014, World Health Organization (2014)
4. Albert, R., Thakar, J.: Boolean modeling: a logic-based dynamic approach for understanding signaling and regulatory networks and for making useful predictions. Wiley Interdisc. Rev. Syst. Biol. Med. **6**(5), 353–369 (2014)
5. Alon, U.: An Introduction to Systems Biology: Design Principles of Biological Circuits. CRC Press, London (2006)

6. Altrock, P.M., Liu, L.L., Michor, F.: The mathematics of cancer: integrating quantitative models. Nat. Rev. Cancer **15**(12), 730–745 (2015)
7. Apte, M., Park, S., Phillips, P., Santucci, N., Goldstein, D., Kumar, R., Ramm, G., Buchler, M., Friess, H., McCarroll, J., et al.: Desmoplastic reaction in pancreatic cancer: role of pancreatic stellate cells. Pancreas **29**(3), 179–187 (2004)
8. Bardeesy, N., DePinho, R.A.: Pancreatic cancer biology and genetics. Nat. Rev. Cancer **2**(12), 897–909 (2002)
9. Bensaid, M., Tahiri-Jouti, N., Cambillau, C., Viguerie, N., Colas, B., Vidal, C., Tauber, J., Esteve, J., Susini, C., Vaysse, N.: Basic fibroblast growth factor induces proliferation of a rat pancreatic cancer cell line: inhibition by somatostatin. Int. J. Cancer **50**(5), 796–799 (1992)
10. Danos, V., Feret, J., Fontana, W., Harmer, R., Krivine, J.: Rule-based modelling of cellular signalling. In: Caires, L., Vasconcelos, V.T. (eds.) CONCUR 2007. LNCS, vol. 4703, pp. 17–41. Springer, Heidelberg (2007)
11. Dunér, S., Lindman, J.L., Ansari, D., Gundewar, C., Andersson, R.: Pancreatic cancer: the role of pancreatic stellate cells in tumor progression. Pancreatology **10**(6), 673–681 (2011)
12. Erkan, M., Hausmann, S., Michalski, C.W., Fingerle, A.A., Dobritz, M., Kleeff, J., Friess, H.: The role of stroma in pancreatic cancer: diagnostic and therapeutic implications. Nat. Rev. Gastroenterol. Hepatol. **9**(8), 454–467 (2012)
13. Erkan, M., Reiser-Erkan, C., Michalski, C., Kleeff, J.: Tumor microenvironment and progression of pancreatic cancer. Exp. Oncol. **32**(3), 128–131 (2010)
14. Faeder, J.R., Blinov, M.L., Hlavacek, W.S.: Rule-based modeling of biochemical systems with bionetgen. In: Maly, I.V. (ed.) Systems Biology. MMB, vol. 500, pp. 113–167. Springer, Heidelberg (2009)
15. Farrow, B., Albo, D., Berger, D.H.: The role of the tumor microenvironment in the progression of pancreatic cancer. J. Surg. Res. **149**(2), 319–328 (2008)
16. Feig, C., Gopinathan, A., Neesse, A., Chan, D.S., Cook, N., Tuveson, D.A.: The pancreas cancer microenvironment. Clin. Cancer Res. **18**(16), 4266–4276 (2012)
17. Gong, H.: Analysis of intercellular signal transduction in the tumor microenvironment. BMC Syst. Biol. **7**(Suppl. 3), S5 (2013)
18. Gong, H., Wang, Q., Zuliani, P., Faeder, J.R., Lotze, M., Clarke, E.: Symbolic model checking of signaling pathways in pancreatic cancer. In: 3rd International Conference on Bioinformatics and Computational Biology, p. 245 (2011)
19. Gong, H., Zuliani, P., Wang, Q., Clarke, E.M.: Formal analysis for logical models of pancreatic cancer. In: 50th IEEE Conference on Decision and Control and European Control Conference, pp. 4855–4860 (2011)
20. Haber, P.S., Keogh, G.W., Apte, M.V., Moran, C.S., Stewart, N.L., Crawford, D.H., Pirola, R.C., McCaughan, G.W., Ramm, G.A., Wilson, J.S.: Activation of pancreatic stellate cells in human and experimental pancreatic fibrosis. Am. J. Pathol. **155**(4), 1087–1095 (1999)
21. Hippert, M.M., O'Toole, P.S., Thorburn, A.: Autophagy in cancer: good, bad, or both? Cancer Res. **66**(19), 9349–9351 (2006)
22. Hurwitz, H., Uppal, N., Wagner, S., Bendell, J., Beck, J., Wade, S., Nemunaitis, J., Stella, P., Pipas, J., Wainberg, Z., et al.: A randomized double-blind phase 2 study of ruxolitinib (RUX) or placebo (PBO) with capecitabine (CAPE) as second-line therapy in patients (pts) with metastatic pancreatic cancer (mPC). J. Clin. Oncol. **32**, 55 (2014)
23. Jaster, R.: Molecular regulation of pancreatic stellate cell function. Mol. Cancer **3**(1), 26 (2004)

24. Jha, S.K., Clarke, E.M., Langmead, C.J., Legay, A., Platzer, A., Zuliani, P.: A Bayesian approach to model checking biological systems. In: Degano, P., Gorrieri, R. (eds.) CMSB 2009. LNCS, vol. 5688, pp. 218–234. Springer, Heidelberg (2009)
25. Kleeff, J., Beckhove, P., Esposito, I., Herzig, S., Huber, P.E., Löhr, J.M., Friess, H.: Pancreatic cancer microenvironment. Int. J. Cancer 121(4), 699–705 (2007)
26. Kondo, Y., Kanzawa, T., Sawaya, R., Kondo, S.: The role of autophagy in cancer development and response to therapy. Nat. Rev. Cancer 5(9), 726–734 (2005)
27. Mahadevan, D., Von Hoff, D.D.: Tumor-stroma interactions in pancreatic ductal adenocarcinoma. Mol. Cancer Ther. 6(4), 1186–1197 (2007)
28. Mariño, G., Niso-Santano, M., Baehrecke, E.H., Kroemer, G.: Self-consumption: the interplay of autophagy and apoptosis. Nat. Rev. Mol. Cell Biol. 15(2), 81–94 (2014)
29. Masamune, A., Satoh, M., Kikuta, K., Suzuki, N., Satoh, K., Shimosegawa, T.: Ellagic acid blocks activation of pancreatic stellate cells. Biochem. Pharmacol. 70(6), 869–878 (2005)
30. Maus, C., Rybacki, S., Uhrmacher, A.M.: Rule-based multi-level modeling of cell biological systems. BMC Syst. Biol. 5(1), 166 (2011)
31. Muilenburg, D., Parsons, C., Coates, J., Virudachalam, S., Bold, R.J.: Role of autophagy in apoptotic regulation by Akt in pancreatic cancer. Anticancer Res. 34(2), 631–637 (2014)
32. Murphy, L., Cluck, M., Lovas, S., Ötvös, F., Murphy, R., Schally, A., Permert, J., Larsson, J., Knezetic, J., Adrian, T.: Pancreatic cancer cells require an EGF receptor-mediated autocrine pathway for proliferation in serum-free conditions. Br. J. Cancer 84(7), 926 (2001)
33. Phillips, P., Wu, M., Kumar, R., Doherty, E., McCarroll, J., Park, S., Pirola, R.C., Wilson, J., Apte, M.: Cell migration: a novel aspect of pancreatic stellate cell biology. Gut 52(5), 677–682 (2003)
34. Siegel, P.M., Massagué, J.: Cytostatic and apoptotic actions of TGF-β in homeostasis and cancer. Nat. Rev. Cancer 3(11), 807–820 (2003)
35. Sneddon, M.W., Faeder, J.R., Emonet, T.: Efficient modeling, simulation and coarse-graining of biological complexity with NFsim. Nat. Methods 8(2), 177–183 (2011)
36. Thomas, R., Thieffry, D., Kaufman, M.: Dynamical behaviour of biological regulatory networks - I. Biological role of feedback loops and practical use of the concept of the loop-characteristic state. Bull. Math. Biol. 57(2), 247–276 (1995)
37. Vardi, M.Y.: Automatic verification of probabilistic concurrent finite state programs. In: IEEE 26th Annual Symposium on Foundations of Computer Science, pp. 327–338 (1985)
38. Von Hoff, D.D., Ervin, T., Arena, F.P., Chiorean, E.G., Infante, J., Moore, M., Seay, T., Tjulandin, S.A., Ma, W.W., Saleh, M.N., et al.: Increased survival in pancreatic cancer with nab-paclitaxel plus gemcitabine. N. Engl. J. Med. 369(18), 1691–1703 (2013)
39. Vonlaufen, A., Joshi, S., Qu, C., Phillips, P.A., Xu, Z., Parker, N.R., Toi, C.S., Pirola, R.C., Wilson, J.S., Goldstein, D., et al.: Pancreatic stellate cells: partners in crime with pancreatic cancer cells. Cancer Res. 68(7), 2085–2093 (2008)

Tool Papers

ASSA-PBN 2.0: A Software Tool for Probabilistic Boolean Networks

Andrzej Mizera[1], Jun Pang[1,2], and Qixia Yuan[1(✉)]

[1] Faculty of Science, Technology and Communication, University of Luxembourg,
Luxembourg City, Luxembourg
{andrzej.mizera,jun.pang,qixia.yuan}@uni.lu
[2] Interdisciplinary Centre for Security, Reliability and Trust,
University of Luxembourg, Luxembourg City, Luxembourg

Abstract. We present a major new release of ASSA-PBN, a software tool for modelling, simulation, and analysis of probabilistic Boolean networks (PBNs). PBNs are a widely used computational framework for modelling biological systems. The steady-state dynamics of a PBN is of special interest and obtaining it poses a significant challenge due to the state space explosion problem which often arises in the case of large biological systems. In its previous version, ASSA-PBN applied efficient statistical methods to approximately compute steady-state probabilities of large PBNs. In this newly released version, ASSA-PBN not only speeds up the computation of steady-state probabilities with three different realisations of parallel computing, but also implements parameter estimation and techniques for in-depth analysis of PBNs, i.e., influence and sensitivity analysis of PBNs. In addition, a graphical user interface (GUI) is provided for the convenience of users.

1 Introduction

Probabilistic Boolean networks (PBNs) [9,12] are a modelling framework widely used to model gene regulatory networks (GRNs). A PBN model is capable to cope with uncertainties both on the data and model selection levels, allowing for systematic analysis of global network dynamics in the context of discrete-time Markov chains (DTMCs). It also provides means for quantifying relative influences and sensitivities of genes in their interactions and for characterising long-run behaviours of the whole genetic network. All these analyses are based on the steady-state probability distribution of the associated DTMC. Therefore, efficient computation of steady-state probabilities is a crucial foundation for analysing a PBN. Due to restricted computational resources, statistical methods are practically the only viable way to deal with large PBNs. However, the applicability of existing methods/tools for computing steady-state probabilities of PBNs are still limited by the network size, e.g., less than 100 nodes [11]. Moreover, to make the PBN framework generally accepted for mathematical modelling of biological systems, a user-friendly tool plays an important role but currently is still missing.

© Springer International Publishing AG 2016
E. Bartocci et al. (Eds.): CMSB 2016, LNBI 9859, pp. 309–315, 2016.
DOI: 10.1007/978-3-319-45177-0_19

In this paper, we present a new release of ASSA-PBN, a tool designed for modelling, simulation and analysis of PBNs. ASSA-PBN in its previous version [3] has overcome the above mentioned network size limitation with the implementation of an efficient simulator and state-of-the-art techniques for steady-state analysis, e.g., the two-state Markov chain approach. The newly released version of ASSA-PBN contributes mainly in three aspects. First, it speeds up the computation of steady-state probabilities of a PBN by using either multiple CPU/GPU core based parallelisation or structure-based parallelisation. Second, it implements parameter estimation and it supports in-depth analysis of PBNs, i.e., long-run influence and sensitivity analysis of PBNs. Third, it provides a graphical user interface (GUI) to ease user interactions with the tool.

A brief overview of the current functionality of ASSA-PBN is presented below. Items highlighted in bold are the new features added in version 2.0, and the tool is publicly available at http://satoss.uni.lu/software/ASSA-PBN/:

- modelling of PBNs in **high-level** ASSA-PBN format and converting a model from Matlab-PBN-toolbox format to ASSA-PBN format;
- random generation of PBNs;
- **efficient simulation** of a PBN;
- computation of steady-state probabilities of a PBN with either numerical methods or statistical methods (the two-state Markov chain approach, the Skart method, and the perfect-simulation method) [1,4];
- **parallel computation** of steady-state probabilities of a PBN with either the two-state Markov chain approach or the Skart method;
- **parameter estimation** of a PBN;
- **long-run influence and sensitivity analysis** of a PBN;
- a command-line tool and a **GUI**.

2   Preliminaries

We briefly introduce the concept of PBN in this section. PBN models a system such as a GRN with binary-valued nodes. For each node there is a certain number of Boolean functions, known as *predictor functions*, which determine the value of the node at the next time step. The selection of the predictor function is governed by a probability distribution: a selection probability parameter is associated with each predictor function of the node. Two variants of PBNs are considered: *instantaneous PBNs* and *context-sensitive PBNs*. In the former variant, the selection of a predictor function is performed for each node at each time step. In the latter variant, the PBN evolves in accordance with selected predictor functions and new selection is performed only if indicated by an additional random variable which is updated at each time step. Moreover, the so-called PBNs with perturbations allow the system to transit to the next state due to random perturbations that are governed by a perturbation rate parameter. We focus on synchronous PBNs, where the values of all the nodes are updated simultaneously, while in asynchronous PBNs only one node is updated at a time step. The dynamics of a PBN can be viewed as a DTMC. In the case of PBNs with perturbations, the underlying DTMC is *ergodic*, thus having a unique *stationary distribution*, the so-called *steady-state* (or *limiting*) *distribution*, which governs the long-run behaviour of the system. Due to the space limitation, we refer to [10] and [12, page 4] for a detailed description of PBNs.

3   Tool Architecture and New Features

The architecture of ASSA-PBN consists of three main modules, i.e., a modeller, a simulator, and an analyser. The modeller provides a simple way to construct a PBN model of a real-life biological system, e.g., a GRN. The simulator takes the PBN constructed in the modeller and performs simulation to produce trajectory samples. Based on the constructed model and generated samples, the analyser performs basic and in-depth analysis of the PBN. The analysis results can be used either to interpret the original system or to optimise the fitting of the system to experimental measurements. Figure 1 depicts the architecture of ASSA-PBN. The analyser requires different input files depending on the analysis task. While simulator and analyser rely on modeller as input, the simulation and analysis results can be used to optimise model fitting.

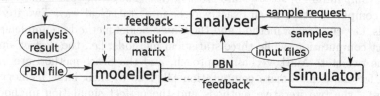

Fig. 1. Architecture of ASSA-PBN.

The newly released ASSA-PBN preserves the original architecture while making improvements to all the three modules. We proceed with describing the three modules in more details, while focusing on the newly implemented features.

Modeller. The PBN modeller can either load a PBN from a specification file or generate a random PBN (e.g., for benchmarking and testing purposes) complying with a given parametrisation [3]. In ASSA-PBN 2.0, a high-level PBN definition format is provided and visualisation of a PBN is supported in the GUI. The high-level PBN definition format provides a way to define Boolean function directly via its semantics instead of its truth table. The visualisation allows the user to check the details of each function and to interactively change the values of a predictor function. The modified PBN can then be used for the *long-run sensitivity with respect to function perturbation* analysis.

Simulator. Statistical approaches are practically the only viable option for the analysis of large PBNs. However, applications of such methods necessitate generation of trajectories of significant length. In order to make the analysis to execute in a reasonable amount of time, ASSA-PBN in its previous version applied the *alias method* [13] to sample the predictor function in each node. In this new version, the simulation process is sped up either with the use of the multiple CPU/GPU core parallelisation technique or with the structure-based parallelisation technique. The multiple core based technique [6] parallelises the simulation with the use of more cores while the structure-based parallelisation [5] achieves the same goal with the use of more memory. In order to parallelise the sample generation process, the algorithms for computing steady-state probabilities in the analyser have to be adjusted. More details are provided in the analyser section.

Moreover, visualisation of the simulation result is supported in ASSA-PBN 2.0 as well. Time-course evolution of selected node values can be displayed.

Analyser. The analyser of ASSA-PBN provides three main functionalities: basic computation of steady-state probabilities of a PBN, in-depth computation of long-run influences and sensitivities of a PBN, and parameter estimation of a PBN. In ASSA-PBN 2.0, the basic computation is largely improved with different parallelisation and multi-core techniques, while the other two are newly implemented.

Parallel computation of steady-state probabilities. The steady-state distributions can be computed either in an exact way or in a statistical way. Two iterative methods, i.e., the Jacobi method and the Gauss-Seidel method are implemented for exact computation; while three statistical methods, i.e., the perfect simulation, the two-state Markov chain approach, and the Skart method are implemented for the approximate computation [3]. Due to their large memory and time costs, the two iterative methods and the perfect simulation method are only suitable for analysing small-size PBNs [3].

Based on incremental sampling, the two-state Markov chain approach and the Skart method are capable of computing steady-state probabilities for large PBNs. In ASSA-PBN 2.0, we provide two types of techniques to speedup steady-state probability computation. Firstly, we implement our approach [6] of combining the Gelman & Rubin method with the two incremental methods to parallelise steady-state probability computation with multiple CPU/GPU cores. The Gelman & Rubin method is used to monitor that all the simulated chains have approximately converged to the steady-state distribution while the two-state Markov chain approach and the Skart method are used to determine the sample size required for computing the steady-state probabilities with specified precision. For a given precision, the lengths of trajectories used to estimate steady-state probabilities in the parallel approach are virtually the same as in the original two incremental methods. However, since the samples are generated with multiple cores in parallel, the processing time is significantly reduced. Details on the combined algorithms can be found in [6]. Secondly, we apply our structure-based parallel technique [5] to speedup the computation. This technique contributes in two aspects: reducing the network size by removing irrelevant nodes and by grouping nodes via merging their predictor functions. The key idea of this technique is to gain faster simulation speed with the use of larger memory.

Long-run influence and sensitivity. In a GRN, it is often important to distinguish which parent gene plays a major role in regulating a target gene and how sensitive the system is with respect to certain changes. PBNs feature quantification of the importances (formally known as long-run influences) and sensitivities [8–10]. ASSA-PBN 2.0 supports computation of long-run influences and sensitivities, i.e., the long-run influence of a gene on a specified predictor function, the long-run influence of a gene on another gene, the long-run sensitivity of a gene in a PBN, the long-run sensitivity of a gene with respect to function perturbation, and the long-run sensitivity of a gene with respect to selection probability perturbation. All these functionalities require the computation of several steady-state probabilities. For each probability, a trajectory of certain length has to be generated. Note that ASSA-PBN does not store the generated trajectory for the sake of memory saving. Instead, ASSA-PBN verifies whether the next sampled state of the PBN belongs to the set of states of interest and stores this information only. Therefore, a new trajectory is required when computing the steady-state probability for a new set of states of interest. ASSA-PBN 2.0 implements computation of steady-state probabilities of several sets of states in parallel with the two-state Markov chain approach [6], allowing the reuse of a generated trajectory. The crucial idea is that each time the next state of the PBN is generated, it is processed for all state sets of interest simultaneously. Different sets require trajectories of different lengths and the lengths are determined dynamically through an iteration process. Whenever the trajectory is long enough for computing the steady-state probability of a certain set of states, the steady-state probability of this set will be computed and this set will not be considered in future iterations.

Parameter estimation. A key challenge in constructing a PBN model is the determination of the model parameters which make the model match the behaviour of the real system. A few algorithms [2,7] have been proposed in literature for parameter estimation of biological systems. ASSA-PBN 2.0 applies the *particle swarm optimisation* algorithm for estimating parameters for a PBN. This algorithm is an iterative process for finding an optimal set of parameters. A set of parameters is called a particle and its fit to the experimental data is verified through a cost function. ASSA-PBN uses the *mean square error* (MSE) function, i.e., $MSE = \frac{1}{n}\sum_{i=1}^{n}(\hat{Y}_i - Y_i)^2$, where n is the number of measurement data, $\hat{Y}_i$ is the ith measurement data point value, and Y_i is the computed steady-state probability corresponding to the ith data point. In each iteration, all the particles are updated and verified using the cost function. The particle that results in the minimum cost function value is the optimal particle. Particle values are updated based on the current values and the current best optimal particle values so that each particle is moving towards the direction of the current best optimal particle. Normally, the verification of the cost function for each particle requires the computation of steady-state probabilities of several sets of states. To make the computation as fast as possible, ASSA-PBN 2.0 provides the support for computing several steady-state probabilities in parallel see [6] for details).

4 Evaluation and Case Studies

As shown in [3], the first version of ASSA-PBN has shown a significant advantage in terms of speed compared to the Matlab tool optPBN [12]. We proceed to compare the performance of ASSA-PBN 2.0 with its previous version. This comparison is done both on randomly generated PBNs and a PBN constructed for a real-life biological system. The newly released version supports three types of parallelised computation of steady-state probabilities of a PBN. We show in [6] that for the multiple CPU core based parallelisation, the speed-up is approximately linear with the number of cores in our hardware environment (CPU cores up to 40). For the multiple GPU core based parallelisation, ASSA-PBN 2.0 can approximately achieve a speed-up of 200; while the structure-based parallelisation shows a promising performance for sparse networks with large percentage of leaves (for more details refer to [5]).

Moreover, we compared the performance of the three types of parallelisations with the sequential approach on an analysis of a 96-node PBN of apoptosis using the two-state Markov chain approach. In [4], the sequential version of the two-state Markov chain approach has been used for the long-run influence analysis on complex2 from each of its parent nodes. Seven steady-state probabilities are required to be computed in order to perform the analysis. We re-perform this computation with the three parallelised versions of the two-state Markov chain approach. Speed-ups of approximately 200 (GPU), 20 (multiple CPUs) and 10 (structure-based parallelisation) are obtained with the use of the three different parallel computations. Detailed comparison of both the random networks and the real-life case-study can be found at http://satoss.uni.lu/software/ASSA-PBN/benchmark.

5 Future Developments

First, we plan to implement a suite of parameter estimation algorithms, e.g., genetic algorithms. Second, user-friendly improvements, such as support for the standard Systems Biology Markup Language (SBML) and graphical editing and visualisation of PBN models, will be introduced in the future releases of ASSA-PBN versions.

Acknowledgement. Qixia Yuan is supported by the National Research Fund, Luxembourg (grant 7814267). The authors also want to thank Gary Cornelius for his work on ASSA-PBN.

References

1. El Rabih, D., Pekergin, N.: Statistical model checking using perfect simulation. In: Liu, Z., Ravn, A.P. (eds.) ATVA 2009. LNCS, vol. 5799, pp. 120–134. Springer, Heidelberg (2009)
2. Kennedy, J., Eberhart, R.: Particle swarm optimization. In: Proceedings of IEEE International Conference on Neural Networks, pp. 1942–1948 (1995)

3. Mizera, A., Pang, J., Yuan, Q.: ASSA-PBN: an approximate steady-state analyser of probabilistic Boolean networks. In: Finkbeiner, B., Pu, G., Zhang, L. (eds.) ATVA 2015. LNCS, vol. 9364, pp. 214–220. Springer, Heidelberg (2015). doi:10.1007/978-3-319-24953-7_16

4. Mizera, A., Pang, J., Yuan, Q.: Reviving the two-state Markov chain approach (Technical report) (2015). http://arxiv.org/abs/1501.01779

5. Mizera, A., Pang, J., Yuan, Q.: Fast simulation of probabilistic Boolean networks. In: Bartocci, E., et al. (eds.) Proceedings of 14th International Conference on Computational Methods in Systems Biology. LNCS, vol. 9859, pp. 216–231. Springer, Heidelberg (2016)

6. Mizera, A., Pang, J., Yuan, Q.: Parallel approximate steady-state analysis of large probabilistic Boolean networks. In: Proceedings of 31st ACM Symposium on Applied Computing, pp. 1–8. ACM Press (2016)

7. Moles, C.G., Mendes, P., Banga, J.R.: Parameter estimation in biochemical pathways: a comparison of global optimization methods. Genome Res. **13**(11), 2467–2474 (2003)

8. Qian, X., Dougherty, E.R.: On the long-run sensitivity of probabilistic Boolean networks. J. Theor. Biol. **257**(4), 560–577 (2009)

9. Shmulevich, I., Dougherty, E., Zhang, W.: From Boolean to probabilistic Boolean networks as models of genetic regulatory networks. Proc. IEEE **90**(11), 1778–1792 (2002)

10. Shmulevich, I., Dougherty, E.R.: Probabilistic Boolean Networks: The Modeling and Control of Gene Regulatory Networks. SIAM Press, New York (2010)

11. Trairatphisan, P., Mizera, A., Pang, J., Tantar, A.A., Sauter, T.: optPBN: an optimisation toolbox for probabilistic Boolean networks. PLOS ONE **9**(7), e98001 (2014)

12. Trairatphisan, P., Mizera, A., Pang, J., Tantar, A.A., Schneider, J., Sauter, T.: Recent development and biomedical applications of probabilistic Boolean networks. Cell Commun. Signaling **11**, 46 (2013)

13. Walker, A.: An efficient method for generating discrete random variables with general distributions. ACM Trans. Math. Softw. **3**(3), 253–256 (1977)

E-Cyanobacterium.org: A Web-Based Platform for Systems Biology of Cyanobacteria

Matej Troják[1], David Šafránek[1(✉)], Jakub Hrabec[1], Jakub Šalagovič[1], Františka Romanovská[1], and Jan Červený[2]

[1] Faculty of Informatics, Masaryk University, Brno, Czech Republic
safranek@fi.muni.cz
[2] Global Change Research Centre AS CR, v. v. i., Brno, Czech Republic

Abstract. E-cyanobacterium.org is an online platform providing tools for public sharing, annotation, analysis, and visualization of dynamical models and wet-lab experiments related to cyanobacteria. The platform is unique in integrating abstract mathematical models with a precise consortium-agreed biochemical description provided in a rule-based formalism. The general aim is to stimulate collaboration between experimental and computational systems biologists to achieve better understanding of cyanobacteria.

1 Introduction

The complexity of dynamical processes occurring in biology is inherent due to the fact that living cells must be responsive to a highly dynamic environment. This applies especially to the family of photosynthetic organisms such as cyanobacteria in which the biophysical processes scale from electron transfers interacting with metabolic biochemistry to genetic regulation and back. Remarkable effort towards a consistent and coherent knowledge-base of cyanobacteria modelling is one of the key activities of CyanoNetwork[1], the international network of top-leading experts on studying these unique bacteria. In cyanobacteria, biophysical processes span in vastly different time scales from femtoseconds to seasons and from individual molecules to aquatic and terrestrial ecosystems.

Such a community-wide modelling effort is notably accelerated by an interactive online platform. In particular, models need to be translated into a unified format, formalised, and uniformly annotated. This allows the models to be fully understood and re-used in the original form, compared with each other, and with wet-lab experiments. To this end, one needs a unified and flexible framework to fully represent partial models and the respective biological knowledge — the involved components as well as their interactions.

An existing resource to inspire further expansion of cyanobacteria modelling are the established web repositories of curated and well-annotated

This work has been supported by the Czech National Infrastructure grant LM2015055 and by the Czech Science Foundation grant GA15-11089S.

[1] http://www.cyanoteam.org.

E. Bartocci et al. (Eds.): CMSB 2016, LNBI 9859, pp. 316–322, 2016.
DOI: 10.1007/978-3-319-45177-0_20

models traversing already through many branches of biology [3,12,17,19]. Unfortunately, cyanobacteria models are strongly under-represented in these repositories, probably because of the natural differences that exist between common bacteria and phototrophic bacteria. There exist several online tools presenting biological networks or genome knowledge [1,4,12]. However, due to enormous complexity of biological processes, it is a challenge to develop tools presenting biological networks alongside with executable or mathematical models. Focusing on domain-specific organisms allows us to integrate the knowledge and present it in a concise and understandable form. This has been already proven on examples of well-known model organisms such as *E. coli* [14] or *C. elegans* [7]. Nevertheless, these resources do not couple the presented knowledge with modelling.

In this tool paper, we present an online platform for cyanobacteria processes — *e-cyanobacterium.org*[2]. The platform integrates several dedicated tools and is distinct in the following aspects: formal rule-based representation of biochemical interactions facilitated by cyanobacteria biochemical entities; repository of kinetic models providing basic analysis tools online (simulation, custom data sets, basic static analysis); integration of models within the rule-based description and export to SBML [10]; storage, maintenance and presentation of experimental data; content visualisation (graphical presentation of models, biochemical space and modelling/experimental data).

The presented release of e-cyanobacterium.org is implemented as an extension of our general database tool-kit — the so-called *comprehensive modelling platform* [15]. Key updates are the formal rule-based language BCSL, wet-lab experiments module, and improved analysis of models. Most importantly, several well-annotated and curated models developed by the consortium are provided within e-cyanobacterium.org including the formalisation of the respective part of biochemical space in BCSL.

2   Web Platform Overview

The platform consists of several dedicated modules (Fig. 1) all connected to a central module – *Biochemical Space* (BCS) [16] – that is the backbone of the platform. BCS provides formal description of the biological problem and it is based on the hierarchy of selected biological processes. It is accompanied with schemas representing relevant biological processes in the context of cyanobacteria. For each process, there are presented relevant models, chemical entities, and rules. Presentation of every process includes detailed information and links to relevant internal and external sources.

2.1   Biochemical Space

Biochemical Space is constructed from hierarchy of entities interacting in rules formally specified in *Biochemical Space Language* (BCSL). The main advantage of BCSL is the adoption of the most important aspects of rule-based features

---

[2] http://www.e-cyanobacterium.org.

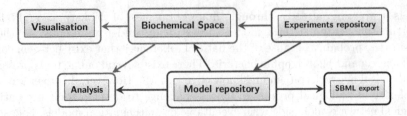

Fig. 1. Scheme of interconnections between modules of E-cyanobacterium.org platform.

while still keeping the syntax human-readable and accessible to communities outside computer science. On the one hand, BCSL has executable semantics that allows basic analysis and consistency checking. On the other hand, the language includes constructs for detailed biological annotation reflecting the known bioinformatics databases and ontologies. BCSL is developed with the consideration of new extensions in SBML level 3. Once the SBML rule-based support becomes to be actively used in tools, our platform will provide relevant export filters.

Rule description provides details of rules, explicit enumeration of substrates and products, and available annotations. Moreover, the rule is schematically visualised (Sect. 3.2) and presented with appropriate links to the process hierarchy.

Entity description provides information about associated models and rules. Links to the process hierarchy and to external sources (UniProt [2], Kegg [12], GeneOnthology [1], etc.) are available.

2.2   Model Repository

Model repository is a collection of implemented mathematical models describing particular parts of biological processes. Every model is represented as a set of ordinary differential equations generated from the model reaction network. Models are integrated within BCS. In particular, each model component should be related to some BCS entity and each model reaction should be related to some BCS rule. Moreover, a model is associated with some parameter value sets (data sets) that enable simulation (Sect. 3.3) in a particular biologically-relevant scenarios. Additionally, several basic non-parameter-specific static analysis techniques (Sect. 3.4) based on model stoichiometry are also provided.

An implemented model then includes complete biological annotation (Sect. 3.1) of all components and reactions that is provided by mapping to BCS. This might help to find connections and overlaps among models. Further, implemented model can be exported to SBML (level 2).

Currently, the repository contains two models describing circadian clock (Miyoshi et al. 2007 [18], Hertel et al. 2013 [8]) and a kinetic model of metabolism (Jablonsky et al. 2014 [11]). Two other models present unpublished results – dynamics of carbon fluxes (Müller et al.) and photosynthesis (Plyusnina et al.).

2.3 Experiments Repository

Experiments repository is a tool for storage and presentation of time-series data from wet-lab experiments. Every experiment is well-grounded by precise description (device, medium, organism, etc.) and appropriate annotations. Experiments are structured – several time series data can be attached to a single experiment. Every time series targets a specific list of measured substances together with time stamps of the individual measurements. Data can be imported/exported in simple text formats. Time series are visualised in a chart (Sect. 3.3).

Registered users can add their own experiments while keeping the selected experiments private or public. For repeated experiments, annotations and details can be cloned from previously inserted experiments.

3 Website Features

E-cyanobacterium.org provides support for modelling and analysis of biological systems. The most important fact is that all these parts are integrated within the Biochemical Space. Therefore it is possible to reveal non-trivial relations between biochemical substances and models.

3.1 Annotation

Annotation is an important task in modelling of biological systems. This is how biological knowledge is mapped to the mathematical description. Our platform considers annotation as an inherent and compulsory part of the modelling procedure. The following aspects of annotation are supported:

– BCS creation and maintenance,
– model annotation by mapping to BCS,
– experiment annotation.

BCS is being continuously extended and revised by the consortium. Researchers supplying their models are required to integrate the models within the current BCS. This gives good feedback to BCS maintainers. In the experiments repository module, emphasis is given to description of conditions under which an experiment has been performed, which is important for interpretation of the data as well as the possibility of reconstructing the measurements.

3.2 Visualisation

Static visualisation is provided by means of schemas showing the process hierarchy with most important objects from BCS. Graphical schemes are supplied with detailed information on the visualised elements of BCS. This is achieved in the information panel below the scheme. An accompanying feature is filtering displayed data to *All* or *Visible* elements. The former option extends the content to all relevant entities and rules that take the part in the displayed process but are not necessarily visualised (e.g., this applies to small "universal" molecules such as ATP, ADP, etc.). Moreover, there is a special visualisation framework for complex networks such as metabolism. This widget allows one to zoom in a specific part of the complex network. This feature is currently employed for metabolic map of cyanobacteria. Owing to requests coming out of the consortium, we have decided to employ manually handled visualisation. In next release, we plan to provide automatically generated SBGN representation of the schemes.

Rule details (Sect. 2.1) are enriched by means of automatically generated visualisation. For every rule, a graphical scheme that displays all its substrates and products is generated.

3.3 Simulation

With every implemented model (Sect. 2.2) it is possible to execute simulations. Registered users can change initial conditions and parameters of the model and set the simulation options (the numerical solver and its parameters). To apply such settings in the simulation, they have to be saved in terms of a *custom dataset* (public or private). The simulation chart is generated for all available datasets and the platform GUI allows one to switch between them. The chart is interactive and allows one to change the axis type, zoom in the selected curves, and show a focused value on a curve.

Simulation data are accessible by exporting them in several data formats. Similar options are also available in charts visualising experimental data in the experiments repository. Additionally, the simulated model including the selected dataset can be downloaded as an Octave file [6]. Additionally, the platform allows one to use different externally called numerical solvers for simulation. In particular, model administrators can select either Octave (default), COPASI [9], or remote simulation with COPASI web services [13].

3.4 Static Analysis

At the current stage, there are three static analysis tasks available. *Matrix analysis* produces incidence matrix, *Conservation analysis* produces mass conservation analysis (moiety conservation), and *Modes analysis* produces elementary flux modes. For these tasks, the well-acclaimed third party tool COPASI [9] is used which allows one to download the task data by means of an SBRML file [5].

4 Conclusion

E-cyanobacterium.org provides several features that contribute to production and presentation of models targeting cyanobacteria. The most principal effort is to interlink biological knowledge with benefits of computational systems biology tools. This is enabled by means of a novel formal notation – Biochemical Space – which allows us to integrate computational models with the biological knowledge and wet-lab experiments.

For future work, we plan to improve the mapping between mathematics and biology, to enhance the website with more analysis tools (currently we are implementing an interface for existing property monitoring and robustness analysis tools such as [20]), and to automatise the comparison of models against experimental data. Moreover, biochemical space of cyanobacteria is continuously being extended and improved with interactive visualisations of reaction networks based on formal description provided in Biochemical Space Language.

References

1. Ashburner, M., et al.: Gene ontology: tool for the unification of biology. Gene Ontology Consortium, Nat. Genet. **25**(1), 25–29 (2000)
2. Bateman, A., et al.: Uniprot: a hub for protein information. Nucleic Acids Res. **43**(D1), D204–D212 (2015)
3. Beard, D.A., et al.: CellML metadata standards, associated tools and repositories. Philos. Trans. R. Soc. Lond. A Math. Phys. Eng. Sci. **367**(1895), 1845–1867 (2009)
4. Croft, D., et al.: The reactome pathway knowledgebase. Nucleic Acids Res. **42**(D1), D472–D477 (2014)
5. Dada, J.O., et al.: SBRML: a markup language for associating systems biology data with models. Bioinformatics **26**(7), 932–938 (2010)
6. Eaton, J.W., et al.: GNU Octave version 3.0.1 manual: a high-level interactive language for numerical computations. CreateSpace Independent Publishing Platform, Seattle (2009)
7. Harris, T.W., et al.: Wormbase: a comprehensive resource for nematode research. Nucleic Acids Res. **38**(suppl 1), D463–D467 (2010)
8. Hertel, S., et al.: Revealing a two-loop transcriptional feedback mechanism in the cyanobacterial circadian clock. PLoS Comput. Biol. **9**(3), 1–16 (2013)
9. Hoops, S., et al.: COPASI - a complex pathway simulator. Bioinformatics **22**(24), 3067–3074 (2006)
10. Hucka, M., et al.: The systems biology markup language (SBML): a medium for representation and exchange of biochemical network models. Bioinformatics **19**(4), 524–531 (2003)
11. Jablonsky, J., et al.: Multi-level kinetic model explaining diverse roles of isozymes in prokaryotes. PLoS ONE **9**(8), 1–8 (2014)
12. Kanehisa, M., et al.: KEGG as a reference resource for gene and protein annotation. Nucleic Acids Res. **44**(D1), D457–D462 (2016)
13. Kent, E., et al.: Condor-COPASI: high-throughput computing for biochemical networks. BMC Syst. Biol. **6**(1), 1–13 (2012)
14. Keseler, I.M., et al.: EcoCyc: fusing model organism databases with systems biology. Nucleic Acids Res. **41**(D1), D605–D612 (2013)
15. Klement, M., et al.: A comprehensive web-based platform for domain-specific biological models. Electron. Notes Theoret. Comput. Sci. **299**, 61–67 (2013)
16. Klement, M., et al.: Biochemical space: a framework for systemic annotation of biological models. In: Proceedings of the 5th International Workshop on Interactions Between Computer Science and Biology (CS2Bio-14). Electronic Notes in Theoretical Computer Science, vol. 306, pp. 31–44 (2014)
17. Le Novère, N., et al.: Biomodels database: a free, centralized database of curated, published, quantitative kinetic models of biochemical and cellular systems. Nucleic Acids Res. **34**, D689–D691 (2006)

18. Miyoshi, F., et al.: A mathematical model for the kai-protein-based chemical oscillator and clock gene expression rhythms in cyanobacteria. J. Biol. Rhythms **22**(1), 69–80 (2007)
19. Olivier, B.G., et al.: Web-based kinetic modelling using jws online. Bioinformatics **20**(13), 2143–2144 (2004)
20. Rizk, A., et al.: A general computational method for robustness analysis with applications to synthetic gene networks. Bioinformatics **25**(12), i169–i178 (2009)

PREMER: Parallel Reverse Engineering of Biological Networks with Information Theory

Alejandro F. Villaverde[1,2,4(✉)], Kolja Becker[3], and Julio R. Banga[4]

[1] Department of Systems and Control Engineering,
University of Vigo, Galicia, Spain
[2] Centre for Biological Engineering, University of Minho,
Braga, Portugal
[3] Modelling of Biological Networks, Institute of Molecular Biology GGmbH,
Mainz, Germany
[4] Bioprocess Engineering Group, Instituto de Investigacións Mariñas (IIM-CSIC),
Vigo, Galicia, Spain
afvillaverde@iim.csic.es

Abstract. A common approach for reverse engineering biological networks from data is to deduce the existence of interactions among nodes from information theoretic measures. Estimating these quantities in a multidimensional space is computationally demanding for large datasets. This hampers the application of elaborate algorithms – which are crucial for discarding spurious interactions and determining causal relationships – to large-scale network inference problems. To alleviate this issue we have developed PREMER, a software tool which can automatically run in parallel and sequential environments, thanks to its implementation of OpenMP directives. It recovers network topology and estimates the strength and causality of interactions using information theoretic criteria, and allowing the incorporation of prior knowledge. A preprocessing module takes care of imputing missing data and correcting outliers if needed. PREMER (https://sites.google.com/site/premertoolbox/) runs on Windows, Linux and OSX, it is implemented in Matlab/Octave and Fortran 90, and it does not require any commercial software.

Keywords: Network inference · Information theory · Parallel computing

1 Introduction

Many biological systems can be meaningfully represented as networks, that is, as a set of nodes (variables) connected by links (interactions). In the context of cellular networks the nodes are molecular entities such as genes, transcription factors, proteins, metabolites, and so on [7]. The network inference problem consists of learning the interconnection structure of the nodes, using as data the values of the variables (e.g. their expression levels or concentrations) at different situations and/or time instants. The concept of mutual information [12] can be

© Springer International Publishing AG 2016
E. Bartocci et al. (Eds.): CMSB 2016, LNBI 9859, pp. 323–329, 2016.
DOI: 10.1007/978-3-319-45177-0_21

used as a statistical measure for estimating the strength of the (possibly non-linear) relations among nodes from a dataset. Indirect interactions, which take place when an entity A exerts an influence in C by means of an intermediate entity B (i.e. $A \rightarrow B \rightarrow C$), are difficult to detect, because a spurious interaction may be deduced (not only $A \rightarrow B$ and $B \rightarrow C$, but also $A \rightarrow C$). The difficulty of discriminating between them increases when dealing with higher-order inter-actions, involving four or more entities. Although a few methods can cope with this issue, their application to large-scale problems is computationally costly [6], especially when dealing with time-series data. One such method, MIDER [13], is a general purpose network inference tool which takes into account time delays. It distinguishes between direct and indirect interactions using entropy reduction [9] and assigns directionality to the predicted links using transfer entropy [11]. It is implemented in Matlab, a widely used programming environment which nevertheless has some drawbacks, mainly (i) the need of buying commercial licenses, and (ii) low computational efficiency compared to other languages.

Here we present PREMER (Parallel Reverse Engineering with Mutual information & Entropy Reduction), a tool that overcomes these issues. It includes an advanced Fortran 90 implementation of the MIDER procedures, which allows for faster computations than Matlab. Additionally, the use of OpenMP directives enable it to run seamlessly in parallel environments, thus allowing for further speedups in performance. Results obtained on different datasets show that PREMER can be orders of magnitude faster than MIDER. Additionally, PREMER's Matlab code is fully compatible with the free Octave environment. Furthermore, PREMER offers two important additional capabilities. One is the ability to take prior knowledge into account, allowing to specify if a particular interaction is known to be non-existent. This is of particular importance in applications such as gene regulatory network (GRN) inference, where only a subset of the genes — the transcription factors, TFs — can regulate other genes. The second one is the ability to handle datasets with missing values and/or outliers, using statistical techniques to impute new values which are coherent with the latent structure of the data. PREMER's work-flow is depicted in Fig. 1. More details about the methodology are given in the supplementary information (user's manual).

2   Implementation and Availability

PREMER is provided as a set of Matlab/Octave scripts and an executable file which carries out the core computations. It has a number of options, which can be tuned by editing the main file, `runPremer`. Executable files are provided for Windows, Linux and OSX, and also as source code in Fortran (F90), which can be compiled to run on any operating system. The executable can also be invoked from the command line, thus avoiding the need for Matlab/Octave. A key feature of PREMER is its ability to run sequentially or in parallel. Parallelization has been implemented using OpenMP directives [3] and is entirely transparent to the user, who only needs to specify the number of threads in the main file. Mutual information and multidimensional entropies are estimated using an adaptive

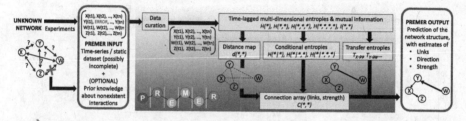

Fig. 1. Work-flow of the PREMER algorithm. First, a data curation module imputes missing data [4] and detects and corrects outliers, thus allowing the use of faulty datasets. Then PREMER calculates the distance between every possible pair of variables $d(X, Y)$ for several time delays. To this end it estimates the entropies of all variables $H(*)$, as well as the joint entropies $H(*, *)$ and the mutual information $I(*, *)$ of all pairs of variables. The user can choose to estimate also the multi-dimensional joint entropies of 3 and 4 variables ($H(*, *, *)$, $H(*, *, *, *)$), in order to use them in the subsequent entropy reduction step. The aim of this step is to determine whether all the variation in a variable Y can be explained by the variation in another variable X or, more generally, in a set of variables $\mathbf{X}$ [9]. By iterating through cycles of adding a variable X that reduces $H(Y|X, \mathbf{X})$ until no further reductions are obtained, the entropy reduction step yields the complete set of variables that control the variation in Y. Finally, directions are assigned to the links using transfer entropy, $T_{X \rightarrow Y}$, a non-symmetric measure of causality [11] calculated from time-lagged conditional entropies.

partitioning algorithm inspired in [2]. The PREMER toolbox is released under the free and open source GNU GPLv3. It is available at https://sites.google.com/site/premertoolbox/. Its use does not require any commercial software.

3   Selected Experimental Results

We tested PREMER on the same set of seven benchmark problems that was used for assessing the performance of MIDER. It has been shown elsewhere [13] that MIDER performs well compared with other state-of-the-art methods in terms of precision and recall of the inferred networks. We found that PREMER predicts the same networks as MIDER (in examples without missing data or prior information) achieving large reductions in computation times, as shown in Fig. 2. Panel A plots the accelerations obtained with PREMER's sequential implementation (i.e. using only one processor) with respect to MIDER. The most computationally costly problems give rise to the largest speed-ups: for example, for benchmark B7 with 3 entropy reduction rounds the computation time decreases from 42 h to roughly 1.5 h. This improvement is obtained using a single processor; additional speed-ups can be achieved in a parallel environment, as shown in panel B. The combined effect of code acceleration and parallel speed-up results in very significant reductions in computation time. For example, using a current 12-core desktop PC (hardware detailed in the caption of Fig. 2), PREMER runs up to 170 times faster than MIDER.

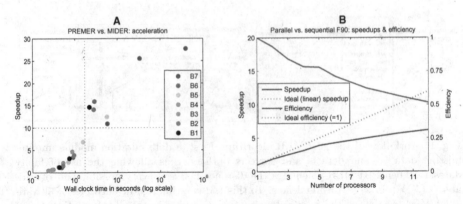

Fig. 2. [**A**]: Accelerations achieved by PREMER w.r.t. MIDER, for benchmarks B1–B7 of [13]. For every benchmark three points are plotted, depending on the number of entropy reduction rounds performed: 1, 2, or 3. [**B**]: Speed-up and efficiency of the parallel vs sequential versions of PREMER (benchmark B7, 3 entropy reduction rounds). Results obtained in a multi-core PC running Windows 7 64-bit with 16 GB RAM and 12 cores, 2 processors/core, Intel Xeon 2.30 GHz. (Color figure online)

By going through several entropy reduction rounds it is possible to discover additional links, but in this process errors may appear: since the accuracy of every network inference method is limited by the information content of the data, some of the extra links can in fact be false positives. Therefore in many cases there is a trade-off between precision and recall: increasing the number of entropy reduction rounds leads to increased recall and decreased precision, and vice versa. Table 1 shows this trade-off for the average of the seven benchmark problems considered in [13].

Table 1. Trade-offs between precision and recall for different numbers of entropy reduction rounds. The values shown are the averages of the seven benchmark problems (B1–B7) considered in [13].

Entropy reduction rounds	1	2	3
Average precision (B1–B7)	0.7676	0.6958	0.6311
Average recall (B1–B7)	0.5267	0.5676	0.5819

Finally, we illustrate the performance improvement that can be obtained by taking prior knowledge into account. With this aim we create a benchmark network with GeneNetWeaver [10] consisting of 18 genes, out of which only 11 are considered transcription factors, and we generate time course data of the expression of each gene at 24 different time points. We evaluate the performance of PREMER using two different modes of including information (removing interactions *a priori* or *a posteriori*) and we compare it to other network inference methods

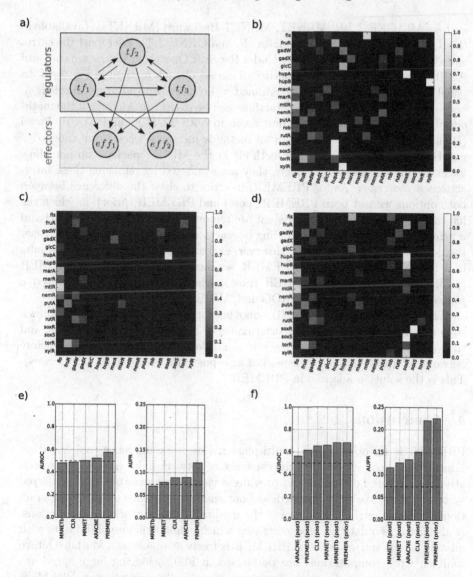

Fig. 3. Incorporating prior knowledge into network inference algorithms: **(a)** Reasoning behind excluding interactions from gene regulatory networks. Only transcription factors effectively serve as regulators in the network, hence interactions from the effector genes to transcription factors or to other effectors can be excluded. **(b)** Regulatory interaction matrix returned by PREMER without incorporating prior knowledge. The heat map scale represents the strength of the predicted interaction (0 = no interaction, 1 = strongly predicted interaction). **(c)** Regulatory interaction matrix returned by PREMER incorporating prior knowledge by removing excluded interactions *a posteriori.* **(d)** Regulatory interaction matrix returned by PREMER incorporating prior knowledge by removing excluded interactions *a priori.* **(e)** Comparison of several network inference methods without incorporating prior knowledge. **(f)** Comparison of network inference methods incorporating prior knowledge either *a posteriori* (post) or *a priori* (prior). In (e) and (f) horizontal dashed lines indicate the theoretical performance of a random classifier.

such as ARACNE, CLR, MRNET, MRNET Backward (MRNETB) (available in the R package MINET [8]), Inferelator [1], and GENIE3 [5]. We report the corresponding values of AUROC (Area Under Receiver Operating Characteristic) and AUPR (Area Under the Precision-Recall curve) of each method in Fig. 3.e–f. In panel (e) the set of regulators is assumed to be unknown (no prior knowledge), and consequently none of the interactions can be excluded. Among all the methods, PREMER achieves the highest score in both AUROC and AUPR. Panel (f) shows that the performance of all methods increases when prior knowledge is taken into account. Since the methods in the MINET package do not allow for excluding interactions *a priori*, they are evaluated by removing these interactions *a posteriori*. As for PREMER, in order to show the difference between both options we test both PREMER (post) and PREMER (prior), in which the interactions known to be non-existent are removed respectively *a posteriori* and *a priori*. It can be seen that excluding interactions *a posteriori* already increases the performance of all methods. However, excluding interactions *a priori* results in a further improvement of PREMER, as shown by the fact that PREMER (prior) is outperforming PREMER (post), where interactions are removed *a posteriori*, both in terms of AUROC and AUPR (Fig. 3.f).

Therefore, we conclude that (i) removing interactions based on prior knowledge is a way of increasing the performance of network inference methods, and that (ii) the improvement is bigger if this information is incorporated before network inference (*a priori*) instead of as a post-processing step (*a posteriori*). This is the solution adopted in PREMER.

4   Conclusions

PREMER is an open-source, multi-platform network inference tool based on information theory. It predicts the existence of network links, estimates their relative strength and direction, and provides a visual representation of the inferred system. It can take prior knowledge about the non-existence of specific interactions into account, which improves the quality of the network reconstructions. It also features a data preprocessing step which enables the use of datasets with missing values and/or outliers. PREMER is freely available as a Matlab/Octave toolbox. Core computations are performed in F90, achieving large speed-ups which can be increased further if working on a parallel environment. PREMER is geared towards ease of use, requiring minimum input from the user.

Acknowledgements. AFV acknowledges funding from the Galician government (Xunta de Galiza) through the I2C fellowship ED481B2014/133-0. KB was supported by the German Federal Ministry of Research and Education (BMBF, OncoPath consortium). JRB acknowledges funding from the Spanish government (MINECO) and the European Regional Development Fund (ERDF) through the project "SYNBIOFAC-TORY" (grant number DPI2014-55276-C5-2-R). This project has received funding from the European Unions Horizon 2020 research and innovation programme under grant agreement No 686282 (CanPathPro). We thank David R. Penas and David Henriques for assistance with the implementation.

References

1. Bonneau, R., Reiss, D.J., Shannon, P., Facciotti, M., Hood, L., Baliga, N.S., Thorsson, V.: The inferelator: an algorithm for learning parsimonious regulatory networks from systems-biology data sets de novo. Genome Biol. **7**(5), R36 (2006)
2. Cellucci, C., Albano, A., Rapp, P.: Statistical validation of mutual information calculations: comparison of alternative numerical algorithms. Phys. Rev. E: Stat. Nonlin. Soft Matter Phys. **71**(6), 066208 (2005)
3. Dagum, L., Menon, R.: OpenMP: an industry standard API for shared-memory programming. IEEE Comput. Sci. Eng. **5**(1), 46–55 (1998)
4. Folch-Fortuny, A., Villaverde, A.F., Ferrer, A., Banga, J.R.: Enabling network inference methods to handle missing data and outliers. BMC Bioinform. **16**(1), 283 (2015)
5. Huynh-Thu, V.A., Irrthum, A., Wehenkel, L., Saeys, Y., Geurts, P.: Inferring regulatory networks from expression data using tree-based methods. PLOS ONE **5**(9), e12776 (2010)
6. Jang, I., Margolin, A., Califano, A.: hARACNe: improving the accuracy of regulatory model reverse engineering via higher-order data processing inequality tests. Interface Focus **3**(4), 20130011 (2013)
7. Le Novère, N.: Quantitative and logic modelling of molecular and gene networks. Nat. Rev. Genet. **16**, 146–158 (2015)
8. Meyer, P., Lafitte, F., Bontempi, G.: minet: A R/Bioconductor package for inferring large transcriptional networks using mutual information. BMC Bioinform. **9**(1), 461 (2008)
9. Samoilov, M., Arkin, A., Ross, J.: On the deduction of chemical reaction pathways from measurements of time series of concentrations. Chaos **11**(1), 108–114 (2001)
10. Schaffter, T., Marbach, D., Floreano, D.: GeneNetWeaver: in silico benchmark generation and performance profiling of network inference methods. Bioinformatics **27**(16), 2263–2270 (2011)
11. Schreiber, T.: Measuring information transfer. Phys. Rev. Lett. **85**(2), 461 (2000)
12. Shannon, C.: A mathematical theory of communication. Bell Syst. Tech. J. **27**, 379–423 (1948)
13. Villaverde, A.F., Ross, J., Morán, F., Banga, J.R.: MIDER: network inference with mutual information distance and entropy reduction. PLOS ONE **9**(5), e96732 (2014)

Abstracts

Morphisms of Reaction Networks

Luca Cardelli[1,2]([✉])

[1] Microsoft Research, Cambridge, UK
[2] Department of Computer Science, University of Oxford, Oxford, UK
luca@microsoft.com

Abstract. The mechanisms underlying complex biological systems are routinely represented as networks, and their kinetics is widely studied. It turns out that relationships between network structures can reveal similarity of mechanism. We define morphisms (mappings) between reaction networks that establish structural connections between them. Some morphisms imply kinetic similarity, and yet their properties can be checked statically on the structure of the networks. In particular we can determine statically that a complex network will emulate a simpler network: it will reproduce its kinetics for all corresponding choices of reaction rates and initial conditions. We use this property to relate the kinetics of many common biological networks of different sizes, also relating them to a fundamental population algorithm. Thus, structural similarity between reaction networks can be revealed by network morphisms, elucidating mechanistic and functional aspects of complex networks in terms of simpler networks In recent joint work, we established a correspondence between network emulation and a notion of backward bisimulation for continuous systems. An emulation morphism establishes a bisimulation relation over the union of two networks, and a bisimulation relation over a network can be seen as an emulation morphism from the full network to the reduced network of its equivalence classes. Along this correspondence, we obtain minimization algorithms for chemical reaction networks, which are of interest for model execution, and algorithms to discover morphisms between networks, which are of interest for model understanding.

© Springer International Publishing AG 2016
E. Bartocci et al. (Eds.): CMSB 2016, LNBI 9859, p. 333, 2016.
DOI: 10.1007/978-3-319-45177-0

From Artificial to Biological Neural Networks

Radu Grosu[✉]

TU Wien, Vienna, Austria
radu.grosu@tuwien.ac.at

Abstract. In this talk I will present our journey during the past few years within the very exciting area of neural networks. I will first start with artificial neurones and neural networks and explain why we got interested in such networks. I will then turn to the mathematical modelling of biological neurones and neural networks. We will distinguish between non-spiking neurones, such as the ones found in C. Elegans, and spiking neurones, as found in most of the other larger organisms. We will discuss our work in analysing the behaviour of such networks, the use in control, and the challenges and opportunities in this area.

© Springer International Publishing AG 2016
E. Bartocci et al. (Eds.): CMSB 2016, LNBI 9859, p. 334, 2016.
DOI: 10.1007/978-3-319-45177-0

Embedding Machine Learning in Formal Stochastic Models of Biological Processes

Jane Hillston[✉]

Laboratory for Foundations of Computer Science,
University of Edinburgh, Edinburgh, Scotland
Jane.Hillston@ed.ac.uk

Abstract. Formal modelling languages such as process algebras are effective tools in computational biological modelling. However, handling data and uncertainty in these representations in a statistically meaningful way is an open problem, severely hampering the usefulness of these elegant tools in many real biological applications. I will present ProPPA, a process algebra which incorporates uncertainty in the model description, supporting the use of Machine Learning techniques to integrate observational data in the modelling. I will explain how this is given a semantics in terms of a generalisation of Constraint Markov Chains, and demonstrate how this can be used to perform inference over biological models.

© Springer International Publishing AG 2016
E. Bartocci et al. (Eds.): CMSB 2016, LNBI 9859, p. 335, 2016.
DOI: 10.1007/978-3-319-45177-0

Posters

Modeling Peptide Adsorption
on Inorganic Surfaces

Priya Anand[(✉)], Monika Borkowska-Panek, Florian Gußmann,
Karin Fink, and Wolfgang Wenzel

Institute of Nanotechnology, Karlsruhe Institute of Technology,
Karlsruhe, Germany
{priya.anand,monika.borkowska-panek,florian.gussmann,
karin.fink,wolfgang.wenzel}@kit.edu

Abstract. The interaction of proteins and peptides with inorganic surfaces is relevant in a wide array of technological applications, yet it is difficult to characterize these interactions with high resolution experimentally. For this purpose we have developed a computational protocol for efficient *in silico* evaluation of the binding affinity of peptides for different inorganic surfaces. With EISM we are able to reproduce and predict experimental results for Au (111) binding peptides. EISM in future can propose peptides sequences functionalizing the surfaces in a desired way.

Keywords: EISM · Peptide binding affinity · SIMONA

1 Introduction and Results

The interaction of proteins/peptides with inorganic surfaces is center to a wide spectrum of biological and chemical phenomena in nature. Bio-selection techniques such as cell/phage-display have been able to isolate peptides sequences that can bind to different inorganic surfaces with high affinity. Unfortunately these experiments involve multistep synthesis protocols and lack peptide specificity information, thus making it a challenging task. Empirical force field based molecular modeling approaches can help in principle to elucidate interaction mechanisms and to design peptides with specific absorption profiles for relevant surfaces. However computational efforts are complicated in part by the same complexities that limit experimental investigations for example; the molecular structures of many inorganic surfaces are not known or may be process-dependent. In addition interactions with the solvent and electrolytes can lead to a strong modification of the surface properties which adds substantial additional complexity. We have developed a new computational protocol, EISM, Effective Implicit Surface Model, for fast and efficient in silico evaluation of the binding affinity of peptides with inorganic surfaces. The energy term includes: E_{INT}: Energy of the peptide, E_{SLIM}: SIMONA layered implicit membrane [1], E_{SASA}: Solvent Accessible Surface Area and E_{SLJ}: Peptide-Surface Lennard-Jones interactions. E_{SASA} corresponds to the attractive and the repulsive interactions between the amino acid (aa) and surface and is proportional to the SASA of the aa with a residue-specific surface tension, denoted as γ_i (where i = 20aa's). The γ_i values are fitted to the experimental or

© Springer International Publishing AG 2016
E. Bartocci et al. (Eds.): CMSB 2016, LNBI 9859, pp. 339–340, 2016.
DOI: 10.1007/978-3-319-45177-0

theoretical data for a given surface (for Au(111) [3]) and conditions like temperature and pH. EISM model has been tested and validated against 12 Au(111) binding peptides and Ag (111). All simulations are done using SIMONA [2]. The calculated free energy of the peptides is then compared with the experimental data available in literature [4]. Results suggest that EISM can differentiate between the strong and weak peptide binders. Despite its simplicity, the EISM model is a powerful tool and it will open new perspective for investigating a large number of different sequences and analyzing whether and how the mutations and the length of the peptide sequence are crucial for binding.

Acknowledgements. We thank Dr. Martin Brieg for model development and funding from BMBF (Project no:031A173A).

References

1. Setzler, J., Seith, C., Brieg, M., Wenzel, W.: SLIM: an improved generalized Born implicit membrane model. J. Comput. Chem. **35**, 2027–2039 (2014)
2. Strunk, T., Wolf, M., Brieg, M., Klenin, K., Biewer, A., Tristram, F., Ernst, M., Kleine, P.J., Heilmann, N., Kondov, I., Wenzel, W.: SIMONA 1.0: an efficient and versatile framework for stochastic simulations of molecular and nanoscale systems. J. Comput. Chem. **33**, 2602–2613 (2012)
3. Palafox-Hernandez, J.P., Tang, Z., Hughes, Z.E., Li, Y., Swihart, M.T., Prasad, P.N., Walsh, T.R., Knecht, M.R.: Comparative study of materials-binding peptide interactions with gold and silver surfaces and nanostructures: a thermodynamic basis for biological selectivity of inorganic materials. Chem. Mater. **26**(17), 4960–4969 (2014)
4. Tang, Z., Palafox-Hernandez, J.P., Law, W.C., Hughes, Z.E., Swihart, M.T., Prasad, P.N., Knecht, M.R., Walsh, T.R.: Biomolecular recognition principles for bionanocombinatorics: an integrated approach to elucidate enthalpic and entropic factors. ACS Nano. **7**(11), 9632–9646 (2013)

Temperature Dependence of Leakiness of Transcription Repression Mechanisms of *Escherichia coli*

Nadia Goncalves[1], Samuel M.D. Oliveira[1], Vinodh K. Kandavalli[1],
Jose M. Fonseca[2], and Andre S. Ribeiro[1(✉)]

[1] Department of Signal Processing,
Tampere University of Technology, 33101 Tampere, Finland
andre.ribeiro@tut.fi
[2] UNINOVA, Campus FCT-UNL, 2829-516 Caparica, Portugal

Abstract. In *E. coli*, transcription repression is essential in cellular functioning. However, its failure rates are non-negligible. We measured the leakiness rate of *lacO3O1* promoter with single RNA sensitivity and its temperature dependence in live cells. After finding strong temperature dependence, we dissected the causes. While RNA polymerase numbers and k_t, the rate of active transcription, vary weakly with temperature, the repression strength (dependent on number of repressors and binding and unbinding rates of repressors to the promoter) is heavily temperature dependent. We conclude that the *lacO3O1* leakiness at low temperatures increases as the repression mechanism's efficiency hampers.

Keywords: Transcription · Repression · Leakiness · lacO3O1 · MS2-GFP RNA detection · Time-lapse confocal microscopy

To investigate *Escherichia coli*'s *lacO3O1* leakiness from *in vivo* single-cell, single-RNA measurements of transcription dynamics, we inserted a single-copy F-plasmid, with *lacO3O1* promoter controlling the expression of a synthetic RNA with multiple binding sites for MS2-GFP and a multi-copy plasmid coding for MS2-GFP on *E. coli* strain BW25113, which expresses the repressor for *lacO3O1*, LacI, in similar quantities to the natural system [1]. Example images are shown in Fig. 1(left).

First, we compared *lacO3O1* transcription kinetics when non-induced in live individual cells at 24 °C and 37 °C (Fig. 1, right). From this data, we find that the rate of *leaky* (non-induced) RNA production, (λ^{Rep}) is ~ 8.5 times higher at 24 °C, allowing concluding that the 'efficiency' of repression by LacI is temperature dependent, being much weaker at lower temperatures.

To search for causes for the temperature-dependence of leakiness, we assume the following models of repression (1) and transcription (2), respectively:

$$\text{Pro}_{OFF}.\text{Rep} \underset{k_{off}}{\overset{k_{on}}{\rightleftharpoons}} \text{Pro}_{ON} + \text{Rep} \tag{1}$$

$$\text{Pro}_{ON} + \text{RNAp} \xrightarrow{k_t} \text{Pro}_{ON} + \text{RNAp} + \text{RNA} \tag{2}$$

© Springer International Publishing AG 2016
E. Bartocci et al. (Eds.): CMSB 2016, LNBI 9859, pp. 341–342, 2016.
DOI: 10.1007/978-3-319-45177-0

where 'Pro' is a promoter and 'Rep' a repressor (LacI tetramer) [2], k_{on} and k_{off} are the rates of binding and unbinding of the repressor, and k_t is the rate at which an RNA polymerase (RNAp) finds the promoter and produces an RNA. From (1) and (2):

$$\lambda_{RNA} = \left[k_{on} \times \left(Rep \times k_{off} + k_{on} \right)^{-1} \right] \times RNAp \times k_t \qquad (3)$$

Next, by confocal microscopy, we measured relative RNAp concentrations of cells expressing fluorescently tagged RpoC [3]. These concentrations are 1.19 times larger at 24 °C than 37 °C. This difference is too small to explain the differences in leakiness.

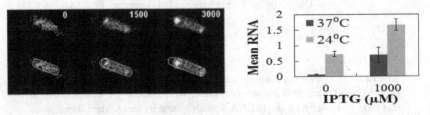

Fig. 1. (Left) MS2-GFP tagged RNAs in a cell over time. Unprocessed (top) and segmented cells and RNA spots (bottom). (Right) RNA numbers resulting from the transcription activity of lacO3O1 promoter as a function of IPTG concentration and temperature, 2.5 h after activation of the production of MS2-GFP reporter molecules.

Next, from the RNA production rates under full induction (Fig. 1), we estimated k_t. We find it to be only 1.92 times faster at 37 °C, also not explaining the higher leakiness at 24 °C. Finally, from the rate of RNA production in the absence of activators, λ^{Rep}, we calculated the change in repression strength, $\beta = \left(Rep \times k_{off} + k_{on} \right)/k_{on}$. We find β to be 13.8 times higher at 37 °C, from which we conclude that the hampering of the repression mechanism at low temperatures is the main cause for the increased leakiness. Given this, we also expect the temperature-dependence of leakiness to vary widely between promoters, as their repressions mechanisms also differ widely.

In the future, we aim to study the effects of low temperatures on the repression mechanisms of various promoters. We expect greater robustness in those associated to *E. coli*'s responses to low temperatures. Also, we aim to explore how the temperature-dependence in leakiness affects gene circuits' robustness to temperature changes.

References

1. Penumetcha, P., Lau, K., Zhu, X., Davis, K., Eckdahl, T.T., Campbell, A.M.: Improving the lac system for synthetic biology. BIOS **81**, 7–15 (2010)
2. Rutkauskas, D., Zhan, H., Matthews, K.S., Pavone, F.S., Vanzi, F.: Tetramer opening in laci-mediated DNA looping. Proc. Natl. Acad. Sci. **106**, 16627–16632 (2009)
3. Bratton, B.P., Mooney, R.A., Weisshaar, J.C.: Spatial distribution and diffusive motion of RNA polymerase in live escherichia coli. J. Bacteriol. **193**, 5138–5146 (2011)

GPU-Accelerated Steady-State Analysis of Probabilistic Boolean Networks

Andrzej Mizera[1(✉)], Jun Pang[1,2], and Qixia Yuan[1]

[1] Faculty of Science, Technology and Communication, University of Luxembourg,
Luxembourg City, Luxembourg
{andrzej.mizera,jun.pang,qixia.yuan}@uni.lu
[2] Interdisciplinary Centre for Security, Reliability and Trust,
University of Luxembourg, Luxembourg City, Luxembourg

Problem Statement. Steady-state computation is important for analysing biological systems modelled as probabilistic Boolean networks (PBNs). Since the state-space is exponential in the number of nodes, the use of statistical methods and Monte Carlo methods remain the only feasible approach to address the problem for large PBNs (e.g., with more than 50 nodes) [2, 5]. Such methods usually rely on long simulations of a PBN; hence the simulation speed becomes critical. For large PBNs with high precision requirements, a slow simulation speed becomes an obstacle of computing the steady-state probabilities. Intuitively, parallelising the simulation process can be an ideal way to accelerate the computation process.

Our Approach. We propose to parallel the simulation of PBNs using multiple graphics processing unit (GPU) cores. A GPU usually contains hundreds or thousands of cores. It uses data parallelism, i.e., the same instruction is run in different cores with different data. The memories provided by GPU can be divided into two types based on the access speed: *fast-memory* and *slow-memory*. Accessing fast-memory is highly efficient, but the size of fast-memory is very limited. A GPU program is executed in parallel by the GPU threads. Usually thousands of threads are launched in parallel to hide latency. Due to the specific architecture of GPUs, parallelising a process in a GPU has to be treated carefully. A discussion of two particular issues follows.

Firstly, synchronisation between cores is very time expensive in a GPU. To avoid it, we let each GPU core handle all the nodes in a PBN. Instead of simulating one trajectory, we simulate multiple trajectories in parallel. Samples from multiple trajectories can be combined to compute steady-state probabilities using a combination of the two-state Markov chain approach with the Gelman &Rubin method [1, 3].

Secondly, the performance of a GPU is highly related to how well the latency is hidden. Latency can be hidden via the use of more threads, more blocks, and/or fast-memory. More threads/blocks require more fast-memory, but the size of fast-memory is very limited. Therefore, a trade-off between the number of threads/blocks and the use of fast-memory is required. We first optimize our

Q. Yuan—Supported by the National Research Fund, Luxembourg (grant 7814267).

© Springer International Publishing AG 2016
E. Bartocci et al. (Eds.): CMSB 2016, LNBI 9859, pp. 343–345, 2016.
DOI: 10.1007/978-3-319-45177-0

data structure to minimize the use of memory and then follow the rule that the frequently accessed information should be put in fast-memory whenever possible since the latency caused by accessing slow-memory is relatively large. To better understand the optimization, we briefly review what a PBN is. A PBN is composed of a set of binary-valued nodes, each of which has a certain number of Boolean functions. The process of simulating a PBN consists of selecting a Boolean function for each node and updating its value in accordance the selected function (see [4, 6] for details). The state (value of all nodes) and the Boolean functions (stored as a truth table) are repeatedly used in the simulation process and require much memory to store, hence we optimize the data structure to represent the state of a PBN and the truth table. We use the primitive integer type (32 bits) instead of Boolean arrays to store a state of a PBN. The integer type is used due to the following two reasons: (1) it reduces the memory usage by 4 times comparing to Boolean arrays; (2) operations on 32 bits data are faster than or equal to those on non-32 bits data in our GPU architecture. The truth table is optimized similarly as the state, i.e., it is also stored using integers. After optimization, we store the state in *registers* (fast-memory), if possible. In the cases that a PBN is extremely large and registers are not enough, the slow global memory is used. However, accessing this slow global memory is reduced by 32 times using an intermediate register. The truth table as well as other frequently accessed arrays (e.g., the selection probabilities) are stored in *shared memory* (fast-memory). Frequently accessed single variables are stored in registers. After arranging all variables in memory, we compute the optimal number of threads and blocks to be launched based on the usage of fast-memory to hide latency as much as possible.

Table 1. GPU speedup for computing four steady-state probabilities.

node #	CPU time (s)	GPU time (s)	speedup
100	635.27	1.84	345×
200	424.18	1.84	231×
500	1567.77	5.80	270×
91	905.10	3.54	256×

Preliminary Results. We have evaluated the proposed GPU-based simulation of PBNs for computing steady-state probabilities of both randomly generated networks and of a real biological network using the approach in [3]. On randomly generated networks, our proposed GPU-based parallelised approach showed more than two orders of magnitude speedups compared to the sequential CPU version. The evaluation on a real biological network was performed by analysing an apoptosis network with 91 nodes [2]. The speedups for computing steady-state probabilities for 3 randomly generated networks and the real 91-node network are shown in Table 1. All experiments were conducted on a high-performance computing (HPC) node, which contained Intel Xeon E5-2680 v3 @2.5 GHz and a NVIDIA Tesla K80 Graphic Card with 2496 cores @824MHz.

References

1. Gelman, A., Rubin, D.: Inference from iterative simulation using multiple sequences. Stat. Sci. **7**(4), 457–472 (1992)
2. Mizera, A., Pang, J., Yuan, Q.: Reviving the two-state markov chain approach Technical report (2015). http://arxiv.org/abs/1501.01779
3. Mizera, A., Pang, J., Yuan, Q.: Parallel approximate steady-state analysis of large probabilistic Boolean networks. In: Proceedings of the 31st ACM Symposium on Applied Computing, pp. 1–8 (2016)
4. Shmulevich, I., Dougherty, E., Zhang, W.: From Boolean to probabilistic Boolean networks as models of genetic regulatory networks. Proc. IEEE **90**(11), 1778–1792 (2002)
5. Trairatphisan, P., Mizera, A., Pang, J., Tantar, A.A., Sauter, T.: optPBN: an optimisation toolbox for probabilistic boolean networks. PLOS ONE **9**(7) (2014)
6. Trairatphisan, P., Mizera, A., Pang, J., Tantar, A.A., Schneider, J., Sauter, T.: Recent development and biomedical applications of probabilistic Boolean networks. Cell Commun. Signal. **11**, 46 (2013)

PINT: A Static Analyzer for Dynamics of Automata Networks

Loïc Paulevé[(✉)]

LRI UMR 8623, Univ. Paris-Sud – CNRS, Université Paris-Saclay,
Saint-aubin, France
`loic.pauleve@lri.fr`

The software PINT[1] is devoted to the formal analysis of the transient dynamics of automata networks, which encompasses Boolean and logical networks. Its main application domain is in systems biology for addressing models of signalling networks and gene regulatory networks. PINT implements formal approximations of transient reachability-related properties, such as cut sets and model reduction which preserves all the traces that lead to a given goal (state). PINT has been applied to numerous large biological networks. It has been experimentally shown that it can address networks with hundreds and thousands of interacting components, which are often intractable with standard approaches.

Input Formalism. PINT takes as input Asynchronous Automata Networks (ANs) [4]. ANs are close to Boolean and multi-valued networks, with the major difference that ANs rely on explicit transitions rules, instead of function-centered specifications. Any logical network can be encoded in AN. The tool `logicalmodel`[2] can export SBML-qual models [2] to the PINT format:

```
java -jar LogicalModel.jar sbml:an model.sbml model.an.
```

Static Analyses for Transient Dynamics. PINT implements various analysis related to *reachability* properties. It relies on static analysis by abstract interpretation which provides algorithms with a low complexity and guaranteed results.

Formal approximations of reachability Necessary and sufficient conditions for reachability have been derived in [4, 8]. They can be efficiently verified on large ANs. The following command line checks those conditions for the reachability of active BCL6 in a TCell differentiation model from [1]. Such a model is too large for exact model-checking, but in this case, the sufficient condition is satisfied (computation took around 1s):

```
$ pint-reach -i TCell-d.an BCL6=1
True
```

Cut-sets are set of component states which predict mutations that should prevent a given reachability to occur. PINT implements a highly scalable formal approximation of cut sets: all identified cut sets are correct, but some may be missed or be non-minimal [7]. The following command computes cut sets with at most 4 components which are not in the initial state for the transient

---

[1] http://loicpauleve.name/pint.
[2] https://github.com/colomoto/logicalmodel.

© Springer International Publishing AG 2016
E. Bartocci et al. (Eds.): CMSB 2016, LNBI 9859, pp. 346–347, 2016.
DOI: 10.1007/978-3-319-45177-0

reachability of BCL6 within the TCell differentiation model. Computation took less than 0.05 s. Given the results, knocking down CD28 and IL6R should prevent the transient activation of BCL6.

```
$ pint-reach --cutsets 4 --no-init-cutsets -i TCell-d.an BCL6=1
"GP130"=1
"STAT3"=1
"CD28"=1,"IL6R"=1
...
"IL6RA"=1,"TCR"=1
```

Goal-oriented model reduction removes transitions that do not contribute to the reachability of the supplied goal state. The reduction can truncate significantly the reachable state space, whereas it preserves all the (minimal) traces to the goal. The following command line reduces the TCell differentiation model for the reachability of BCL6 before using NuSMV to verify if active IL2RA and IL6RA form a cut set for BCL6 (CTL property not E [IL2RA!=1 and IL6RA!=1 U BCL6=1]). Without the model reduction, the model-checking of this property was impossible [6].

```
$ pint-export -i TCell-d.an --reduce-for-goal BCL6=1 \
  | pint-nusmv --is-cutset"IL2RA=1,IL6RA=1" BCL6=1
```

Interaction with Other Softwares. Bridges with other standard tools for dynamical models have been developed, in particular for model-checking (NuSMV [3], ITS [5], Mole [9]). It allows to take advantage of the static analyses of PINT beforehand further exact dynamical analyses.

References

1. Abou-Jaoudé, W., Monteiro, P.T., Naldi, A., Grandclaudon, M., et al.: Model checking to assess T-Helper cell plasticity. Front. Bioeng. Biotech. **2** (2015)
2. Chaouiya, C., Bérenguier, D., Keating, S.M., Naldi, A., et al.: SBML qualitative models: a model representation format and infrastructure to foster interactions between qualitative modelling formalisms and tools. BMC Syst. Biol. **7**(1), 135 (2013)
3. Cimatti, A., Clarke, E., Giunchiglia, E., Giunchiglia, F., Pistore, M., Roveri, M., Sebastiani, R., Tacchella, A.: NuSMV 2: an opensource tool for symbolic model checking. In: Brinksma, E., Larsen, K.G. (eds.) CAV 2002. LNCS, vol. 2404, pp. 359–364. Springer, Heidelberg (2002)
4. Folschette, M., Paulevé, L., Magnin, M., Roux, O.: Sufficient conditions for reachability in automata networks with priorities. Theor. Comp. Sci. **608**, 66–83 (2015)
5. LIP6/Move. Its tools. http://ddd.lip6.fr/itstools.php
6. Paulevé, L.: Goal-Oriented Reduction of Automata Networks. Research report, May 2015. https://hal.archives-ouvertes.fr/hal-01149118
7. Paulevé, L., Andrieux, G., Koeppl, H.: Under-approximating cut sets for reachability in large scale automata networks. In: Sharygina, N., Veith, H. (eds.) CAV 2013. LNCS, vol. 8044, pp. 69–84. Springer, Heidelberg (2013)
8. Paulevé, L., Magnin, M., Roux, O.: Static analysis of biological regulatory networks dynamics using abstract interpretation. Math. Struct. in Comp. Sci. **22**(04), 651–685 (2012)
9. Schwoon, S.: Mole. http://www.lsv.ens-cachan.fr/~schwoon/tools/mole/

Linear Temporal Logic for Biologists in BMA

Benjamin A. Hall[1(✉)], Nir Piterman[2], and Jasmin Fisher[1,3]

[1] University of Cambridge, Cambridge, UK
bh418@mrc-cu.cam.ac.uk
[2] University of Leicester, Leicester, UK
[3] Microsoft Research, Cambridge, UK

The BioModelAnalyzer (BMA[1]) is a web based tool for the development of discrete models of biological systems. Through a graphical user interface, it allows rapid development of complex models of gene and protein interaction networks and stability analysis without requiring users to be proficient computer programmers [1, 2]. Here I will present a new set of tools in the BMA that allow users to perform complex queries over models in linear temporal logic, allowing biologists to test specifications based on the dynamics observed in simulations. In keeping with the core objective of tool, queries are constructed graphically and results are presented to the users with examples of simulations. Alongside stability analysis, this new tool allows biologists to verify complex specifications to validate executable biological models.

Linear temporal logic queries are substantially more complex than stability testing due to the fundamental requirement for users to construct the query and select a path length. Biologists specifically face further problems; biological models typically have many variables that may be included in a query, they may not be familiar with complex operator precedence issues, and they must balance parentheses. Whilst this is handled by NuSMV in other tools [3, 4], a design principle of BMA is that computing proficiency is not required so necessarily this must be achieved in the GUI.

We address these issues in the tool through a two-stage workflow (Fig. 1). Users define *LTL states*; large conjunctions of variable assignments that may be

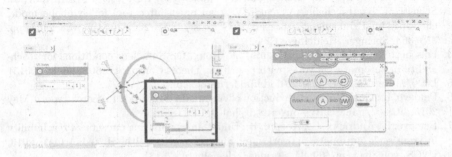

Fig. 1. The LTL state editor and the LTL query editor.

---

[1] http://biomodelanalyzer.research.microsoft.com/.

© Springer International Publishing AG 2016
E. Bartocci et al. (Eds.): CMSB 2016, LNBI 9859, pp. 348–350, 2016.
DOI: 10.1007/978-3-319-45177-0

created by dragging and dropping from the model canvas, or through a drop-down menu inspired by file browsers. These states are tansformed into an LTL query in a second temporal and logical layer. Operators in this layer carry a number of "sockets" into which operands (i.e. LTL states) or other operators may be dragged and dropped. As such complex queries can be developed through repeatedly nesting operators.

To aid users several default states are included. In addition to "True", users may also search for fixpoints or cycles in the system. These are valuable for studying biological models as they allow users to study developmental end-points specifically. The description of these states through other operators would be difficult. For example, a self loop state is characterised by the formula $\bigwedge_{v \in V} v = Xv$. To the best of our knowledge, most LTL tools do not support such a direct comparison between the value of a variable and its value in the next state.

User testing indicated that the LTL operator "until" confused unfamiliar users, as the first operand need not hold. To address this we supplemented the list of operators with the non-standard operator "upto", which carries a similar meaning in English but ensures that both operands hold (A upto B corresponds to A and next A until B).

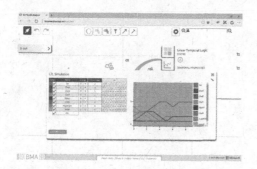

Fig. 2. An example trace from an LTL query.

On clicking the test button both the query and its negation are checked. This produces three types of result- always true, never true, and true for some. The user can then choose to see examples of simulations that satisfy, or fail to satisfy the query (Fig. 2).

References

1. Benque, D., Bourton, S., Cockerton, C., Cook, B., Fisher, J., Ishtiaq, S., Piterman, N., Taylor, A., Vardi, M.Y.: Bma: visual tool for modeling and analyzing biological networks. In: Madhusudan, P., Seshia, S.A. (eds.) CAV 2012. LNCS, vol. 7358, pp. 686–692. Springer, Heidelberg (2012)

2. Chuang, R., Hall, B., Benque, D., Cook, B., Ishtiaq, S., Piterman, N., Taylor, A., Vardi, M., Koschmieder, S., Gottgens, B., Fisher, J.: Drug target optimization in chronic myeloid leukemia using innovative computational platform. Sci. Rep. **5**, 8190 (2015)
3. Naldi, A., Thieffry, D., Chaouiya, C.: Decision diagrams for the representation and analysis of logical models of genetic networks. In: Calder, M., Gilmore, S. (eds.) CMSB 2007. LNCS (LNBI), vol. 4695, pp. 233–247. Springer, Heidelberg (2007)
4. Bean, D., Heimbach, J., Ficorella, L., Micklem, G., Oliver, S., Favrin, G.: esyN: network building, sharing, and publishing. PLoS ONE **9**, e106035 (2014)

Deregulation of Osmotic Regulation Machinery Explains and Predicts Cellular Transformation in Cancer and Disease

David Shorthouse[✉], Angela Riedel, Jacqueline Shields, and Benjamin A. Hall

MRC Cancer Unit, University of Cambridge, Cambridge, UK

Osmotic regulation is a hugely important homeostatic system in all cells. Cells respond to osmotic stresses by activating or upregulating proteins involved in the transportation of charged ions, primarily Chlorine, Potassium, Sodium, and Calcium. Additionally, the movement of ions and osmotically obliged water are necessary for many of the cellular hallmarks exhibited in the transformations associated with disease states such as cancer. In particular, the aberrant expression of ion channels are hallmarks for increased proliferative and invasive behaviours [1, 2]. We present a formal model of the osmotic regulation machinery within a mammalian cell. The model can provide a mechanistic explanation for the behavioural changes observed in highly diverse cellular systems of murine premetastatic Lymph Node stromal cells, and Lung Cancer Fibroblasts. The model explains phenotypic transformations within each cell types, and predicts behaviour from datasets not involved in its generation. Furthermore, we use the model to predict key proteins involved in each transformation, and propose experiments to alter the behaviour of cells in controllable ways.

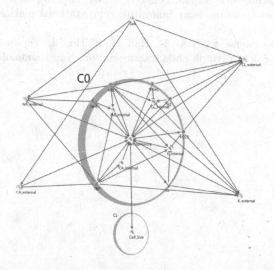

Fig. 1. The model of osmoregulation as rendered by BMA. Phenotype nodes not shown for clarity.

© Springer International Publishing AG 2016
E. Bartocci et al. (Eds.): CMSB 2016, LNBI 9859, pp. 351–352, 2016.
DOI: 10.1007/978-3-319-45177-0

A qualitative network of key channels, ions, and transporters was constructed using the BioModelAnalzer (http://biomodelanalyzer.research.microsoft.com/, [3, 4]). As osmoregulation achieves a homeostasis, the model was verified both through stability analysis (which proves the existence of a global attractor) and simulation. Initially a specification was constructed from the literature, and then refined against microarray data from resting fibroblast reticular cells (FRCs) in the lymph node [5].

To model the response of the FRCs to upstream tumuors at different time-points, a subset of the ion channels and transporters were deregulated. This in turn caused wide-spread, coordinated changes in other channels through osmotic pressure alterations, and subsequent changes in cellular phenotypes. The model was found to accurately predict the changes observed in the FRCs, and subsequent validation of expression changes supported the model findings.

References

1. Prevarskaya, N., Skryma, R., Shuba, Y.: Ion channels and the hallmarks of cancer. Trends Mol. Med. **16**, 107–21 (2010)
2. Djamgoz, M., Coombes, R., Schwab, A.: Ion transport and cancer: from initiation to metastasis. In: Philosophical Transactions of the Royal Society of London. Series B, Biological Sciences, vol. 369 (2014)
3. Benque, D., Bourton, S., Cockerton, C., Cook, B., Fisher, J., Ishtiaq, S., Piterman, N., Taylor, A., Vardi, M.Y.: BMA: visual tool for modeling and analyzing biological networks. In: Madhusudan, P., Seshia, S.A. (eds.) CAV 2012. LNCS, vol. 7358, pp. 686–692. Springer, Heidelberg (2012)
4. Chuang, R., Hall, B., Benque, D., Cook, B., Ishtiaq, S., Piterman, N., Taylor, A., Vardi, M., Koschmieder, S., Gottgens, B., Fisher, J.: Drug target optimization in chronic myeloid leukemia using innovative computational platform. Sci. Rep. **5**, 8190 (2015)
5. Riedel, A., Shorthouse, S., Haas, L., Hall, B., Shields, J.: Tumor-induced stromal reprogramming drives lymph node transformation. Nat. Immunol. **17** (2016)

Game Theoretic Consideration of Transgenic Bacteria in the Human Gut Microbiota Converting Omega-6 to Omega-3 Fats

Ahmed M. Ibrahim[1]($\boxtimes$) and James Smith[2,3,4]

[1] 44 El-Geish St, Mansoura, Dakahlia, Egypt
wetawdt@gmail.com

[2] Department of Applied Mathematics and Theoretical Physics,
Centre for Mathematical Sciences, Cambridge Computational Biology Institute,
Wilberforce Rd, Cambridge CB3 0WA, UK

[3] MRC Elsie Widdowson Laboratory, MRC Human Nutrition Research,
University of Cambridge, 120 Fulbourn Rd, Cambridge CB1 9NL, UK

[4] School of Food Science and Nutrition,
Faculty of Mathematics and Physical Sciences,
University of Leeds, Leeds LS2 9JT, UK
j.smith252@leeds.ac.uk

Abstract. Prophylactic use of functional foods and the design of nutraceuticals has a far-reaching public health benefit. Understanding the phenotypic manipulations needed to take advantage of gut microbial ecology is fundamental to bioengineering and the food, diet and health industries. This work considers a hypothetical adjustment of gut microbiota by an introduced transgenic bacterial strain that contributes to increased exposure of essential omega-3 (n-3) poly-unsaturated fatty acids, the so-called *fish oils*. Absorption of the essential poly-unsaturated fats from food is dominated by the omega-6 (n-6) fats over the omega-3 (n-3) fats. Unfortunately, long-term depleted levels of n-3-containing lipids in blood plasma is a high-risk indicator for outcomes such as metabolic syndrome, cardiovascular disease and diabetes-related conditions.

In our vignette, a genetically modified strain converts excessive dietary n-6 into bioavailable n-3 in the gut. Maintaining a long-term co-existence between indigenous gut bacteria and the transgenic strain is the challenge. Game theory is an appropriate formalism for exploring such conflicts. We show that long-term co-existence is predicted if the two forms of bacteria engage in the Snowdrift game. Our model explores putative mechanisms for addressing metabolic syndrome and related conditions by locally increasing n-3 production by the transgenic gut bacteria. Our model suggests long-term therapeutic supplementation by a functional probiotic food is possible without detriment to the indigenous bacteria.

Keywords: Game theory · Snow drift game · Prisoners' dilemma · Non-linear behaviour · Gut microbiome · Fat metabolism

© Springer International Publishing AG 2016
E. Bartocci et al. (Eds.): CMSB 2016, LNBI 9859, p. 353, 2016.
DOI: 10.1007/978-3-319-45177-0

Revealing Biomarker Mixtures in Lipid Pools from Large-Scale Lipidomics

Kai Loell[1], Albert Koulman[2], and James Smith[1,2,3]($\boxtimes$)

[1] Department of Applied Mathematics and Theoretical Physics,
Centre for Mathematical Sciences, Cambridge Computational Biology Institute,
University of Cambridge, Wilberforce Rd, Cambridge CB3 0WA, UK
10311kai@gmail.com
[2] MRC Elsie Widdowson Laboratory, MRC Human Nutrition Research,
120 Fulbourn Rd, Cambridge CB1 9NL, UK
ak675@cam.ac.uk
[3] Faculty of Mathematics and Physical Sciences,
School of Food Science and Nutrition, University of Leeds, Leeds LS2 9JT, UK
j.smith252@leeds.ac.uk

Abstract. Lipids are key structural elements, energy sources, and components for intracellular signalling and metabolic processes. Their constituents are a small number of fatty acids (FAs), indicators of metabolic health and nutrition and biomarkers for disease risk. Lipids contain singlet, doublet and triplet combinations of FAs. Different combinations of FAs can have equivalent configurations that in aggregate are observed as lipid mixture pools. Traditionally, the lipid pools have been considered to be biomarkers, however it is now recognised that sub-populations of explicit lipid species are more informative. Lipid biomarkers for metabolic states, so far, come from the latent (hidden) structure of sub-populations in the pools and this needs to be addressed.

Epidemiological high-resolution lipidomics data is required to derive the *mixtures* of lipid species contributing to lipid pools. FA signals are acquired using gas chromatography and lipid profiling performed by direct infusion high resolution mass spectrometry. Profiling identifies lipid mixture pools as spectral peaks separated by their m/z ratio. However, not all constituent lipid sub-populations are easily distinguished. Furthermore, the data generated is *compositional* with the signals of FAs and lipid pools normalised separately.

Our approach to this problem uses both lipid pool and FA data. Lipid data is re-scaled to account for the combinations of FAs required in each observed pool. A linear algebra Gauss-Jordan reduction algorithm is applied to the stoichiometry of FAs incorporated in the explicit lipid species and the combinations of lipid species in every pool. The method solves the contributing lipid species sub-populations, that is, the representative combinations of FAs that form the pools. Abundances of explicit FA combinations not only improve lipid biomarker identification but also provide a more detailed picture of metabolic responses.

Keywords: Compositional mixture modelling · Optimisation · Biomarkers · Metabolic states · Big data · Lipidomics

© Springer International Publishing AG 2016
E. Bartocci et al. (Eds.): CMSB 2016, LNBI 9859, p. 354, 2016.
DOI: 10.1007/978-3-319-45177-0

Author Index

Printed in the United States
By Bookmasters